Indian Reservations in the United States

UNIVERSITY OF CHICAGO GEOGRAPHY RESEARCH PAPER NO. 242

Titles published in the Geography Research Papers series prior to 1992 and still in print are now distributed by the University of Chicago Press. For a list of available titles, see the end of the book. The University of Chicago Press commenced publication of the Geography Research Papers series in 1992 with no. 233.

Indian Reservations in the United States

Territory, Sovereignty, and Socioeconomic Change

KLAUS FRANTZ

The University of Chicago Press • Chicago and London

Klaus Frantz is currently research professor at the Institute for Urban and Regional Research of the Austrian Academy of Sciences in Vienna and professor of geography at the University of Innsbruck in Austria.

The University of Chicago Press, Chicago 60637
The University of Chicago Press, Ltd., London

08 07 06 05 04 03 02 01 00   2 3 4 5

ISBN: 0-226-26089-5 (paper)

Originally published as *Die Indianerreservationen in den USA*. ©1993, Franz Steiner Verlag, Stuttgart.

The translation of this publication was sponsored by the Austrian Ministry for Sciences and Arts; the University of Innsbruck; and Swarovski, the worldwide leading manufacturer of full cut crystal.

Library of Congress Cataloging-in-Publication Data

Frantz, Klaus.
[Indianerreservationen in den USA. English]
Indian reservations in the United States : territory, sovereignty, and socioeconomic change / Klaus Frantz.
p. cm. — (University of Chicago geography research paper ; no. 242)
Includes bibliographical references.
ISBN 0-226-26089-5 (paper : alk. paper)
1. Indian reservations—United States. 2. Indians of North America—Social conditions. 3. Indians of North America—Economic conditions. I. Title. II. Series.
E93.F7313 1999
330.973'09'08997—dc21   99-13353
CIP

∞ The paper used in this publication meets the minimum requirements of the American National Standard for Information Sciences—Permanence of Paper for Printed Library Materials, ANSI Z39.48-1992.

For Erica, Nicola, and Lisa

Contents

Figures

Tables

Abbreviations

AAAG	Annals of the Association of American Geographers
AHFC	Arizona Historical Foundation Collection (Hayden Library, Arizona State University, Tempe, Arizona)
AILR	American Indian Law Review
ARCIA	Annual Report of the Commissioner of Indian Affairs
ARIL	U.S. Department of the Interior, BIA, Office of Trust Responsibilities (yearly issues, 1974 to 1985, Annual Report of Indian Lands
AzR	The Arizona Republic
BAR	Branch of Acknowledgement and Research (BIA)
BC	Bureau of the Census
BIA	Bureau of Indian Affairs
BLM	Bureau of Land Management
CAP	Central Arizona Water Project
CERT	Council of Energy Resource Tribes
CRTF	Colorado River Tribal Farm
ED	U.S. Department of the Interior, BIA, 1986: Report of the Task Force on Indian Economic Development
EDA	Economic Development Administration
EPA	Environmental Protection Agency
FATCo	Fort Apache Timber Company
GAA	General Allotment Act
GPO	U.S. Government Printing Office
GR	Geographische Rundschau
GRev	Geographical Review
HIP	Housing Improvement Program
HO	Human Organization
HUD	Department for Housing and Urban Development
I.	Indians
ICC	Indian Claims Commission
IHS	Indian Health Service
I.R.	Indian Reservation
IRA	Indian Reorganization Act

JES	Journal of Ethnic Studies
MICA	Maricopa Indian Cooperative Association
NCHS	National Center for Health Studies
OEO	Office of Economic Opportunities
OSCC	Oglala Sioux Community College
R.I.	Reservation Indians
SBA	Small Business Administration
SRP	Salt River Project
USGS	U.S. Geological Service

Preface

I first became interested in American Indians as an Austrian graduate student on a Fulbright scholarship to the United States in the mid '70s. I was studying geography at the University of Wisconsin in Madison at the time and took advantage of semester breaks to travel throughout the United States and Canada. On this trip I first saw and became fascinated by Indian reservations in the Northwest, the Southwest, and the Dakotas. It was not until the summer of 1982, however, that I felt inspired to contemplate writing this book. During a one-month stay at the John F. Kennedy Institute for North American Studies at the Free University of Berlin, funded by the German Academic Exchange Service, I had the opportunity to search for relevant literature at their fine library and to have constructive talks with Burkhard Hofmeister (Technical University of Berlin), Karl Lenz (John F. Kennedy Institute, Free University Berlin), and Horst Hartmann (Section of Indians of North America, Museum of Ethnology). It was in Berlin that I first realized that it would be rewarding from a geographer's point of view to study American Indian reservations.

The following spring, an invitation to give a paper at a conference, organized and sponsored by the University of New Orleans, gave me the opportunity to stay on in the United States for six weeks and go on a fact-finding trip that included visits to the Bureau of Indian Affairs, the Racial Statistics Branch of the Bureau of the Census, the Smithsonian Institution (all in Washington, DC), the Newberry Library in Chicago, and several university libraries throughout the country. I also had the chance to seek out and contact a number of geographers, historians, anthropologists, and other specialists, whose names I had come across while searching for literature on American Indians at the John F. Kennedy Institute in Berlin. A great many scholars, such as James Goodman (University of Oklahoma, Norman), Leslie Hewes and David Wishart (University of Nebraska, Lincoln), Elliot

McIntire and Bea Medicine (Northridge), Imre Sutton and Robert Young (Fullerton), George van Otten (Flagstaff), Emroy Sekaquaptewa (Tucson), Alfonso Ortiz (University of New Mexico, Albuquerque), David Woodward (University of Wisconsin, Madison), and the late Wilcomb Washburn (Smithsonian Institution) willingly shared their ideas with me, which was most rewarding and illuminating. They also provided reprints of their publications that were useful for my work. It was the advice and encouragement from all these scholars which finally persuaded me to begin my project notwithstanding my Austrian nationality. It was also during this initial exploratory trip that I decided, after preliminary inquiries on a number of different Indian reservations, to concentrate my fieldwork in the Southwest.

Before work on this project could begin I had to secure financial assistance to meet all the expenses of a prolonged stay in the United States. I am therefore particularly grateful to the Austrian Society for American Studies for kindly supporting my application for a six-month scholarship from the American Council of Learned Societies. Further much-needed support came from the Ford and Mellon Foundations. I am greatly indebted to all these institutions for financial aid which enabled me to stay for almost fourteen months, beginning in early February 1985. I also wish to express my gratitude to the Austrian Fulbright Commission for covering my travel expenses, and to the University of Innsbruck for granting me a one-year leave.

The first two months in the United States were spent in Washington, DC, primarily at the Bureau of Indian Affairs (BIA), the Bureau of the Census (BC), the Indian Health Service (IHS), the National Center for Health and Statistics (NCHS), and the Administration for Native Americans. I am indebted to those many people in Washington, DC, too numerous to mention, who helped and directed me to the information I needed, despite their initial surprise that an Austrian was interested in Indian affairs. In particular, I would like to extend special thanks to Edna Paisano, until recently associated with the Racial Statistic Branch (BC), and to Tony D'Angelo (IHS) and Jeff Maurer (NCHS), who shared census data with me and have kept me updated with American Indian statistics to this day. I am also grateful to Bud Shapard, former chief of the Branch of Acknowledgement and Research (BIA), who placed an office at my disposal for two months, set me in contact with several branch chiefs within the BIA and other federal agencies, and gave me access to many useful materials and Indian policy-related insights.

I visited several other specialists in the field throughout the country to obtain additional material, and undertook a one-month information-gathering tour to several reservations in the Southwest before moving to the Arizona State University at Tempe, where I was allowed to set up headquarters for the remainder of my stay. On the basis of this preliminary research, I decided to concentrate my fieldwork in Arizona—a state three and a half times the size of Austria—27% of which is Indian land. This large land base required a considerable degree of mobility, and the fact that Indian reservations, with a few exceptions, have no accommodation for visitors also posed a problem. The consequence was that during my fieldwork in Arizona I was probably the only "nomad" among the reservation Indians, spending the nights and a considerable part of the days in an old pickup truck, registering an additional 30,000 miles before it was sold again in March 1986.

My fieldwork on Indian lands required me to come into contact with a great number of people both on and off the reservations. It was usually possible to establish such contacts fairly quickly and to find a satisfactory basis for interviews and productive conversations. This was in part due to the fact that as a non-American I had the advantage of being seen as exotic, a bonus which opened many doors for me. As I had little to offer in return, I was grateful to be given the opportunity to show slides of Austria and to talk about Alpine landscapes and customs to tribal and federal agencies, schools, religious communities, and veterans organizations. Many Indian and non-Indian families also invited me as a guest into their homes and I was even asked to introduce them to Austrian cuisine. Such activities kept many doors open and facilitated the conduct of my research.

Lack of space makes it impossible to name all the people who provided both professional and personal support during my fieldwork on reservation lands. I would, however, like to single out and extend my warm thanks to Carla Alchesay, Robert Brauchli, Gloriana Dayaye, and Floyd Massey (Fort Apache Indian Reservation), Dorothy Hallock, Colleen Moya, and Nick Sunn (Gila River Indian Reservation), Amelia Flores, Joycelyn Martinez, and Ron Moore (Colorado River Indian Reservation), Earl Havatone, Robert McNichols, Herb Voigt, and Lucilla Whatahomigie (Hualapai Indian Reservation), Wayne Sinyella (Havasupai Indian Reservation), and Dick Jeffries, Bob Jones, and Fred Malroy (BIA-Phoenix Area Office), all of whom provided countless hours of their time to patiently answer my questions and provided access to source materials.

While in Tempe, I also received much help from Malcolm Comeaux, Ray Henkel, and the late Mayland Parker, all geography professors at Arizona State University, who also continued to keep me informed about Indian-related matters after I returned to Austria. These colleagues and their families were most hospitable, made us feel at home in Arizona, and did a great deal to make conditions more agreeable for my wife and our little daughter during the frequent periods while I was away conducting fieldwork. My heartfelt thanks also go to Donna Webber, a librarian of the Massachusetts

Institute of Technology and longtime friend, who helped me to obtain documents and bibliographical sources that would otherwise have been hard to come by.

I would also like to express my gratitude to professors Jürgen Bähr (University of Kiel), Peter Meusburger (University of Heidelberg), Christoph Stadel (University of Salzburg), Hans Wilhelm Windhorst (Hochschule Vechta), and Eugen Wirth (University of Erlangen–Nürnberg) for reading preliminary drafts of particular chapters. I owe an especial debt of gratitude to Hans Becker (University of Bamberg), Burkhard Hofmeister (Technical University of Berlin), Peter Schwarzbauer (University of Natural Resources–Vienna and founding member of the Indians of North America "working circle," an Austrian branch of the Society for Threatened Peoples), and Guido Weigend (Arizona State University), who read the whole manuscript. The criticism and encouragement I received from these individuals was most helpful.

The German manuscript, first submitted as a thesis for the certificate of habilitation, the Austrian precondition for gaining tenure, was completed in April 1991 and published two years later in the geographic research paper series "Erdkundliches Wissen" by Franz Steiner Verlag in Stuttgart. A slightly altered second edition followed in 1995. At about the same time I was asked by the University of Chicago Press whether I would be willing to prepare an English translation of the book for publication, an honor which I gladly accepted.

The original study was designed for the German-speaking world, which generally has a keen interest in North American Indians, but little knowledge. I therefore decided to give an overall view of American Indian reservations and their residents in the conterminous United States. This necessarily includes a broad range of disparate topics, enriched by regional examples, that might help a non-American readership to understand American Indians and reservation lands. I do not pretend to be an authority on all these topics, nor was my intention to write a comprehensive survey of such a huge and complex subject, but I hope that, as a European, I might be able to view some Indian-related aspects from a different perspective.

The English translation represents a revised version of the second edition of the German book. Besides leaving out quite a few details that would have been self-evident for an American reader, changing some passages, and restructuring the original table of contents, maps have had to be redrawn and some formal aspects altered in order to meet the requirements of the University of Chicago Press.

At the end of the 1980s I collected regional and national data from various sources, analyzed them, and linked them together, and the results of these inquiries were published in the original German version of this book. These data were drawn from fieldwork on the reservations and the 1980 Census, as well as census material from other federal agencies and information regarding the situation of U.S. American Indians, collected in Washington, DC, and other places. For the English version of this book it was unfortunately not possible to redo the fieldwork and research, so I have refrained from updating parts of the original version with new census data and information from the 1990s lest I destroy the coherence of the initial text. Instead I decided to rewrite the final chapter, where I summarized the findings, to update some of the results with 1990 Census material and incorporate new issues of American Indian politics, the welfare of reservation Indians, and the state of the reservation economy. I also expanded the bibliography with some of the more recent work on American Indians published since the late 1980s. The updating of the book remains, however, far from complete, both for lack of space and also because, living in Austria, I could not remain fully informed about all the latest trends in Indian affairs in spite of a number of shorter visits to America in the early 1990s.

Without the assistance and generosity of a number of people and institutions this book would not now be published in English. In addition to my duties at the University of Innsbruck, I have been working as a research professor at the Institute for Urban and Regional Research at the Austrian Academy of Sciences in Vienna from 1997 to 1999. I am grateful to these institutions for their forbearance in allowing me to devote considerable time to this project. Financial support to have the maps redrawn and the German text translated into English came from the Austrian Ministry for Sciences and Arts, Prof. Dr. Hans Moser, Chancellor of the University of Innsbruck, and Dipl. Ing. Helmut Swarovski (Swarovski Corporation, Wattens). I would like to acknowledge a debt of gratitude to them. Helmut Heinz-Erian (Innsbruck) assisted me with the cartographic work, and the main part of the initial draft of the translation was done by Dr. Greeley Stahl (Lüneburg). I then elaborated on this version, going through it meticulously and polishing wherever necessary.

I also wish to extend my warm thanks to geography professors and friends Michael Conzen (University of Chicago), William Loy (University of Oregon, Eugene), and Robert Sauder (University of New Orleans), who took the time to read and correct draft sections from the initial translation and also those parts I added subsequently. I am particularly grateful to Professor Dick Winchell (Eastern Washington University, Spokane), who carefully read through the entire English manuscript, polishing and correcting the rough copy, verifying the geographical and specialist terms, and generally raising the standard of English. His efforts regarding the style and tone of the English version of the book were invaluable, far exceeding what any author has a right to expect. For me all the above mentioned exemplify the way friendship and academic criticism should function, if only we lived in a perfect world.

I am grateful to the staff of the University of Chicago Press for their role in bringing this project to fruition. Further thanks are due to Michael Conzen for helping to convince me that the German book was worthy of publication in English, and Penelope Kaiserlian, associate director of the University of Chicago Press, for her role in expediting the project. My thanks also go to Carol Saller, the copy editor, whose skills are reflected in every page, and Mike Brehm, who created the book design.

Finally my heartfelt thanks go to my wife Erica for her encouragement, patience, and companionship, and her tremendous help in getting the translation completed. Without her support this book would never have become a reality. It is to her and to our two daughters Nicola and Lisa that I dedicate this book.

1 Introduction

AMERICAN INDIANS IN GEOGRAPHIC AND RELATED RESEARCH: A SUMMARY SURVEY

The study of American Indians and their reservations opens up to the geographer a wide spectrum of possibilities for research that encompasses many disciplines in the field. It is therefore a rather surprising fact that geographers, whether of the German- or English-speaking world, have generally paid little attention to this subject (see Haas 1925; Ballas 1966, Carlson 1972; Lazewski 1973, etc.), as a systematic examination of specialist journals[1] and other geographical literature proves. The following pages will provide a brief introduction to selected literature on the subject.

Prior to 1960, the geographic research on American Indians was generally limited to the work of Carl Sauer (1936, 1944, 1954) and his students (Brand 1935; Sauer and Brand 1930; Hewes 1942a, b, c; Rostlund 1951; Aschmann 1970, et al.) as well as Jonas W. Hoover (1929a, b; 1930, 1931a, b). The geographers from Berkeley, working in close cooperation with Berkeley anthropologists (Kroeber 1939, etc.), concentrated exclusively on historical and cultural questions,[2] while Hoover (see 1929b, 1931a, etc.), at the Arizona State Teachers College in Tempe, was already preoccupied with the structure of specific reservations and their residents. Hoover's work gives scant information about prevailing socioeconomic conditions, however, and his approach is very largely descriptive.

Beginning in the mid-1960s there was an increase in the number of geographers who concentrated on American Indians over a longer period of

1. Besides the standard geographical journals, the following yearbooks and periodicals in English were examined: Antipode, Australian Geographer, Bulletin of the Illinois Geographical Society, Canadian Geographer, Focus, Geographical Analysis, Geographic Perspectives, Great Plains Rocky Mountain Journal, Journal of Historical Geography, Institute of British Geographers, Journal of Geography, Journal of Cultural Geography, Landscape, Papers and Proceedings of Applied Geography Conferences, Geographical Journal, Geographical Survey, Professional Geographer, and the yearbooks of the California, Eastern Lakes, Kansas, Mississippi, Oregon, Pacific Coast, Southeastern and Virginia Geographers.

2. Most of these studies, however, were carried out in northern Mexico or in Central or South America.

time, instead of limiting themselves to a dissertation or one or two papers, a trend which resulted in the founding of the Native American Specialty Group within the Association of American Geographers in 1977. Yet here, too, historical topics remained in the foreground in the work of Stephen Jett (1977, 1978, 1981) and Elliot McIntire (1967b, 1971, etc.), whose research on the Navajos and the Hopis emphasized the traditional forms of tribal homes and settlements as well as the beginnings of their agriculture. William Denevan (1976) sought to calculate the extent of the Native population in prehistoric times and David Wishart (1976) examined the Indian fur trade. Wishart (1979, 1985) also studied the loss of land suffered by the Pawnee Indians in Nebraska as well as their living conditions during the nineteenth century. Donald Ballas (1960, 1973a, b, 1985) gave his attention to the early years on the Eastern Cherokee Indian Reservation in North Carolina and the Teton Dakota in South Dakota.

Few geographers, however, have concentrated on the present-day situation of American Indians, a surprising fact in view of the number of geographically significant questions involved. Some of those who have, Harvey Flad (1972), Bryan Higgins (1982), Tony Lazewski (1976, 1982) and Elaine Neils (1971), for example, have done research exclusively on American Indians living in large cities and Indian migration to urban areas. Compared to this urban research, relatively few studies have focused on the present-day structure of the Indian reservations. These studies include James Goodman (1982a, 1982b, etc.), Goodman and G. Thompson (1975), Jesse McKee (1971, 1987), McKee and Steve Murray (1986), Imre Sutton (1967, 1975, 1985), and the works of Donald Ballas (1960, 1973, 1985), who also considered current issues. Goodman described the Navajo Indian Reservation in an atlas (1982a) and related research (1982b; Goodman and Thompson 1975) while McKee (1971, 1987; McKee and Murray 1986) concentrated on the Choctaw of Mississippi. Imre Sutton has published two books (1975, 1985) and several articles (1967, 1976, etc.), from a broader perspective which examines questions of territory and sovereignty of Indian lands across the country.[3]

The marginal role that U.S. geographers have given to American Indians as the "First Americans" can also be seen in the textbooks on human geography and regional geographies of the United States. As a rule American Indians in these texts are dealt with as if they had only a past, with no present and no future. This applies to J. Wreford Watson (1979, 20–48), Paul Knox et al. (1988, 51–53), and also to Wilbur Zelinsky (1973, 14–17). The

3. Almost half of the geographers mentioned here are either directly or indirectly associated with Carl Sauer, both in their research on the American Indians and in the historical and cultural direction of their studies. Denevan studied in Berkeley; Ballas and Wishart were students of Hewes, and Homer Aschmann was Sutton's teacher.

British geographer John H. Paterson (1984, 44–45), in his much read regional geography of North America, devoted only one and a half pages of a very general character to the indigenous population of the continent, which does not come as a surprise, in view of the lack of geographical research on the present conditions of American Indians.

There is understandably even less geographical literature in German concerning American Indians. Except for the inevitably brief accounts in the various regional geographies,[4] only Dietrich Fliedner (1974, 1975, etc.), Ingo Mose (1990), Rainer Vollmar (1982) and, above all, Burkhard Hofmeister (1975, 1978, 1979, 1980, etc.) have done studies in this field. Of all these authors only Fliedner (1974) and Hofmeister (1975) have carried out extensive fieldwork. With only a few exceptions, such as the Finnish geographer Eino Siuruainen (1979, 1980), hardly any work has been done on American Indians in other European countries.

Geographers who seek more extensive information about American Indians must therefore turn to the relevant specialist literature of historians, anthropologists, sociologists, and economists.[5] In sifting through this very extensive material, one is struck by the fact that there is a conspicuous lack of literature linked to the present day, and that socioeconomic questions are very largely neglected. This is probably to be explained by the apparent division of labor in studies regarding racial and ethnic minorities in the United States,[6] with sociology and the socioeconomic subdisciplines of regional science and economics on the one hand, and the research of anthropologists on the other. A careful investigation shows that remarkably few sociologists, economists, and regional scientists concern themselves with American Indians. This small group of scholars includes Americans Howard Bahr, Bruce Chadwick, and Robert Day (1972), Kent Gilbreath

4. See Blume (1979, 273–277; and 1987, 136–139), Boesch (1973, 27–28), Dietrich (1933, 50), Friese and Hofmeister (1983, 110–111), Hahn (1990, 339–344), Hofmeister (1988, 83–97), Kramm and Brunner (1977, 67–68), Pohl and Zepp (1975, 21–22 and 119–120), Schmieder (1963, 37–41, 226–228, 340–341 and 352–359).

5. Further information may also be obtained from articles by journalists (see Biegert 1983a and 1983b) and members of the numerous organizations that support the cause of American Indians (see Schwarzbauer 1986 and 1989). In the German-speaking world alone there are more than a dozen such organizations.

6. To support this statement a number of relevant monographs and anthologies (see Bibliography) as well as the following English- and German-language journals were examined: American Journal of Public Health; American Sociological Review; Angewandte Sozialforschung; Annals of the Academy of Political and Social Science; British Journal of Sociology; Environment; Ethnic and Racial Studies; Ethnicity; Growth and Change; Journal of Economic Issues; Journal of Ethnic Studies; Kölner Zeitschrift für Soziologie; Österreichische Zeitschrift für Soziologie; Plateau; Schweizer Zeitschrift für Soziologie; Social Forces; Social Research; Social Science Journal; Soziale Welt; SWS-Rundschau. Of all these periodicals, the Journal of Ethnic Studies, an interdisciplinary journal, seems to be the most rewarding.

(1973) and Alan Sorkin (1973), as well as the German sociologist René König (1980), who has written about the Navajo Indians.

There are many anthropologists, on the other hand, whose studies are mainly, though not exclusively, concerned with the cultural traditions of American Indians, as Hofmeister has remarked (see 1975, 209),[7] and the number of historians who have studied the history of Indian-white relations is only slightly lower (see Prucha 1977, 1982). Anyone interested in American Indian affairs will find a plentiful source of information and a valuable stimulus to further thought in many of these studies.[8] This comprehensive literature cannot, however, obscure the fact that research on the demographic and socioeconomic situation of present-day Indians is only beginning. Geographers can also make their contribution by filling some of these gaps in research.

THE TOPIC SELECTION AND THE PROBLEM OF COLLECTING DATA ON INDIAN RESERVATIONS

In this study, I have attempted to provide a comprehensive picture of certain aspects of Indian reservations in the United States. For this purpose, I collected official tribal and federal agency data as well as archival and statistical material in Washington, DC, and at a number of other locations. Combined with numerous interviews and informative conversations, this material served as a basis for a general survey of the subject. Such a general description undoubtedly gains in value when it is supported with detailed studies of particular regions. Because of the size of the country and the limited length of time at my disposal, however, it was not possible to carry out such detailed studies in a number of different states at the same

7. In addition to relevant monographs, anthologies and the publication series by university departments and museums (see the bibliography), the following journals were examined: American Antiquity; American Anthropologist; American Indian Culture and Research Journal; American Indian Quarterly; Current Anthropology; Ethnohistory; Ethnicity; Human Organization; Kiva (Arizona Archaeological and Historical Society); Southwestern Journal of Anthropology; European Review of Native American Studies (interdisciplinary); Review of Ethnology (Vienna); Wiener Ethnohistorische Blätter; Anthropos; EAZ; Tribus; Ethnologische Zeitschrift (Zürich). Among these publications, Human Organization (The Journal of the Society for Applied Anthropology) with its cultural and anthropological orientation deserves particular attention. This journal has numerous articles about the present-day American Indians, and is particularly rewarding for the attention it gives to socioeconomic questions.

8. Among the authors of these studies only the historians Francis P. Prucha and Wilcomb Washburn as well as the ethnographers and anthropologists Henry Dobyns, Christian Feest, Wolfgang Lindig, Joseph Jorgensen and Edward Spicer are mentioned here. Their publications encompass a very broad spectrum of research concerning American Indians. For additional literature see the bibliography.

time. Most of the fieldwork was therefore concentrated on the Indian reservations in Arizona.

I chose Arizona, among other reasons, because reservation Indians play such an important role in the state, owing to their large numbers and the extensive land area they control. One-third of all reservation Indians in the United States live in Arizona spread over twenty-one reservations, which comprise approximately 37% of the total U.S. reservation land. This state, along with only a few other U.S. regions, plays a special role regarding ethnic diversity and the Indian sense of culture and tradition.

Even in Arizona, however, the size of the state and organizational restrictions make a fully comprehensive study impossible. It was therefore necessary, after an exploratory journey to all the reservations in Arizona and adjacent areas in Colorado, California, New Mexico, and Utah, to select a smaller number of tribes to be investigated.

The original idea was to pick out regional examples that would reflect the heterogeneous character of the Indian reservations of Arizona. It followed, therefore, that I should consider such issues as the traditional foundation of the way of life of the respective tribe, the territorial and socioeconomic development of the reservation in question, the natural resources available, and also the geographical location. An additional objective was to study reservations which previous research had generally ignored. This explains why I did not select the Navajo and Hopi Indian Reservations for regional fieldwork, since these reservations, as well as the Pine Ridge and Rosebud Indian Reservations in South Dakota, are among those which have been studied most frequently (see Correll et al. 1969, 1973; Iverson 1976; Laird 1977; Marken and Hoover 1980).[9]

Along with these considerations, other factors played a role in the choice of the reservations to be investigated, factors over which the individual researcher has little control. Since Vine Deloria's article entitled "Anthropologists and Other Friends" was published, in which the author, a Sioux and champion of all reservation Indians, attacked the profession of anthropology (see Deloria 1972, 83–104), non-Indian researchers have gradually begun to understand that American Indians who are the subject of their research are increasingly disinclined to be used for this purpose or to tolerate the misuse of the reservations as "living laboratories" (Feest 1976, 335). I also encountered this reserved attitude toward researchers when in the spring of 1983—without having read Deloria's article—I visited a number of Pueblos in the Southwest of the United States and had preliminary,

9. Laird's Hopi bibliography alone includes 2,935 titles, yet by far the largest part of this literature deals with archaeological, historical, and anthropological questions.

exploratory talks with tribal leaders.[10] This experience made it clear, first of all, that fieldwork would meet with strong opposition among the Pueblo Indians which would be difficult to overcome, and secondly, that without advance notice in writing and an explanation of one's purposes, it would not be advisable or productive to begin interviews in Indian country or to be seen doing fieldwork. Today it is necessary to make a request in writing in order to obtain permission for such work and to have the official authorization of the tribal council, which usually takes a long time and is often refused. This may partly explain why geographers have written so little about present-day conditions on Indian reservations.

Before making the final selection of the regions to be examined, I addressed inquiries to all tribal governments and Bureau of Indian Affairs (BIA) agencies in the Southwest, and followed up by talking with the responsible authorities. On three reservations—the Payson, San Carlos, and the Papago (which has since changed its name to Tohono O'odham, "Country People")—inquiries were strictly forbidden. On several other reservations restrictions were imposed, often after time-consuming interviews with the tribal chairman or members of the tribal council. These restrictions prevented access to the tribal enrollment lists and to the registers of land use and livestock associations.

On all of the reservations, I found that informal conversations and participant observation were the most productive ways to collect data. The questions for such conversations were not rigidly determined, and I made no attempt to obtain well-defined answers of the kind that is demanded in clear-cut standardized interviews (see König 1974, 144). In an effort to avoid understandable mistrust on the part of reservation Indians, I did not take notes during the interviews, but waited until later to record my memories of them. While it is true, of course, that such information cannot be used for a mathematical, statistical evaluation or for rigorous scientific assessments, more precise data are not available, and the fact remains that this kind of qualitative research is invaluable in helping the outsider to see the reservation from the viewpoint of local residents.

In the end, I decided to concentrate the fieldwork on three very different reservations, also taking into account case studies on other reservations for particular issues. These three reservations fulfill the requirement of heterogeneity regarding their geographical locations, territorial development, and the nature of their environment and resources as well as the prevailing socioeconomic conditions, as previously postulated. The reservations selected are the Gila River Indian Reservation, ideally located for

10. Among the reservations I visited at that time were the Hopi Indian Reservation in Arizona and the San Ildefonso, San Juan, Laguna, and Zuni Indian Reservations in New Mexico.

development because it borders on the metropolitan area of Phoenix; the Hualapai Indian Reservation, poor in natural resources; and the Fort Apache Indian Reservation, which has an abundance of timber and grazing land, and access to water. The latter two reservations are located in isolated areas of Arizona. These three reservations, along with other specific case studies, provide a representative cross-section of the reservations in Arizona, and their study makes it possible to obtain a generally reliable impression of Indian reservations across the United States, although one must be wary of generalizations in view of the individual character of each tribe and its territory.

PURPOSE AND STRUCTURE OF THE PRESENT STUDY

At first glance the cultural landscape of the United States appears to the European observer to be strikingly uniform, which also seems to be reflected in the behavior of the country's inhabitants. Behind this apparent cultural and social melting pot one suspects there exists a "standardized" civilization, whose social and economic mechanisms, apparently uninfluenced by the diversity of the landscape or the varying ancestry of the inhabitants, have resulted in urban and rural development patterns that look very much alike throughout the country.

Upon closer examination, however, one can find regions in the United States which contrast sharply with this apparently standardized American landscape, such as the Spanish-influenced borderlands of the Southwest (Bolton 1921). There are also cultural islands with distinct regional ways of life and unique features in the landscape that might not be easily perceptible to the outsider, even though these unique cultural regions are in many respects strikingly independent. To understand and describe these enclaves, to discover their territorial and cultural development throughout history and to identify the "inner forces" which lay the foundations for their social and economic structures would appear to be of topical interest to a geographer with a pluralistic attitude toward the subject.

The Indian reservations of the United States are, in fact, excellent examples of such regional and cultural entities. At first glance most of these reservations do not appear to differ very significantly from the surrounding non-Indian environment as far as the cultural landscape is concerned. American Indians have generally been able to preserve their cultural identity in the midst of the Anglo-American world down to the present day, however, despite the fact that from their first contact with American Indians, white American society has constantly subjected them to pressure to assimilate, pressure that has often been very strong. It is the purpose of this study to call attention to certain aspects of this identity, depicting American Indian people and their lands in both general terms and with reference to particular reservations.

The study opens with a brief survey of the essential phases of U.S. Indian policy, beginning with the concept of Indian Territory and continuing to the present federal policy of self-determination. First I shall discuss the rapid reduction of the Indian country and the brief periods in which some tribes were able to regain some tracts of land, in accordance with the respective federal policies over time. I shall also examine the present-day distribution of Indian lands, showing the extent of these territorial adjustments by means of detailed maps, based on the annual reports of the Commissioner of Indian Affairs.

Chapter 3 explains the different procedures by which reservations have been established, and the present-day conditions of landownership on reservations, illustrated by specific examples. By means of two ideal-typical diagrams, I will summarize the remains of traditional sovereignty rights and the territorial development of Indian reservations.

Chapter 4 describes the fluctuation and the distribution of the Indian population in the United States as well as its basic demographic structure, and includes a critical analysis of census data. The diversity of ethnic elements in the reservation population is analyzed in detail using the Hualapai tribe as an example. In chapter 5, I will discuss several aspects of the standard of living, the means of livelihood, and the employment and educational status of the Indians, and compare them with those of other racial and ethnic minorities in the United States. The discussion of these subjects is based for the most part on data found in the available census bureau or BIA statistics, because on the reservations in question it was not possible to carry out the detailed inquiries that would be required to produce representative information.

The main part of this study has to do with the reservation economy, and following a discussion in chapter 6 of the sociocultural framework on reservations which is generally ignored in the economic policies of the federal government, and an outline of other structural economic conditions, the principal topics treated are mining (chapter 7), the problem of "Indian" water rights (chapter 8), and agriculture and forestry on the reservations (chapter 9). The historical development both of the relevant laws and the use of natural resources is always a component of the background for this analysis of present-day conditions on the reservations. Chapter 10 discusses industrial parks on the reservations and the private service sector, and considers the question of the conditions of ownership. Finally, chapter 11 includes more recent data on American Indians. There, I describe new issues in Indian politics as well as the state of the reservation economy, before presenting my conclusions.

All these topics are of importance in the complex of issues regarding American Indians. While not all aspects can be fully dealt with here, they present myriad topics for further studies in human geography, as the American Indian reservations represent a fascinating and unique subject of inquiry.

2 Two Hundred Years of Indian Policy in the United States

THE COLONIAL HERITAGE AND THE BEGINNINGS OF U.S. INDIAN POLICY

Today a considerable number of American Indians live on reservations, separated from the rest of the U.S. population by different systems of law and government, and by their own sense of identity. The course toward a policy of separation was already set during the colonial period. From the mid-seventeenth century on, certain compensations along with small reservations in New England and in Virginia were guaranteed to the Indians by treaty in exchange for title to large tracts of their lands (see Feest 1976, 36; Lurie 1968, 31; Washburn 1975a, quoted in Hofmeister 1980, 69). While the colonial powers of Great Britain and France were still fighting each other for dominance in North America, powerful tribes such as the Iroquois, the Cherokee, and the Creek were sometimes able to play the two opponents against each other and hold individual colonies in check. From the defeat of the French in 1763 until the beginning of the American War of Independence, the British were free to pursue a more far-reaching Indian policy in their colonies. The watershed of the Appalachian Mountains was set as the boundary line between the territory for white settlements and that of American Indians in the Proclamation Act (1763) (see Deloria and Lytle 1983, 59; McNickle 1973, 43; Prucha 1962, 13–20). White colonists were no longer permitted to settle beyond this boundary, and those who had already established themselves there were asked to leave. Thus the concept of a separate territory for Indians began to develop, a concept which the Americans were to take up again in the nineteenth century.

Even at that time, however, it was clear to British legislators that because of the pressures of population and colonization these boundary regulations would be valid for only a short time and that there were not enough local officials and military to check and enforce them. This skeptical attitude was already evident in the wording of the agreement, which made it clear that the British Crown had an interest in containing the activity of colonists in the hinterland. It was further stipulated in the Proclamation Act that no one would be permitted to acquire land either directly or through the concession of a colony. In this way Britain's responsibility for Indian affairs was clearly expressed, an attitude later taken over by the United States federal government which still prevails today.

When the Americans gained independence at the close of the eighteenth century, they took over the fundamental principles of the Indian policy from their former colonial masters, the British. During the first few years following the revolution, for example, representatives of the newly founded U.S. government entered into negotiations with numerous tribes in border areas to the north and west of the white population's sphere of influence as the British had done before them. Peace treaties with American Indians were regarded as an appropriate means for the acquisition of large tracts of land (see Hilliard 1971, 495). At first the respective tribes were granted a legal status similar to that of sovereign nations. Some Indian tribes, such as the Delaware, who originally lived on the Northeast coast, or the Cherokee, whose home was in the Southeast (fig. 2.1), still had the option at this time to decide for themselves whether they wanted to join the Union as a separate state or not (Joyner 1978, 32).[1]

As long as the Americans were still at war with the British, the founding fathers had no interest in useless skirmishes for land with Indian tribes when this land could be acquired through purchase at less expense,[2] so essentially the Proclamation Act was maintained. The old boundary line was preserved for the time being and trade with the Indians was permitted only with a federal license.[3]

After vast territorial gains were made through the Louisiana Purchase in 1803 and the war came to an end in 1814, some states were even less inclined than before to recognize the Old Indian Territory beyond the Appalachian boundary. In 1804 President Jefferson had already suggested offering American Indians east of the Mississippi a land exchange, a proposal, however, which the Congress did not accept (Tyler 1973, 54). By that time Jefferson had realized that the federal government was powerless to protect the Indians in that region from the loss of their territory through seizure by individual states intent on a policy of aggressive expansion. It was also obvious that such a policy would lead to treaty violations, and furthermore, that many people could not understand why the

1. Until the 1830s both parties in Congress accepted in principle the possibility that individual Indian tribes might establish their own states (see Barsh and Henderson 1980, 210).

2. See George Washington to James Duane (1783), quoted in Prucha 1975, 2.

3. Trade and Intercourse Act (1790), quoted in Prucha 1975, 15.

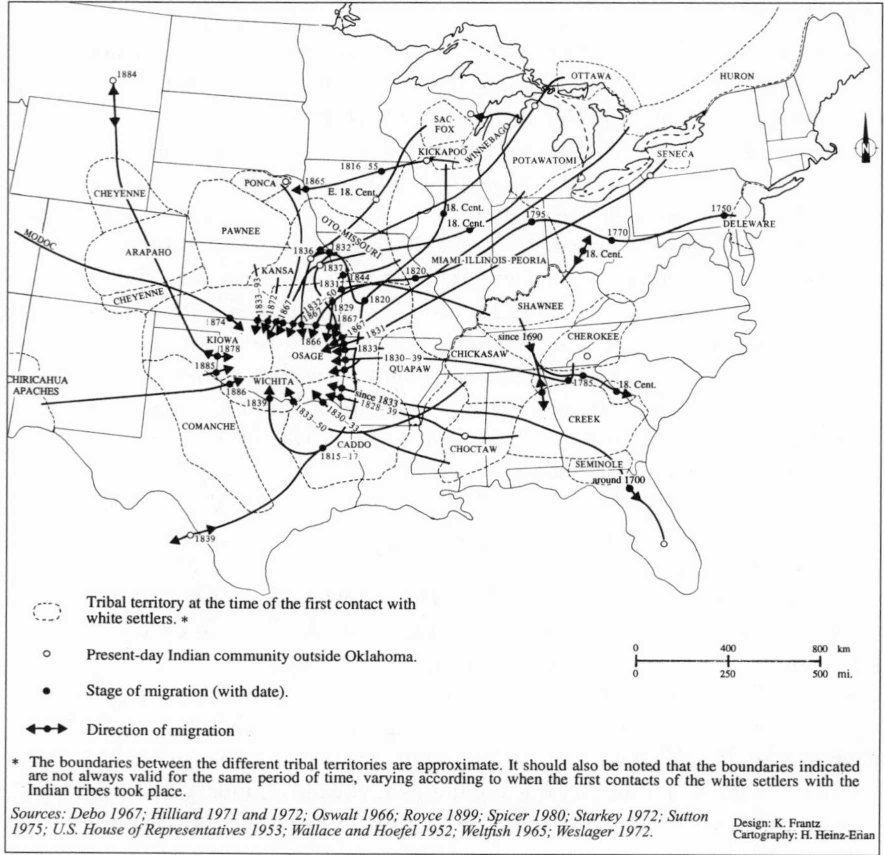

Fig. 2.1: Forced resettlement of Indian tribes in the Indian Territory during the nineteenth century

young republic should tolerate the existence of largely independent Indian nations which, though under the jurisdiction of the federal government, were not subject to the state in which they were located. Both the controversial demarcation line and the claim to sovereignty of many Indian tribes in regions adjacent to white settlements led to constant tensions, which, in the case of the Cherokee, escalated to the point of litigation between the tribe and the State of Georgia (*Cherokee Nation vs. State of Georgia*, 1831, 16–17).

Hard-pressed by colonists in Georgia, who wanted to expropriate even the Indians' remaining land that was guaranteed by treaty, the Cherokee set up a constitution, following the American example, in which they repeatedly stressed their status as a sovereign nation. They also decided not to sell any more land. The State of Georgia simply ignored this claim to sovereignty, and, in a countermove, passed a series of laws which declared the constitution of the Cherokee invalid and divided up their territory into separate counties (Andress and Falkowski 1980, 98–99). As a "foreign nation" the Cherokee appealed for help to the United States as a protecting power and asked the U.S. Supreme Court to clarify the legal situation. In a verdict that has since become famous[4] the Supreme Court acknowledged that the Cherokee and other similar tribes had all the attributes of a foreign nation, adding, however, that the United States could not tolerate any "imperium in imperio." Thus a new legal category was created for tribes as "domestic dependent nations" alongside the sovereign federal state and the individual states. Furthermore, the court expressed the view that the Cherokee "occupy a territory to which we assert a title independent of their will, which must take effect in point of possession, when their right of possession ceases" (*Cherokee Nation vs. State of Georgia* 1831, 17).

It goes without saying that the Cherokee could not be satisfied with this judgment, for it paved the way for the speedy expropriation of more land in the future.

Immediately after the end of the War of 1812 against the British, Jefferson's idea of a land exchange was again revived. The pressure from white settlers in Eastern and Southern states increased and it was only in a halfhearted manner that the federal government fulfilled its obligations to protect the Indians in these regions. A Permanent Indian Territory in the newly acquired territory between the Mississippi and the Rocky Mountains now appeared to be the ideal solution to the Indian problem in the East.

It was argued that the only way to ensure lasting protection of American Indians was to push the Indian boundary farther west once again, which meant that the treaties were violated. In this apparently deserted region,[5] then still known as the Great American Desert and considered an uninhabitable region for white settlers (see Brown 1948, 370–372; Bowden 1969; Hofmeister 1988, 117–118), it was hoped that the newly arrived tribes would give up their "nomadic" way of life and that they would integrate themselves into Western civilization with governmental assistance. Schools, churches, and the necessary farming equipment were to be put at their disposal in order to help them.[6]

4. See *Cherokee Nation vs. Georgia* (1831), quoted in Prucha 1975, 58–59.

5. Indeed, several tribes such as the Kansa, Osage, and Quapaw had their ancestral homes in the new Indian Territory, and their agreement to land cessions had not been obtained (Hilliard 1971, 496).

6. Message of President Monroe on Indian Removal (1835), quoted in Prucha 1975, 39 and 71–72.

These "noble" intentions on the part of the government cannot obscure the fact that whites were really only interested in appropriating Indian lands and natural resources for their own benefit. The epithets "nomad" and "uncivilized" certainly did not apply to the "Five Civilized Tribes" in the Southeast of the United States,[7] sedentary farmers with a high level of civilization (see Lindig and Münzel 1985, 132–140).

The first removal took place in the 1820s, and at first involved only relatively small tribes from the East coast (Deloria and Lytle 1983, 64). Between 1830, the year the Indian Removal Act was drafted, and the period shortly after the Civil War, new territory was allocated to nearly fifty tribes (Hilliard 1971, 496; see fig. 2.1). Under the supervision of the U.S. Army many tribes were compelled to exchange their ancestral homelands partly located in a humid, subtropical climate for new homes in a semiarid land west of the Mississippi, and the conditions of this forced migration alone claimed many deaths.[8] A particularly bizarre case of this removal policy is illustrated by the case of the Kickapoos, Algonquian-speaking forest Indians of the Great Lakes Territory, some of whom ended up as far afield as El Nacimiento, a little village in the Mexican state of Coahuila, eighty miles from the U.S.-Mexican border (Latorre and Latorre 1976).

Five years after the Indian Removal Act, President Jackson stated with satisfaction that with only a few exceptions, the eastern half of the United States between Lake Michigan and Florida would soon be cleared of Indians.[9] Opposition to this policy was successful only in very exceptional cases, including a few splinter groups of the Seminole and Cherokee who managed to hold out on their ancestral land. A few tribes in the Northeast and the Middle West were also successful in their delaying tactics (see fig. 2.2).

Many of the relocated tribes were soon compelled to realize that the land which had been granted to them and their descendants forever—the text of the treaty contained the words "as long as the grass grows"—was again being coveted by non-Indians and becoming a controversial issue. After only a few years Kansas and Nebraska were excluded from the new Indian Territory, and the Indians of Oklahoma were obliged to make room for their companions in misfortune, who were now forced upon new lands once again. Today, 160 years after the beginning of the removal policy, there is

7. The five tribes in question are the Cherokees, Creeks, Choctaws, Chickasaws, and Seminoles, the vast majority of whom were later forcibly settled west of the Mississippi (see fig. 2.1).

8. Thus, for example, thousands of Cherokee died during the Trail of Tears, the march to their new home in Oklahoma in the 1830s.

9. President Jackson on Indian Removal (1835), quoted in Prucha 1975, 71.

only one reservation left in Oklahoma, a territory which was once reserved for American Indians (see fig. 2.3).[10]

By the middle of the nineteenth century it had became clear that the concept of separate territory for American Indians, set apart from the white population's sphere, could no longer be maintained. The "frontier" continued to advance farther west, and the indigenous people of the United States, with their traditional ways of life, were obstacles to this driving expansion. Therefore, beginning in 1851, reservations were established on the ancestral lands of the Indians who had not yet been pacified (see fig. 2.4). These reservations were put under the supervision and protection of the Bureau of Indian Affairs (BIA) and the military. In the period between 1851 and 1856, after either peaceful or military measures, as many as fifty-two treaties (Schmeckbier 1972, 44) were negotiated, leading to the creation of many reservations (see fig. 2.2).

During the period from 1851 to 1887 the U.S. Army built fifty-eight military posts in what is now Arizona (see fig. 2.5). Their purpose was to pacify certain Indian tribes and to protect the white settlers. These military posts, most of which were located on the tribal lands of the nonsedentary Apache, Navajo, and Pai peoples, were often established either in the center of Indian country or at the fringe of expanding settlement areas to protect white settlers. Others were built along important traffic routes and streams or in the vicinity of strategic passes to prevent Indian raids or to deny access to water for hostile Indians (see Comeaux 1981, 125). Except for the fort and camps on the Colorado, Santa Cruz, and Verde Rivers, most of the military posts were in mountainous areas. With the single exception of Fort Huachuca, located in southeastern Arizona, all these posts had lost their military function by the beginning of the twentieth century. Fort Apache, for example, is now used as a BIA boarding school (see fig. 2.6).

Within one generation the Indian tribes west of the Mississippi lost the last bit of land in which they could move around in freedom and tribal members were confined to reservations under the control of the BIA and the military. This marked the end of their previous status as sovereign nations, at least in name, and established tribal "trust" relations in which the federal government became the manager in control of tribal lands and resources.

THE SEARCH FOR A FINAL SOLUTION

The last quarter of the nineteenth century marked the beginning of a period of Indian policy which, from the viewpoint of American Indians, certainly

10. In Oklahoma, which means "red people" *(humma okla)* in the Choctaw language, only the Osage Indian Reservation remains today.

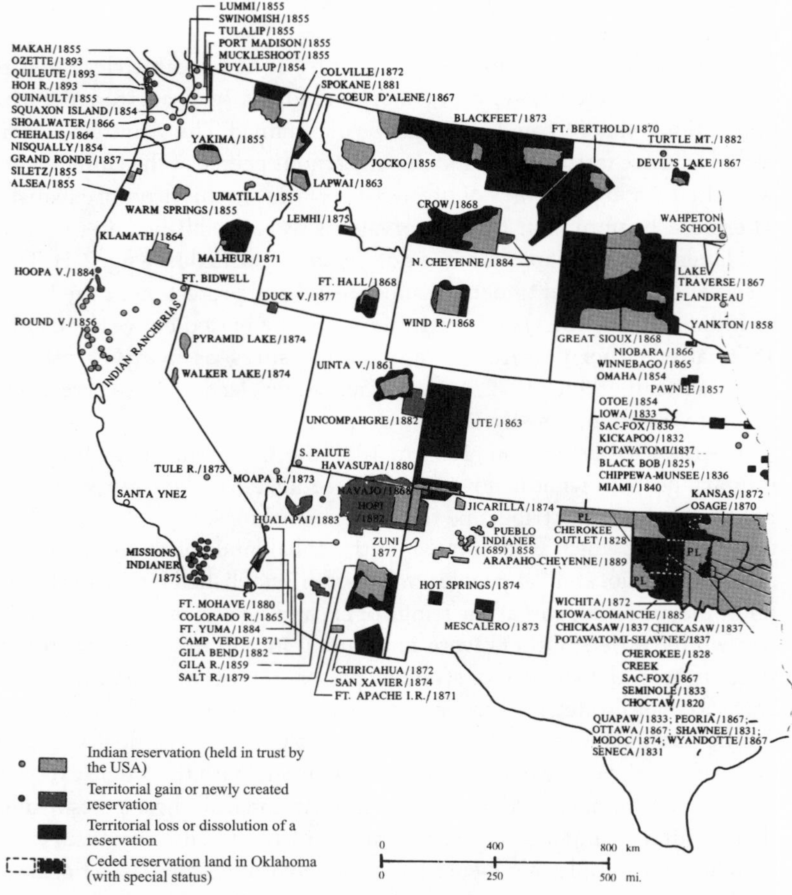

Sources: Data found in the U.S. Department of the Interior (BIA), the Library of Congress, and the National Archives in Washington, D.C.; ARCIA 1874-1901; Hilliard 1972; Pipenstem and Rice 1978; Royce 1899.

Fig. 2.2: Territorial gains and losses of Indian reservations in the United States, 1875–1902

brought the most painful encroachments on their culture and way of life. The time when the United States had to consider the Indian tribes as a threat to its territorial expansions and its settlement policies was over, and all the tribes had been defeated and forcibly settled on reservations. Warfare and epidemics had decimated and weakened Indian populations, and within a short period of time formerly independent peoples had been degraded to such an extent that they became dependent on the goodwill of their former enemies. Their "white brothers" no longer had any reason to negotiate formal treaties with the Indians as though they were still sovereign nations. In 1871 therefore,

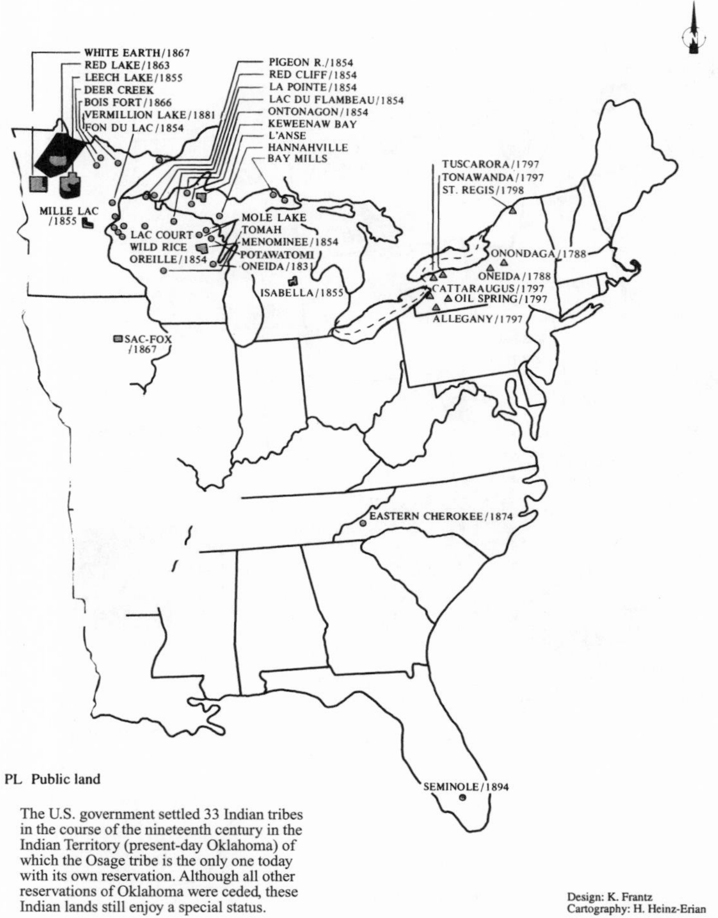

The U.S. government settled 33 Indian tribes in the course of the nineteenth century in the Indian Territory (present-day Oklahoma) of which the Osage tribe is the only one today with its own reservation. Although all other reservations of Oklahoma were ceded, these Indian lands still enjoy a special status.

after 370 treaties between the United States and individual Indian tribes had been signed, Congress abruptly put an end to this diplomatic procedure. The U.S. government no longer regarded the Indians as negotiating partners with equal rights but as wards of the BIA.

An era now began in which the American Indian was deprived of political, economic, and cultural autonomy and subjected to very strong pressures to assimilate. "Kill the Indian in him and save the man"[11] was the slogan, and any means to this end seemed acceptable.

11. Captain R. H. Pratt, first superintendent of the Carlisle Indian Boarding School in Pennsylvania run by the BIA, quoted in Wilkinson and Biggs 1977, 143.

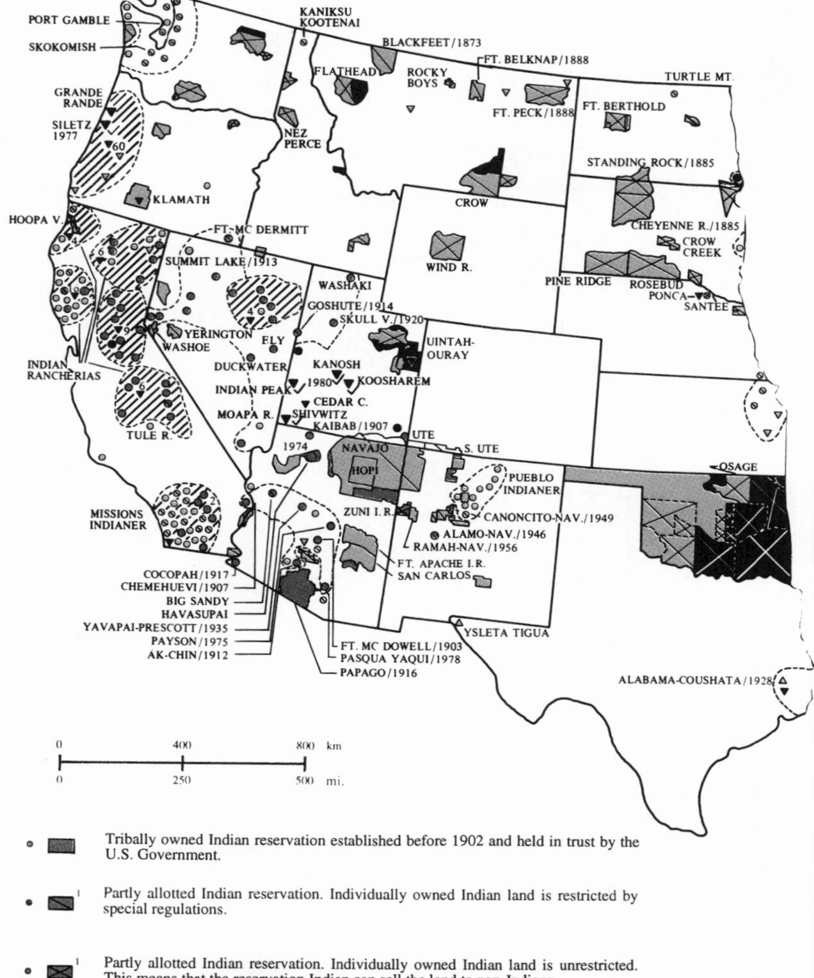

Tribally owned Indian reservation established before 1902 and held in trust by the U.S. Government.

Partly allotted Indian reservation. Individually owned Indian land is restricted by special regulations.[1]

Partly allotted Indian reservation. Individually owned Indian land is unrestricted. This means that the reservation Indian can sell the land to non-Indians.[1]

▲ State reservation

▼ Group federally not recognized.

Sources: Data gathered at the U.S. Department of the Interior (Bia), the Library of Congress, and the National Archives in Washington, D.C.

Fig. 2.3: Territorial gains and losses of Indian reservations in the United States since 1902

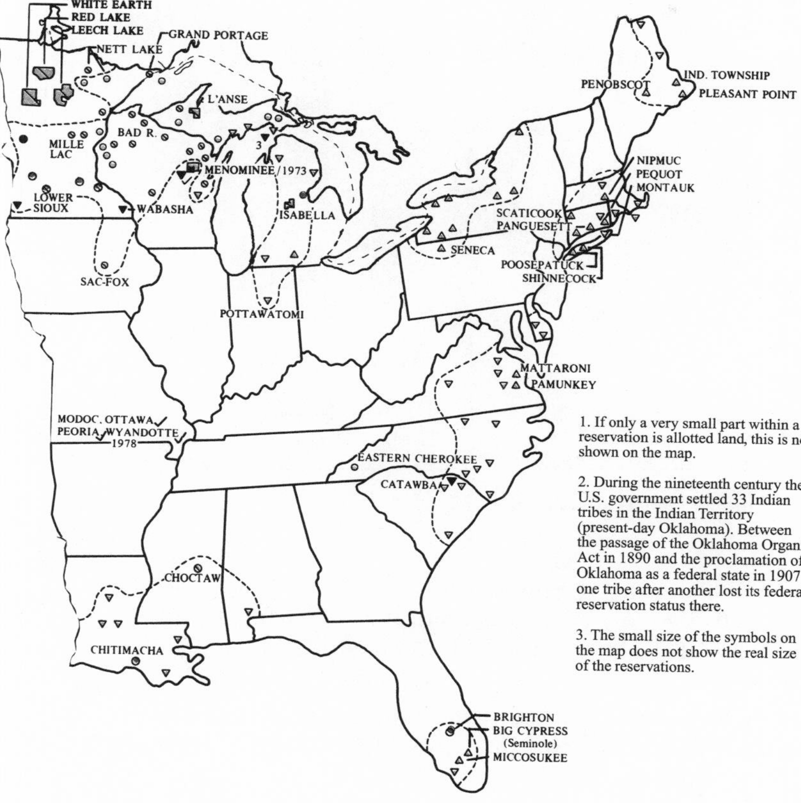

1. If only a very small part within a reservation is allotted land, this is not shown on the map.

2. During the nineteenth century the U.S. government settled 33 Indian tribes in the Indian Territory (present-day Oklahoma). Between the passage of the Oklahoma Organic Act in 1890 and the proclamation of Oklahoma as a federal state in 1907 one tribe after another lost its federal reservation status there.

3. The small size of the symbols on the map does not show the real size of the reservations.

Territory with numerous little Indian reservations, rancherias, or missions. [3]

Territorial gain or newly created reservation.

Territorial loss or closing of a reservation. (Termination policy not shown here).

Ceded reservation land in Oklahoma (with special status). [2]

Reservation status abrogated by Eisenhower's termination policy.

Termination of reservation status was reversed.

Region with numerous small Indian reservations, rancherias, and missions, some still in existence, some terminated. The figure inserted gives the number of terminated reservations. [3]

Design: K. Frantz
Cartography: H. Heinz-Erian

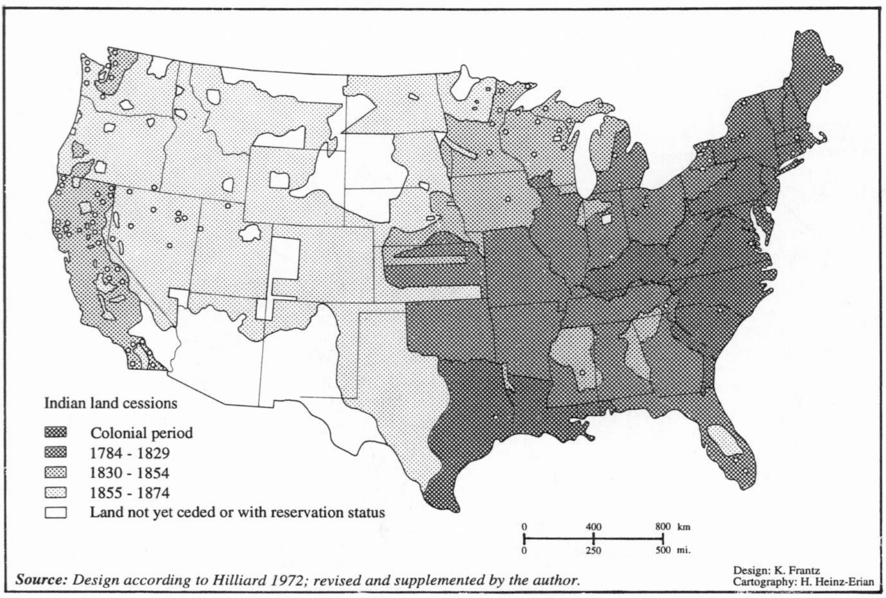

Source: *Design according to Hilliard 1972; revised and supplemented by the author.*

Fig. 2.4: Indian land cessions up to 1875

The reservations at this time were like prison camps which American Indians could leave only with the authorization of the BIA agent responsible (see fig. 2.7). The military strictly forbade them to leave the reservation, although this would have been necessary for Indian tribal members to maintain their traditional subsistence economy. Thus tribal members were made dependent on the food rations that were distributed at the military posts at irregular intervals.

If these regulations were not followed the rations could be withheld at any time. Should parents refuse to send their children to boarding schools hundreds of miles from their homes, where they would be subjected to a systematic reeducation program for several years, for example, they could expect to have their rations stopped.[12] The task of educating the students in these schools was less important than the task of assimilating them, the aim being to deprive them of their cultural identity.[13] The students, upon entering the school, were not only given military haircuts and uniforms

12. Appropriation Act (1893), quoted in Cohen 1982, 14.

13. Besides the 52 missionary schools already in existence—34 of which were Catholic (ARCIA 1901, 26–28)—the BIA established boarding schools throughout the country after 1879. Some of these schools were in or near large cities, while others were located on reservations (see fig. 5.13). In 1900, which marked the heyday of this BIA school policy, there were 25 off-reservation boarding schools and 88 on-reservation boarding schools (ARCIA 1901, 20–23). It was only after the Second World War that the number of these schools rapidly de-

(see figs. 2.8, 2.9), but they were also given Anglo-American names and were assigned to a Christian religious denomination.[14] It was further decreed that for all Indian school children "barbarous dialects" had to give way to English (ARCIA 1887, 19–21). Children who did not comply with this regulation were imprisoned—a simple matter, as many of the school buildings had formerly been military barracks.[15]

In 1870, when it appeared that neither the military nor the BIA had had much success in their attempts to "civilize" American Indians, President Grant handed the job of assimilating the Indians over to the broad spectrum of Christian denominations in the United States.[16] In this policy the constitutional principle of the separation of church and state was ignored,[17] and all the reservations were divided up and assigned to the different religious denominations. The pattern of the distribution of tribes to church groups shows the relative influence of the respective denominations (see fig. 2.10).

The lion's share of the Indian agencies, and the Indian population, was entrusted to the Quakers (Orthodox and Hickside Friends), Methodists, Presbyterians, Baptists, and Episcopalians. The distribution of denominations varied markedly from region to region (see fig. 2.10). The Quakers claimed tribes in what is now Oklahoma, Kansas, and Nebraska while the Episcopalians were assigned large areas in North and South Dakota. Most of the Northwest tribes were assigned to the Methodists and those in Nevada to the Baptists, while the Presbyterians and the Reformed Dutch Church shared the Southwest. In view of their large number of schools, it is striking how poorly the Catholics fared in the number of reservations entrusted to them, yet perhaps not really surprising in an overwhelmingly Protestant America. The Catholic Church was even compelled to put an end to its missionary work for several tribes (see fig. 2.11). The Lutherans, on the other hand, hoped to have better success in their missionary work in Africa, and it was not until the end of the nineteenth century that they concentrated on the Apaches in Arizona, who had been avoided by all other religious denominations.

In the first decades after President Grant's assimilation mandate to the Christian denominations, the respective religious communities laid claim

---

clined. Toward the end of the 1980s only seven of the large off-reservation boarding schools in cities were left, and these too will probably be closed soon <1>. *[Note: numbers within angled brackets refer to the list of informants on p. 331.]*

14. This policy still prevailed in the 1950s.

15. Even as late as the 1950s school children on the Hualapai Indian Reservation in Arizona and elsewhere were punished with solitary confinement in darkened enclosures if they were caught on school grounds speaking their native language.

16. President Grant's Second Annual Message to Congress (1870), quoted in Prucha 1975, 135.

17. As far back as 1819 the Congress established the Civilization Fund, which supported various religious communities for Indian missions on an annual basis (Act of March 13, 1819, 3 Stat. 516 1846).

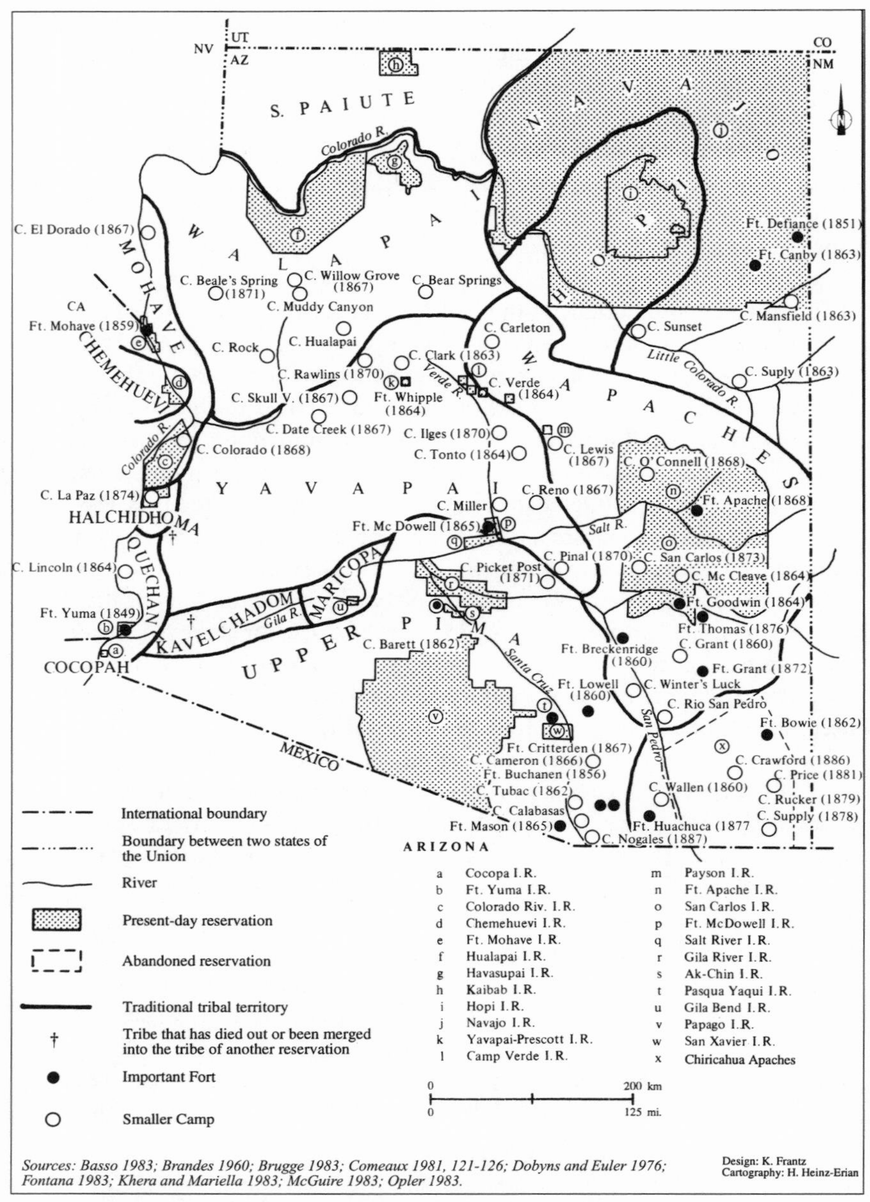

Fig. 2.5: American military posts among the Indian tribes of Arizona in the nineteenth century

to a virtual monopoly of missionary work on their respective reservations. They often constituted the entire staff of the BIA agency, and were able to influence the choice of people to run the trading posts and those who were assigned Indian land. In this way they gained absolute control over reservation Indians entrusted to their care so that, according to a decree of the BIA, tribal members even lost their religious freedom (Federal Agencies Task Force 1979, 5–6).

Toward the close of the century, Indian lands which had been set aside as trust land on reservations were divided up and assigned on a large scale to individual tribal members, a major blow to American Indians, with consequences that many tribes to the present day have not been able to reconcile (see chapter 3). The idea of American Indian private property in land dates as far back as 1633 (Kinney 1937, 82) and even President Jefferson saw an appropriate instrument for civilizing the Indians in this policy (Feest 1976, 39). Around the middle of the nineteenth century the idea of the individualization of landholdings was again taken up, for example, in the treaty negotiations with the Chippewa and the Shawnee (Cohen 1982, 98–100). From the viewpoint of American Indians, however, all these experiments were failures, because ultimately it often meant that the commonly held trust land was granted to individual Indians, who in turn very often lost their land.

With the General Allotment Act (GAA) of 1887 the transfer of tribal lands to private ownership through "allotment" to individual Indians became the official policy of the federal government. Influential philanthropic organizations from the East Coast,[18] lobbyists for private land speculators as well as railway and mining companies, representatives of the timber industry, and large cattle ranchers were all behind this legislation. The philanthropists believed that the tribal ideology of collectivism was the basic reason for the general misery of American Indians. In their opinion only a healthy egoism, an "intelligent greed," and the desire for individual landownership could free Indians from their "barbarous" condition (Gates 1896, 11). According to the reformers the reservations were the only obstacles to civilization and progress, and they believed that abolishing them (Abbot 1885, 53–54), as well as giving individual Indians private land instead of maintaining tribal ownership, would teach them personal responsibility and help them to develop into good American citizens. On the other hand, the pragmatic lobby groups, believing that large land reserves were not being used properly, aimed only at their acquisition. Their rational cal-

18. Such as the Board of Indian Commissioners (founded 1869), the Indian Rights Organization (founded 1882) and the Lake Mohonk Conference of Friends of the Indian, which until the Indian Reorganization Act (1934) were largely responsible for official Indian policy (see Prucha 1973; 1975, 131–134; and 1981; Washburn 1968 and 1975b).

Fig. 2.6: BIA boarding school in Fort Apache, Fort Apache Indian Reservation, Arizona, 1985.

Fig. 2.7: Distributing food rations at Camp San Carlos in Arizona, ca. 1875. Photo courtesy AHFC.

Fig. 2.8: Drill corps of the Phoenix Indian High School in Phoenix, Arizona, 1903. Photo courtesy AHFC.

culation was that the Indian, as a farmer or stockman, needed less land than the nonsedentary hunter and food gatherer and that was the reason why they advocated this new policy.

From the viewpoint of whites the General Allotment Act was a great success, for in barely fifty years American Indian tribes lost nearly 60% of their territory, which by that time had already dwindled to a fraction of what it had been originally. Similar to the Homestead Act (1862), the GAA provided that the head of each Indian family living on a reservation should receive 160 acres of land. Additionally, unmarried Indians over 18 years of age and all orphans should receive 80 acres, while single persons under 18 were to have 40 acres. Married women did not receive any land at all.[19] The size of the allotted land could be doubled if it was in a region suitable only for grazing. Young people were often ignored in the allotments, while Indian chiefs were sometimes offered a great deal more land, in an effort to gain their favor.[20]

The choice of land was generally the responsibility of the BIA, although this was not provided for by law. After all allotments of land had been made,

19. General Allotment Act (1887), quoted in Prucha 1975, 173–174.

20. An amendment to the GAA in 1891 cut the land allotments from the original act in half (Cohen 1982, 133).

Fig. 2.9: Issuing of "orders" for girls at the Phoenix Indian High School in Phoenix, Arizona, ca. 1900. Photo courtesy AHFC.

the U.S. government would purchase all surplus land from the tribe and offer it to prospective white buyers. Because the BIA was usually influenced by these buyers, the result was that the best land for agriculture and grazing and the richest timberlands were usually not allotted to the Indians but sold to white people. The proceeds of the compulsory land sales were seldom paid out directly to the Indians, but placed in trust for the respective tribe. The BIA then used the interest from these trust accounts to finance assimilation programs.

In this way, the great Sioux reservation which had been given to the Western Sioux in the Fort Laramie Treaty of 1868 (see fig. 2.2) was reduced by the U.S. government to six smaller reservations after a series of land cessions in 1882 and 1883, although this was contrary to the terms of the treaty. This made an area of 13,905 square miles available for white settlement (Josephy 1977; Ortiz, R. 1980, 5–6). One of these reservations was the Rosebud Indian Reservation in South Dakota, which still covered an area of 4,988 square miles in 1887. Under the GAA 2,878 square miles of the Rosebud Reservation were allotted to individual Indians, 29 square miles stayed with the tribe, and the remaining 2,082 square miles were offered for sale to white settlers as surplus land (ARCIA 1890, 1916–1934). Today the reservation covers an area of 1,475 square miles, of which 816 square miles are tribally owned (ARIL 1984, 5).

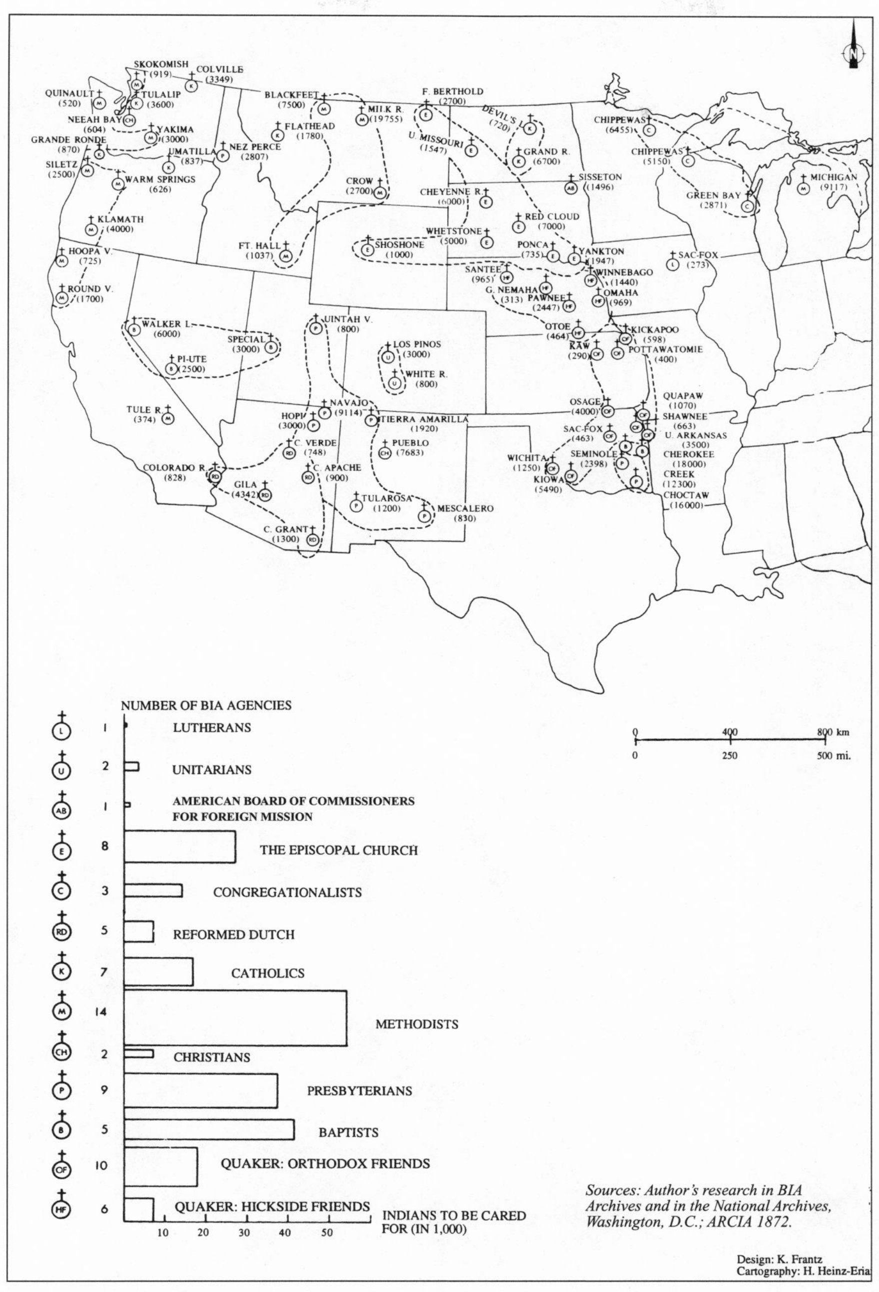

Fig. 2.10: President Grant's assignment of Indian agencies to religious denominations, 1870

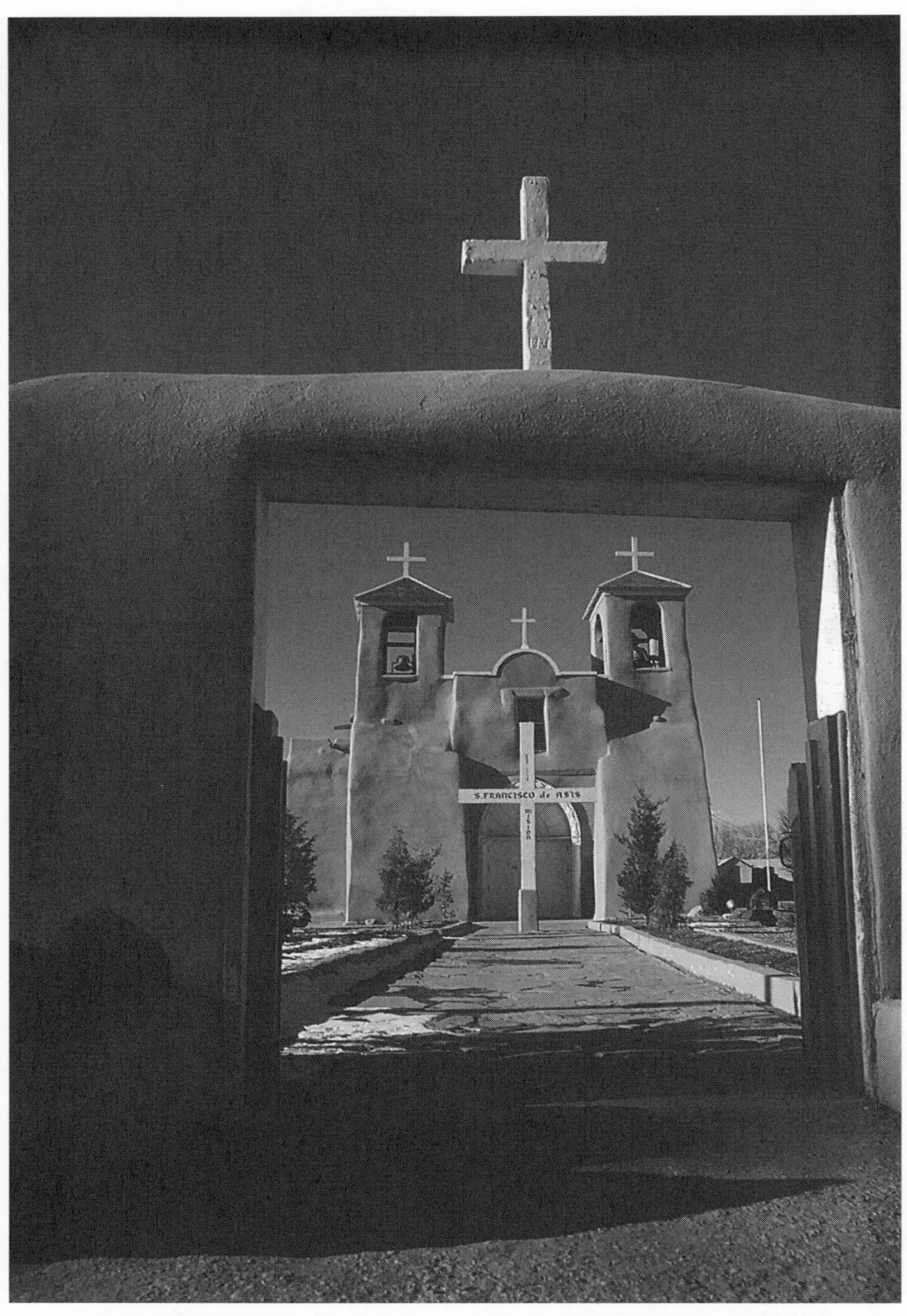

Fig. 2.11: San Francisco de Assisi Mission Church in Ranchos de Taos, New Mexico, 1976

Not all Indian tribes were included in the allotment policy. In Oklahoma and Nebraska allotment was not taken until the Curtis Act of 1898, which finally led to the abolishment of almost all reservations in this area (see fig. 2.3). Other tribes like the Menominee in Wisconsin and the Seneca in New York State were able to avoid the allotment under the GAA. In the arid regions of the West and the Southwest, however, even the legislators could not help but see that their measures to individualize the tribal land base and create Indian farmers would fail (see fig. 2.3).

There was a provision under the GAA to protect the land allotted to individual Indians from seizure by non-Indian settlers for the first twenty-five years, but at the end of this period either the time limit could be extended or the trusteeship restrictions were removed and the Indian allottee received the free right of disposal. This meant that individual Indians could then sell their property, and that they were liable to taxation. In this way de jure they were made American citizens.[21] The right to sell had in fact already been granted in 1906, but in the following three years around 2,700 applicants received the patent-in-fee title to their lands, and 60% of these Indian patentees felt obliged to sell their land to white settlers immediately thereafter (Cohen 1982, 136). In the same period the number of new land allotments was at its peak (see fig. 2.12).

Considered in terms of the official purpose of the GAA, to make American Indians private landowners and farmers, the allotment policy was a miserable failure. In the region where it was carried out most consistently, the dry prairie belt west of the hundredth meridian, it was applied to Indians who had been hunters and gatherers, and whose self-respect and religious viewpoint regarding "Mother Earth" did not allow them to take a plow into their hands. Moreover, the possession of private land as property was incompatible with their traditional way of thinking. The Indians who were willing to take the plunge and become farmers in spite of their traditions received only halfhearted support from the government, whether moral or financial. Moreover, the land was often unsuitable for agriculture. The following generations were left to reap the bitter fruits of the GAA, for they had not been taken into consideration when the land was allotted. When this act was passed the size of the American Indian population was at a low point, but the number of Indians increased considerably during the twentieth century until there was no longer any surplus land for subsequent generations.

21. Many American Indians were not granted citizenship until 1924 under the Indian Citizenship Act in recognition of their voluntary military service during the First World War. In most states this meant that they were also granted the right to vote. In Arizona, however, this right was denied until 1948 because Indians were still considered "persons under guardianship" or "wards of the government" (Washburn 1971, 164). In Utah, where they were not liable to taxation, it was not until 1956 that they were permitted to cast their votes.

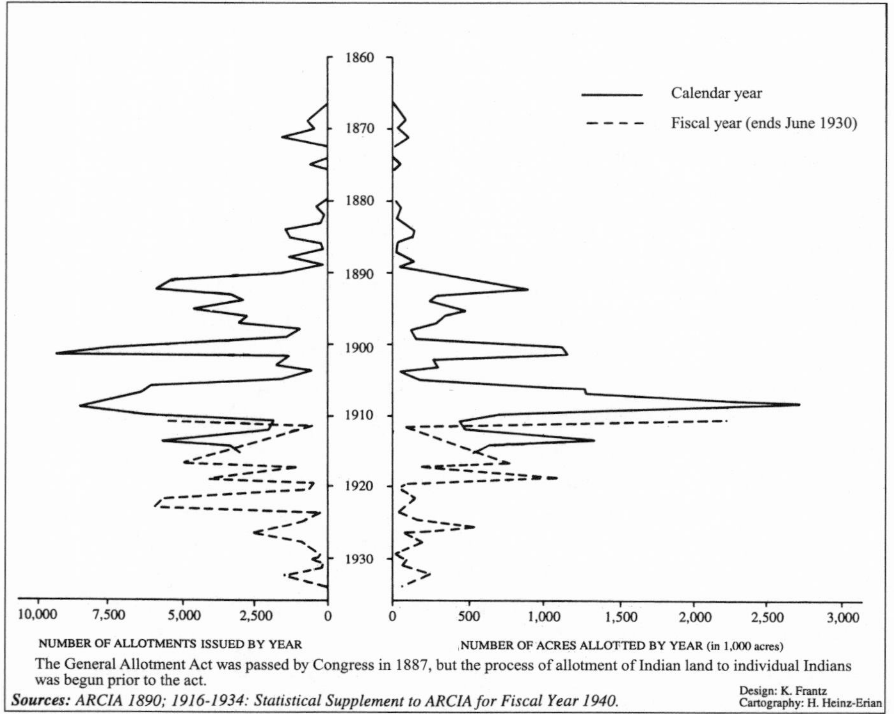

Fig. 2.12: The extent of the General Allotment Act policy, 1871–1934

Furthermore, the legislators had not considered the effects of land inheritance, which divided the land among heirs and led to increasingly smaller tracts of land so that future generations often no longer had any foundation for their existence.

FEDERAL INDIAN POLICY SINCE THE 1930s

After the First World War the government commissioned a survey of the social and economic conditions of reservation Indians.[22] The resulting Meriam Report showed that the assimilation policy previously pursued had led to a great loss of Indian land and an overblown administrative bureaucracy, while at the same time the Indians' economic advancement and their readiness for initiatives was not promoted (Meriam et al. 1928, 41). This report further revealed the deplorable socioeconomic situation of America's indigenous populations: their dire poverty, their high child mortality rate, the generally poor state of their health, their appalling living conditions,

22. It is significant that this survey was not carried out by the BIA but by a private organization, the well-known Brookings Institution in Washington, DC.

and an altogether inadequate level of education. In retrospect this report can be seen as marking a change to a more liberal Indian policy, which was framed in the Indian Reorganization Act (IRA)[23] at the beginning of the Roosevelt era and lasted until shortly after the Second World War.

The IRA, passed in 1934, put an end to the allotment policy, and instituted a series of reforms which gave American Indians more political, economic, and cultural independence. Unlike past Indian policies, there was now a positive attitude, both toward the existence of the tribe and toward the idea of the reservation. After fifty years of a policy aimed at undermining them, the tribes were now offered the possibility to form themselves anew with an elected tribal council, an elected government, and their own constitution, as well as a tribal court and police system. A considerable number of the tribes—including most of the Pueblo Indian communities of the Southwest and, at first, the Navajos—objected to this offer, however, as it would have obligated them to copy the United States model, which was foreign to their culture.[24] Some of the tribes were able to modify the standardized procedure of the U.S. government in accordance with their own way of thinking by integrating their traditional leaders and practices into what was for them a novel form of government.[25] On other reservations a situation developed in which the two forms of government—the traditional and the new—existed side by side, though the new form was only partly accepted. This situation has led to permanent tension down to the present day.[26]

In addition to reforms of the political structure which recognized tribal governments, the IRA stressed economic incentives from the New Deal era which differed sharply from the earlier Indian policy (see ARCIA 1934, 78–83). For the first time in the history of Anglo-Indian relations, American Indians were given money with which they could buy back land. Even if critics of this reparation policy are right in maintaining that it was only a "drop in the ocean," the fact remains that on some reservations this action

23. Indian Reorganization Act (1934), quoted in Prucha 1975, 222–223.

24. Of the 266 federally recognized Indian tribes at that time, a total of 77 voted against a constitution and tribal government to be elected every two years (Witt 1968, 59). Even when the vote was in accordance with what the IRA intended, these tribal constitutions were often a subject of controversy because many who had the right to vote failed to do so. Many of the traditional tribal members simply did not cast votes because, according to their view, they already had a legitimate political structure within their tribe.

25. For example, according to their constitution the traditional chief of the Havasupai Indians in Arizona automatically had a seat in the tribal council. In principle there has been no change in this policy down to the present, yet this seat has not been occupied for a long time owing to controversies <3>.

26. This dualistic political structure is characteristic of the Hopi Indian Reservation in Arizona, for example.

did at least lead to a modest consolidation of the confused conditions of landownership.

With financial assistance and the support of trained BIA personnel the federal government helped to construct tribal enterprises, although at first these were almost exclusively limited to the primary sector. Tribal farms and grazing enterprises were developed which made it possible for American Indians to make use of land that had formerly been leased to whites. In addition, cattle owners on the reservations were encouraged to join together with Indian families that did not yet have any cattle to organize cooperatives.[27] To help families obtain the capital necessary for such efforts, the BIA founded its own credit fund, a necessary arrangement, because the landed property of individual Indians was held in trust by the United States, and private banks could not use the land title as security.

All these arrangements, which still exist on many reservations, show that the federal government was again prepared to tolerate the predominantly collective mentality of American Indians. At the same time, however, it promoted measures to encourage individual initiative, even if under different conditions. The BIA[28] was commissioned to see to it that the tribes could assign land for an unlimited time period to tribal members, who would then have the means to be self-supporting.

The IRA policy also followed new paths in Indian education. On the one hand, it supported the development of the BIA Day Schools on reservations, which served the needs of the local population more than the boarding schools did. On the other hand, the Johnson-O'Malley Act made it possible for education for Indian students to be contracted to public schools adjacent to the reservations, while the individual states were reimbursed for the expenses.[29]

Toward the end of the IRA era in 1946 Congress established the Indian Claims Commission (ICC),[30] a form of compensation court to which federally recognized Indian tribes could appeal with their claims for reparations for land that had been taken from them illegally. During its more than thirty years in existence this commission made 670 judgments, of which almost 60% were in favor of the tribes <4>.[31] Together with cases handled

27. From 1934 to 1947 the amount of Indian-owned livestock more than doubled and the income from agriculture increased twenty-six fold (McNickle 1962, 60).

28. By this time the BIA had begun to lose its monopoly, because some reservations, on their own initiative, established contact with other federal agencies or with private organizations.

29. Johnson O'Malley Act (1934), quoted in Prucha 1975, 221–222.

30. Indian Claims Commission Act (1946), quoted in Prucha 1975, 231–233.

31. Tracing as far back as 1863, Indian tribes have made hundreds of petitions for reparations to the U.S. government. Before the ICC was founded, only 152 claims were authorized and reached the Court of Claims; 104 cases were dismissed and 16 cases were pending

in other law courts up to 1985, the U.S. government made reparation payments of $1.26 billion (Washburn 1985, 24).[32] A number of lawsuits, which concern not only land but also the violation of water, fishing, and hunting rights or the improper BIA management of natural resources on the reservations, remain unsettled. Only in the most exceptional cases did the verdict of the ICC result in the restitution of land, however, and financial compensation was the general rule. Moreover, the amount of this compensation was based on the estimated value of former Indian lands at the time when they were ceded, and not on their current value. Apart from that, Indians could not claim interest that would have accumulated in the meantime. If a tribe accepted compensation payments, it renounced all future claims of any kind to tribal lands. For this reason, a number of tribes decided not to accept money.

With this financial compensation the U.S. government hoped that it could free itself once and for all of the moral burden of the wrongs it had previously done to American Indians. However, according to the commissioner of Indian affairs at the time, another purpose was to prepare a way to justify the BIA's plan to back out of its trust responsibilities as specified by treaty (Dillon Myer 1952, quoted in Josephy 1971, 54). The very fact that the federal government was now willing to give American Indians the right to bring charges against the federal government or a U.S. state in a law court was a sign of great progress when compared with the Indian policy of previous epochs.

There was another about-face in federal Indian policy at the beginning of the 1950s, however. President Eisenhower did not approve of the social experiments of the prewar period which, in his opinion, only perpetuated the illusion of a future for Indians as Indians (McNickle 1973, 104). Therefore, in accordance with the general retreat from the Roosevelt administration's ideology of "big government" there was also a drastic cutback of BIA programs. The questions as to which of the BIA activities could be dropped and whether the federal government should not simply disengage itself from its historic trust relationship with the federally recognized Indians became

when the Indian Claims Commission Act was passed. A mere 32 of all these cases were awarded claims judgment. The court awarded approximately $38 million (Jorgensen 1978, 22) of the $2 billion that had been claimed by Indian tribes.

32. These lawsuits are usually very long and drawn out with a host of costly lawyers, ethnologists, and historians on both sides called in to reconstruct, defend, or dispute, as the case may be, the indigenous people's lawful right to their land. If a tribe wins its case, then according to the law, its legal advisers may retain 10% of the settlement as the fee for their services. Moreover, the federal government may make deductions for services that it has rendered to the tribe in the past that were not included in the original treaty negotiations. At least 20% of the reparation payments must remain with the respective tribe, while the rest may be divided up among the individual tribal members <5>.

critical. The way was thus prepared for the hoped-for "final solution" of the Indian problem in the *termination policy* of 1953.[33] While this legislation proclaimed that American Indians should "emancipate" themselves from the tutelage of the BIA, and that only as "real" Americans would they be able to share the same privileges and responsibilities applicable to other U.S. citizens, it also made it clear that this would be possible only by putting an end to the separate status which the Indians enjoyed under federal law. It was also hoped that the time would come when the reservations would no longer be necessary.[34]

Contrary to the assertions of the legislators, this termination, considered critically, was not really an act of charity. It was not the American Indians who were now freed—in fact, they had not asked for this freedom—but rather the U.S. government, which was thus able to wash its hands of its responsibilities toward Indians as stipulated by treaty. A large number of federal government development programs for Indians in education, welfare, housing assistance, and other social affairs as well as many BIA schools were discontinued as a result of this policy. Additionally, the BIA relinquished all health care responsibilities to the Public Health Service.[35]

Without consulting the tribal governments, the BIA was asked to conduct an investigation to produce a list of the reservations which, in its opinion, would no longer need the assistance of "Big Brother" in the future and be qualified to manage their own affairs. As a result of the data collected, around 120 reservations, generally relatively small, were "terminated" by 1962, and around 3% (approximately 11,500) of all federally recognized American Indians lost their special status (see Cohen 1982, 174; U.S. Commission 1981, 23; Wilkinson and Biggs 1977, 149). Most of these terminated tribes were on the West Coast (see fig. 2.3). The two largest reservations affected by this policy were the Klamath Indian Reservation in Oregon and the Menominee Indian Reservation in Wisconsin, but by 1973, the latter regained its reservation status (see Hofmeister 1976, 509). Of all the selected tribes only the Turtle Mountain Chippewa Indian Reservation in North Dakota and the Eastern Cherokees in North Carolina were successful in evading this termination policy.

The consequences of the new policy were serious for the tribes affected by it. Not only did they lose their autonomous rights as governments, in most cases they also lost a considerable part of their land and control over their natural resources. Moreover, being subject to state laws and taxation,

33. House Concurrent Resolution 108 (1953), quoted in Prucha 1975, 233.

34. Long before Resolution 108 the federal government had removed the trust status of several reservations, to the extent that more tribes' trusteeships were terminated in the period from 1880 to 1915 than after the Second World War (Sutton 1975, 138).

35. In 1954 the Indian Health Service was attached to the U.S. Public Health Service.

Fig. 2.13: "Raising the Flag on Iwo Jima," with the Pima Indian Ira Hayes (Second World War memorial in Washington, DC, 1983)

they had to rely on public relief from the state, for the tribes did not have the revenue needed to finance their own social services or other programs. The Menominee Indian Reservation, for example, once one of the richest Indian communities in the country, soon became the poorest county in Wisconsin (see Lurie 1972). Paradoxically, although the termination policy was also intended to finally abolish the BIA, this agency, now chiefly concerned with putting an end to the Indian reservations, increased both in personnel and the size of its budget.

In connection with the attempt to terminate the trust relationship of Indian reservations and to end the special status of its residents, the U.S. government also developed a policy to strongly encourage American Indians to accept off-reservation jobs and integrate into urban American society through incentives for Indians to migrate to urban areas and assimilate. Until 1890 almost all Indians lived in rural areas, and during the GAA and IRA era there was little change. Of the indigenous population that had become landless, only a small portion migrated to urban areas; and the Voluntary Relocation Program, established in 1931, did not meet with much approval <6>. It was not until the Second World War that there was a real change in this situation. Besides 25,000 volunteers for military service (Josephy 1971, 52; see fig. 2.13) around 40,000 Indians answered the call of the war-related industries, located for the most part in large cities, which offered favorable working conditions to make up for the great shortage of labor. The consequence was that toward the end of the war around 80,000 Indians were living in cities, mostly on the West Coast (Spicer 1980, 110).

Many veterans and defense industry workers returning from the war saw themselves once again confronted with a deplorable socioeconomic situation on the reservations. It is not surprising, therefore, that they often responded favorably when the BIA instituted a new relocation program[36] which helped Indians find jobs in the cities, gave them job training for a limited time, and provided short-term financial assistance. This policy enabled 78,000 Indians to leave the reservation for the city by 1972, although in fact only 48,000 found work (Jorgensen 1978, 65). Forced to face urban poverty and cultural isolation, 35% of those willing to move to the cities eventually returned to their reservations between 1954 and 1972 (Spicer 1980, 111). Those who remained in the cities, however, soon lost the special services status that had been provided by the BIA, and could not expect support from those who stayed behind in poverty on the reservations. Despite the high percentage who returned to the reservation, the federal government was generally satisfied with the results obtained, as the underlying objective of assimilation was apparently realized in many cases.

Toward the end of the 1960s the national conscience was again stirred by the conditions of American Indians. Roused by the protests of a new generation of Indian leaders who forcefully and skillfully opposed the termination policy, and by a number of reports which called attention to the appalling social and economic state of reservation Indians,[37] the government called for a war on Indian poverty. The new slogan was "self-determination,"[38] and during President Nixon's administration this policy was enshrined in the Indian Self-Determination and Education Assistance Act (1975). It finally recognized that despite decades of pressure to assimilate the many Indian tribes, Indians would never completely blend into the American melting pot. With a view to helping tribes in their self-determination, the BIA and all other federal agencies concerned with Indian affairs were authorized to transfer their powers and responsibilities by contract to individual tribal governments, should tribes wish them to do so. The U.S.

36. Commissioner of Indian Affairs, Glen Emmons (1954): Relocation of Indians in Urban Areas, quoted in Prucha 1975, 237–238.

37. Indian Education: A National Tragedy—A National Challenge, Senate Report no. 501 (1969) quoted in Prucha 1975, 253–256.

38. President Nixon, Special Message on Indian Affairs (1970), quoted in Prucha 1975, 256–259.

departments concerned would then finance the respective programs and provide technical assistance to tribes when needed. The paternalistic BIA, which despite the IRA had hitherto treated the tribal governments as nothing more than instruments to carry out its instructions, was now to become an advisory agency, working as a partner. Since the time of the Kennedy administration, the Office of Economic Opportunity and VISTA (Volunteers in Service to America),[39] with its domestic development aid, had already successfully demonstrated this new approach.

The generation of soldiers returning from the war were also better prepared than their fathers had been to use efficiently the instruments of self-government provided by the Roosevelt era. The unrestricted power of individual BIA agents, who always stood firm when dealing with tribal governments and tribal leaders inexperienced in administrative affairs, was a thing of the past. An increasing number of tribal leaders and decision makers had acquired political and administrative skills in their military and work experience. The more prosperous tribes also hired white experts to help them defend tribal interests in Washington directly and without the involvement of the BIA hierarchy. In such cases the BIA superintendent and staff on some reservations became little more than a monitoring body whose task is to control the proper use of federal funds. Today most Indian communities exhibit a dual system of administration with the BIA on one side and the tribal government on the other, which can lead to conflict. The fact remains, however, that the tribes are becoming more and more successful in the administration of tribal programs under tribal control.[40]

Another goal of the self-determination policy was to bring about a change in the structure of the BIA personnel to benefit reservation Indians. This had already been envisaged in the IRA era, yet until after the end of the Second World War the Bureau of Indian Affairs remained a domain of non-Indians. This changed fundamentally in subsequent years, and by 1985, 81% of all BIA employees were American Indians, the product of preferential hiring. American Indians are no longer limited to menial work.

39. The OEO also finances the Head Start programs, by which Indian children of preschool age are prepared for school. A goal of this program is to deliver instruction in the native language. In the past Indian children were often instructed year after year by non-Indian teachers whose language the students did not understand. VISTA has also helped some tribes to realize development projects such as farmers' cooperatives.

40. Today it is the general rule rather than the exception for the police force, the court system, and the health and social services as well as the housing authority to be under tribal administration. The BIA has maintained much of its authority in the management of tribal lands and natural resources, since the Indian country, along with the natural resources, are held in trust by the federal government. The Indians' rights are generally limited to the use of these lands and resources.

Table 2.1. Racial distribution of full-time employees in the BIA headquarters in Washington, DC, classified according to rank (numbers and percentage), 1985

Rank	American Indian		White		African American		Other		Total	
2–7	246	85.4	19	6.6	21	7.3	2	0.7	288	100.0
8–12	146	73.0	43	21.5	6	3.0	5	2.5	200	100.0
13–15	67	56.8	47	39.8	1	0.8	3	2.6	118	100.0
16–18	16	54.6	13	1.9	2	6.5	—	—	31	100.0
Assistant secretary	1	100.0	—	—	—	—	—	—	1	100.0
Total	476	74.6	122	19.1	30	4.7	10	1.6	638	100.0

Sources: Author's analysis of unpublished BIA records

On the contrary, almost all superintendents of BIA agencies located on reservations and all the directors of the twelve BIA area offices are now Indians, and Indians are also well represented at the BIA headquarters in Washington, DC (see table 2.1). BIA superintendents and other executives of the local agencies are rarely members of the tribe where they are stationed, however.

3 From Sovereign Tribal Territory to the Indian Reservations of Today

THE REDUCTION OF INDIAN TERRITORY AND THE GEOGRAPHICAL DISTRIBUTION OF TODAY'S RESERVATIONS

The indigenous peoples of the United States once possessed exclusive tenure to all the lands. Today they are left with only 2.3% of the land which once was their exclusive possession. That is approximately the area of Idaho, or two and a half times the size of Austria. This dramatic loss of land is closely connected to a peculiarly European and American sense of legal tradition, which shaped Indian policy from the very first encounters between whites and American Indians.

At first the colonial powers of Europe justified the seizure of the North American continent by the so-called *principle of discovery*. According to this principle the mere discovery of a stretch of coast was in itself sufficient grounds to lay claim to that land as well as its ill-defined hinterland. The right to make use of their land was at first conceded to the indigenous population, yet, in the case of need, this right could be declared null and void at any time. This interpretation of the conditions of land tenure appeared to legitimize the liberal grants of land from European rulers to those of their subjects who wanted to emigrate. Many Christian denominations approved of this policy, which they combined with a missionary effort to convert the Indians to Christianity (see Joyner 1978; Pierce 1977; Prucha 1962).

The Americans adopted this viewpoint, adding to it the principle of "best use of the land,"[1] by which they justified their policy of unbridled expansion. With this argument agriculturalists who would cultivate the land were given precedence over American Indians who had previously occupied

1. This principle goes back to the Swiss jurist of the eighteenth century, Emmerich von Vattel, who maintained that no nation had the right to claim more land than it settled and cultivated (see Prucha 1962, 143, 227, and 241).

the land as mere hunters and gatherers. According to President Theodore Roosevelt, "the settler and pioneer have at bottom had justice on their side; this great continent could not have been kept as nothing but a game preserve for squalid savages" (cited in Washburn 1971, 38). Many of the Indian tribes that had been dispossessed, however, were not roaming hunter-gatherers, but were settled farmers for whom hunting only complemented their means of livelihood.

The Americans, with the certainty that right was on their side, set about to progressively annex the lands of the native population. Within half a century Indian territory, once so unified, had been dissolved, leaving only tribal enclaves, which were further reduced in size again and again (figs. 2.2, 2.3, 2.4). In 1881, just prior to the adoption of the General Allotment Act, America's indigenous peoples were still left with a land base equal to the size of California and Idaho combined. The introduction of privately owned allotments assigned to individual Indians resulted in another reduction of their land by two-thirds, to a total of 74,160 square miles. The majority of this loss (70%) was caused by the sale of what was deemed surplus land by the Bureau of Indian Affairs (BIA). Many of the American Indian families who became landholders under allotments sold or lost the rest of their land by 1934 (Cohen 1982, 138).

The Indian Reorganization Act (IRA) was the first federal action to halt this accelerating erosion of Indian country. Some tribes at that time were even able to recover a portion of their tribal land base, a total of 5,716 square miles, through purchase or redesignation as a result of the IRA. During the course of the Second World War, however, the Indian reservations lost 772 additional square miles through dam construction and other federal projects. Another change in direction of the federal Indian policy, like the swing of a pendulum, gave rise to the Termination Act of 1953, which cost American Indians more than 3,862 square miles of land by 1957 (Senate Committee 1958, 101). The Klamath Indian Reservation in Oregon, the largest reservation affected by the Termination Act, suffered the greatest losses. The Klamath tribe's best land (70%) was immediately sold off, and in the following years a large part of the remaining area ended up in white hands (Staff of the American . . . 1976).

During the 1970s and 1980s the land situation improved slightly for American Indians. Reparations were made by the Indian Claims Commission (ICC) for lands taken illegally, but these were almost always purely financial. A few tribes, however, were also successful in regaining land. Many tried hard to have sacred land returned to them, but only the Yakima Indian Reservation in Washington and the Pueblo Indians of Taos in New Mexico succeeded, while the Sioux and other tribes were offered cash settlements, which they refused. The demands of all other tribes have been rejected up to the present day. On the sacred lands of American Indians today you can find ski developments, mining, and lands flooded behind dams.[2] In other cases, land areas of considerable size, economic value, and above all symbolic significance were returned to the Havasupais and the San Carlos Apaches in Arizona (figs. 2.3 and 7.1), and the Passamaquoddy and Penobscot Indians of Maine, along with financial compensation.[3] The largest territorial gains, however, were made by the Navajos, whose reservation was created in 1868 (see Goodman 1982a, 57–58; Hofmeister 1980, 77–78) and has since increased four and a half times as a result of measures taken by the president and by Congress. Because of this expansion the Navajo Nation now completely surrounds the Hopi Indian Reservation, much to the Hopis' distress, as they had previously lived apart from the Navajos (see fig. 3.1).

If one adds up the total amount of land returned to all tribes, American Indians have regained 5,948 square miles of their original tribal land since the Second World War (see fig. 3.2). This is almost equal to the combined area of Connecticut and Rhode Island. In most cases public land was returned, while the private landholdings of whites, though often acquired by unlawful means, were almost always exempted.

The geographical distribution of reservation land within the United States shows that no less than 93% of this area can be found in eleven western states as well as South Dakota. East of the Mississippi barely 3% of former tribal lands remain in reservation status. Just over half of the states have no reservations at all (see table 3.1), and this unequal distribution is a direct consequence of past Indian policy, in particular, of the forced migrations and relocations during the nineteenth century.

Much of what is left of reservations is land in which white settlers have long had no particular interest. Much of this land is arid or semiarid territory of little value for either agriculture or pastureland. Because of its generally poor infrastructure and its distance from important urban centers there is little prospect of future economic development (see figs. 3.3, 6.9, 6.10). The irony of history is, however, that some of the land given to American Indians has turned out to be very valuable, containing up to 10% of the petroleum and natural gas deposits, 33% of open-pit mining

2. For example the Hopis and the Navajos oppose skiing on the San Francisco Peaks in Arizona. The Papagos have taken legal action against copper and gold mining in the Baboquivari Mountains in Arizona and the Sioux are trying to regain the Black Hills in South Dakota. Furthermore, the Navajos oppose a dam project in the drainage area of the Colorado River in Arizona and the Cherokees oppose a dam along the Tennessee River in North Carolina.

3. In 1969 more than 772 square miles of land was returned to the San Carlos Apaches. In 1975 the Havasupais received 286 square miles, and in 1980 the Passamaquoddy and Penobscot Indians in Maine were granted approximately 463 square miles of land.

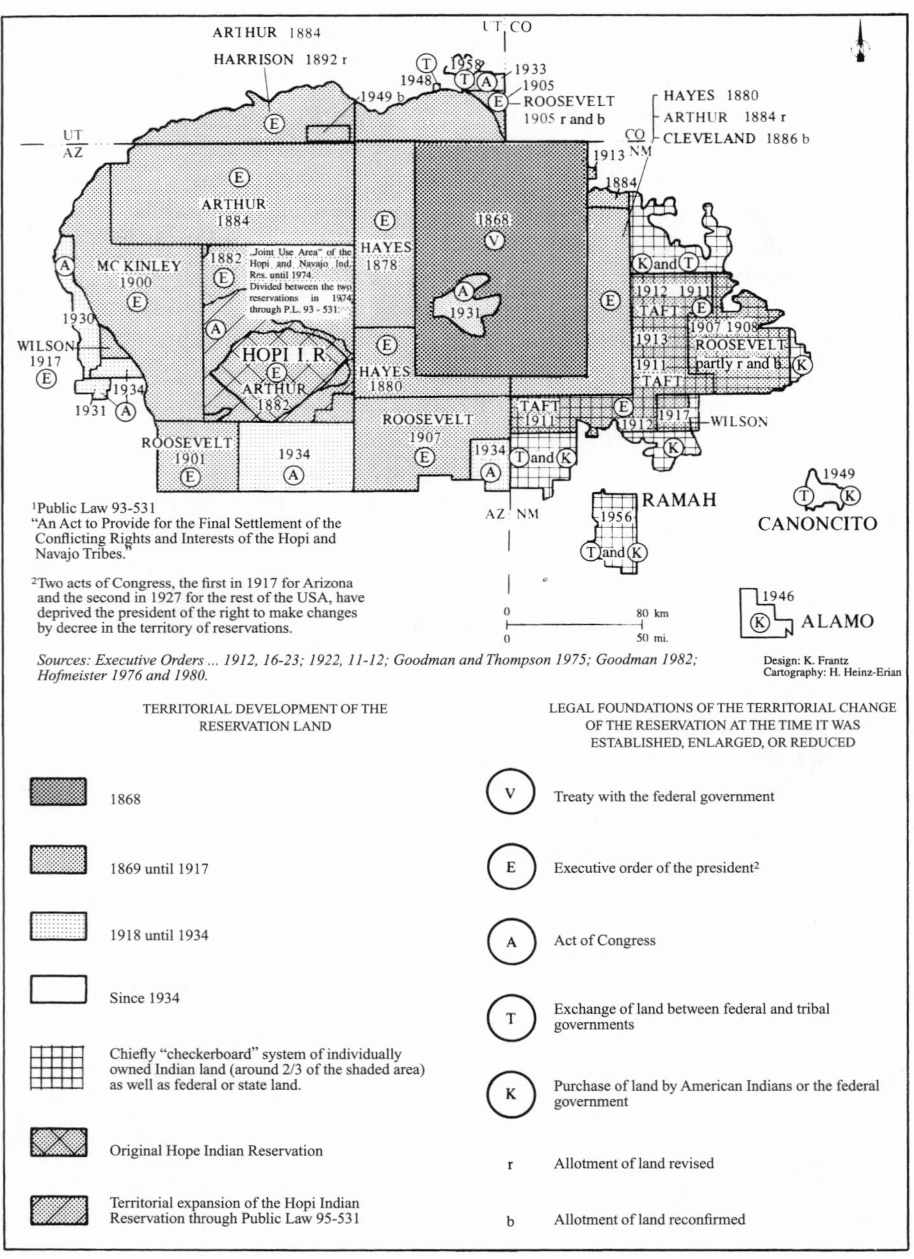
ARTHUR 1884
HARRISON 1892 r
UT CO
1958
1948
1933
1905
1949 b
ROOSEVELT
1905 r and b
HAYES 1880
ARTHUR 1884 r
CLEVELAND 1886 b
UT
AZ
CO
NM
1913
1884
ARTHUR
1884
1868
HAYES
1878
MC KINLEY
1900
1882
"Joint Use Area" of the Hopi and Navajo Ind. Res. until 1974. Divided between the two reservations in 1974 through P.L. 93 - 531.
1931
K and T
1912 1911
TAFT
1907 1908
1913
ROOSEVELT
partly r and b
1911
TAFT
1930
WILSON
1917
HOPI I.R.
ARTHUR
1882
HAYES
1880
1934
1931
TAFT
1911
1912
1917
WILSON
ROOSEVELT
1907
ROOSEVELT
1901
1934
1934
T and K
AZ NM
1949
RAMAH
1956
CANONCITO
T and K
1946
ALAMO
[1]Public Law 93-531
"An Act to Provide for the Final Settlement of the Conflicting Rights and Interests of the Hopi and Navajo Tribes."
[2]Two acts of Congress, the first in 1917 for Arizona and the second in 1927 for the rest of the USA, have deprived the president of the right to make changes by decree in the territory of reservations.
0 80 km
0 50 mi.
Sources: Executive Orders ... 1912, 16-23; 1922, 11-12; Goodman and Thompson 1975; Goodman 1982; Hofmeister 1976 and 1980.
Design: K. Frantz
Cartography: H. Heinz-Erian
TERRITORIAL DEVELOPMENT OF THE RESERVATION LAND
1868
1869 until 1917
1918 until 1934
Since 1934
Chiefly "checkerboard" system of individually owned Indian land (around 2/3 of the shaded area) as well as federal or state land.
Original Hope Indian Reservation
Territorial expansion of the Hopi Indian Reservation through Public Law 95-531
LEGAL FOUNDATIONS OF THE TERRITORIAL CHANGE OF THE RESERVATION AT THE TIME IT WAS ESTABLISHED, ENLARGED, OR REDUCED
V Treaty with the federal government
E Executive order of the president[2]
A Act of Congress
T Exchange of land between federal and tribal governments
K Purchase of land by American Indians or the federal government
r Allotment of land revised
b Allotment of land reconfirmed

Fig. 3.1: The origin of the Navajo Indian Reservation (Arizona, New Mexico, Utah): an example of how a reservation is created by a combination of events

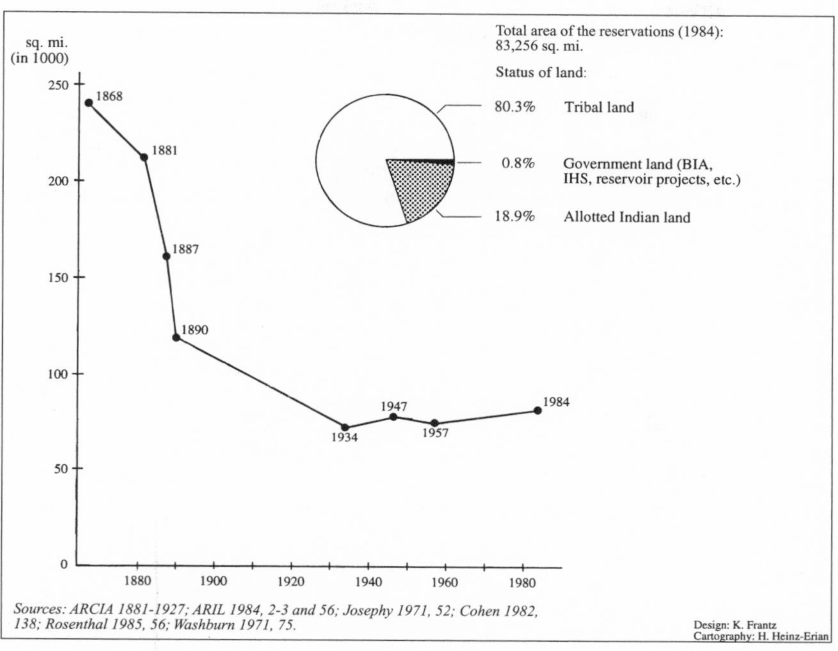

Fig. 3.2: The reduction of Indian country 1868–1984

Fig. 3.3: "Desert Solitaire" in Monument Valley, Navajo Indian Reservation, Arizona, 1985

Table 3.1. The distribution of the ownership and tenancy of reservation lands in the United States, 1984

State	Total reservation land (sq. mi.)	Percentage of total land			% of total state land	% of total reservation lands	% of leased reservation land[d]
		Tribal trust land[a]	Allotted land in trust[a]	Gov't. owned (BIA, IHS, etc.)			
NORTHEAST[b]	348	100.0	—	—	0.2[c]	0.4	—
Connecticut	2	100.0	—	—	< 0.1	< 0.1	—
Maine	346	100.0	—	—	1.0	0.4	—
MIDWEST[b]	11,312	54.7	45.3	< 0.1	1.5[c]	13.4	16.2
Iowa	7	100.0	—	—	< 0.1	< 0.1	24.6
Kansas	47	24.9	74.1	1.1	0.1	0.1	59.5
Michigan	34	57.2	42.7	0.1	0.1	$ 0.1	3.0
Minnesota	1,196	93.3	6.7	—	1.4	1.4	1.7
Nebraska	101	36.3	63.7	—	0.1	0.1	81.6
North Dakota	1,331	25.2	74.7	< 0.1	1.9	1.6	24.5
South Dakota	7,950	52.1	47.8	< 0.1	10.2	9.4	17.2
Wisconsin	646	80.5	19.5	< 0.1	1.1	0.8	0.7
SOUTH[b]	2,110	24.2	75.8	< 0.1	0.2[c]	2.6	54.2
Florida	244	99.8	—	0.2	0.4	0.3	7.7
Louisiana	< 1	100.0	—	—	< 0.1	< 0.1	2.0
Mississippi	28	98.8	0.1	1.1	< 0.1	< 0.1	14.5
North Carolina	88	99.8	—	0.2	0.2	0.1	0.2
Oklahoma	1,750	7.9	91.5	0.6	2.5	2.1	64.5
WEST[b]	70,429	86.0	13.1	0.9	3.9[c]	83.6	11.9
Alaska	1,426	9.9	90.1	—	0.2	1.7	0.2
Arizona	31,309	98.3	1.3	0.4	27.2	37.2	1.1
California	890	88	11.9	0.1	0.6	1.1	5.0
Colorado	1,230	99.6	0.4	—	1.2	1.5	3.5
Idaho	1,288	56.2	39.8	4.0	1.5	1.5	41.8
Montana	8,158	43.5	56.4	0.1	5.5	9.7	81.5
Nevada	1,920	93.2	6.4	0.4	1.7	2.3	0.7
New Mexico	12,442	87.8	8.5	3.7	10.1	14.8	1.0
Oregon	1,201	82.3	17.7	—	1.2	1.4	6.5
Utah	3,625	98.5	1.4	—	4.2	4.3	1.7
Washington	3,989	81.4	18.5	0.1	5.8	4.7	10.7
Wyoming	2,951	94.9	5.0	0.1	3.0	3.5	2.5
Total	84,199	80.2	18.9	0.8	2.3[c]	100.0	13.6

[a]Held in trust by the federal government.

[b]Based on regions identified by the Bureau of the Census (1980).

[c]Percentage of regional totals.

[d]Percentage of surface leases (i.e., agriculture, forestry, businesses, etc.) and subsurface leases (mining). These specific uses often overlap and cannot be determined individually because of lack of precise statistical data.

Source: ARIL 1984, 2–3 and 55–57.

coal of low sulfur content, and 55% of the uranium deposits of the United States <7>.

Most of the reservations—about two-thirds—are very small, covering an area of less than 50 square miles each (see fig. 3.4), while 7% comprise more than 1,000 square miles.[4] These nineteen largest reservations encompass nearly 74% of all Indian territory (see table 3.2). The Navajos have by far the largest amount of land, nearly 24,334 square miles, approximately the size of West Virginia.

THE LEGAL BASIS FOR ESTABLISHING RESERVATIONS

American Indian reservations developed from a variety of legal origins, and a differentiated knowledge of this background is necessary for understanding the reservation status. Although this difference in origin has not so far had a significant effect on the status of the reservations, it could play a decisive role in the future for the various Indian communities concerned.

Until 1871 it was common practice for the U.S. government to enter into treaty negotiations with a tribe as if it were negotiating with a foreign nation. In many cases Indian tribes were allowed to keep a portion of their ancestral lands as a reservation, even if often only temporarily, on condition that they give up the remainder. Under this type of *treaty reservation*, it was conceded that the tribes still retained a part of their ancient legal title to the land.

Beginning around the middle of the nineteenth century American Indians, to a large extent, were denied their decision-making powers, and more or less all Indian territory was declared public land. The U.S. government at that time reserved the right for itself to allocate part of its territory as a reservation for a tribe. This, for instance, was the case with the Crow Indians of Montana and Wyoming, who were granted an ever smaller part of their aboriginal tribal land in three successive treaties (see fig. 3.5a).[5]

In the era of the massive relocation of Indian tribes, when they were

4. The Federal Register in 1985 listed for the United States 309 federally recognized Indian communities (excluding Alaska). Of these only 264 had their own reservation territory (ARIL 1984). These numbers do not, however, correspond with the BIA figures. The population statistics of the U.S. Bureau of the Census in 1980 lists 278 reservations (see chap. 4). The two smallest of the federally recognized Indian communities, the Nooksack Indian Reservation in Washington state and the reservation of the Me-Wuk Indians in California, have an area of only one acre each (ARIL 1984, 26 and 28).

5. In a friendship treaty in 1825 the consequence was simply the recognition of the supremacy of the U.S. government. The Crow Indians in the first Fort Laramie Treaty in 1851 were still able to reserve for themselves around 59,500 square miles of land. By the time of the second Fort Laramie Treaty in 1868 the federal government, however, reduced their land to 12,360 square miles. Between 1882 and 1904 the reservation land was further reduced to its present size of 2,350 square miles by three forced land sales <8>.

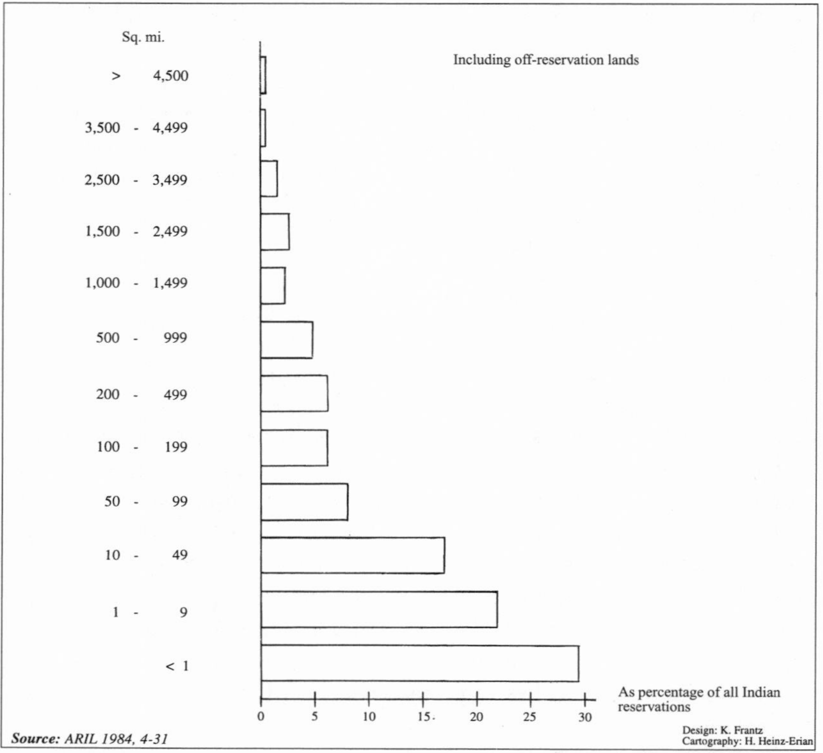

Fig. 3.4: The size of Indian reservations in the United States listed according to their area in square miles and as a percentage of all reservations, 1984

moved far from their lands, partly because of the Indian Removal Act, federal policy prescribed *exchanges of land* in treaty negotiations for the Indians. This enabled them to take control of the areas left behind, which could then be given, often for free, to white settlers. A treaty signed in 1826, for example, provided that the Creek Nation would cede all its territory in the Southeast to the federal government, on the condition that the tribe would receive as compensation territory of approximately the same size in what is now Oklahoma (Royce 1900, 714–715). This newly assigned reservation, however, would later be lost a second time to white settlers (see fig. 3.5b).

While many reservations were established by treaty, others had their origin from a different legal basis. One variation is a *reservation purchased* by American Indians or the federal government, or both parties. These purchases could be made through the treaties described above between the tribes and representatives of the federal government, or they could be made independently from the government by individual Indians or a tribe. This

Table 3.2. The twenty-five largest Indian reservations in the United States (by area), 1984

Reservation	State	Area (sq. mi.)	% reservation that is tribal land	% of total U.S. reservation lands
Navajo[a]	AZ, NM, UT	24,368	93.9	29.2
Papago[b]	AZ	4,462	98.6	5.3
San Carlos	AZ	3,059	95.3	3.6
Wind River	WY	2,951	95.0	3.5
Pine Ridge[c]	SD	2,789	40.0	3.3
Fort Apache	AZ	2,602	100.0	3.2
Hopi	AZ	2,439	99.9	2.9
Crow[c]	MT	2,378	26.8	2.8
Cheyenne River[c]	SD	2,187	68.2	2.6
Yakima[c]	WA	1,800	78.2	2.1
Colville	WA	1,659	96.2	2.0
Uintah-Ouray	UT	1,596	98.6	1.9
Hualapai	AZ	1,552	99.9	1.8
Rosebud[c]	SD	1,492	55.3	1.8
Blackfeet[c]	MT	1,470	27.3	1.7
Fort Peck[c]	MT	1,432	43.1	1.7
Standing Rock[c]	ND, SD	1,322	41.5	1.6
Jicarilla Apache	NM	1,201	100.0	1.4
Warm Springs	OR	1,005	91.9	1.2
Flathead	MT	978	92.4	1.2
Fort Belknap[c]	MT	965	34.7	1.1
Ute Mountain	CO, NM, UT	933	96.4	1.1
Red Lake	MN	892	100.0	1.0
Fort Hall[c]	ID	816	49.8	1.0
Pyramid Lake	UT	745	100.0	0.9
Totals		67,093	83.7	80.0

[a]Includes the Alamo, Canoncito, and Ramah districts of the Navajo Indian Reservation and public lands outside the reservation given to the tribe.

[b]Includes Gila Bend, San Xavier, and Florence districts of the Papago Indian Reservation.

[c]A considerable portion of the reservation consists of allotted private land (not held in trust by the BIA).

Source: ARIL 1984, 4-3.

procedure has continued down to the present day in cases of enlargement of reservation lands. For example, various reservations, one after another, were allocated to the Sauk and Fox Indians by the U.S. government. These reservations, however, were far removed from the original tribal territory, and after only a short time their existence was once again threatened and eventually the reservations were terminated (see figs. 2.1 and 3.5c). In the second half of the nineteenth century, therefore, individual members of the

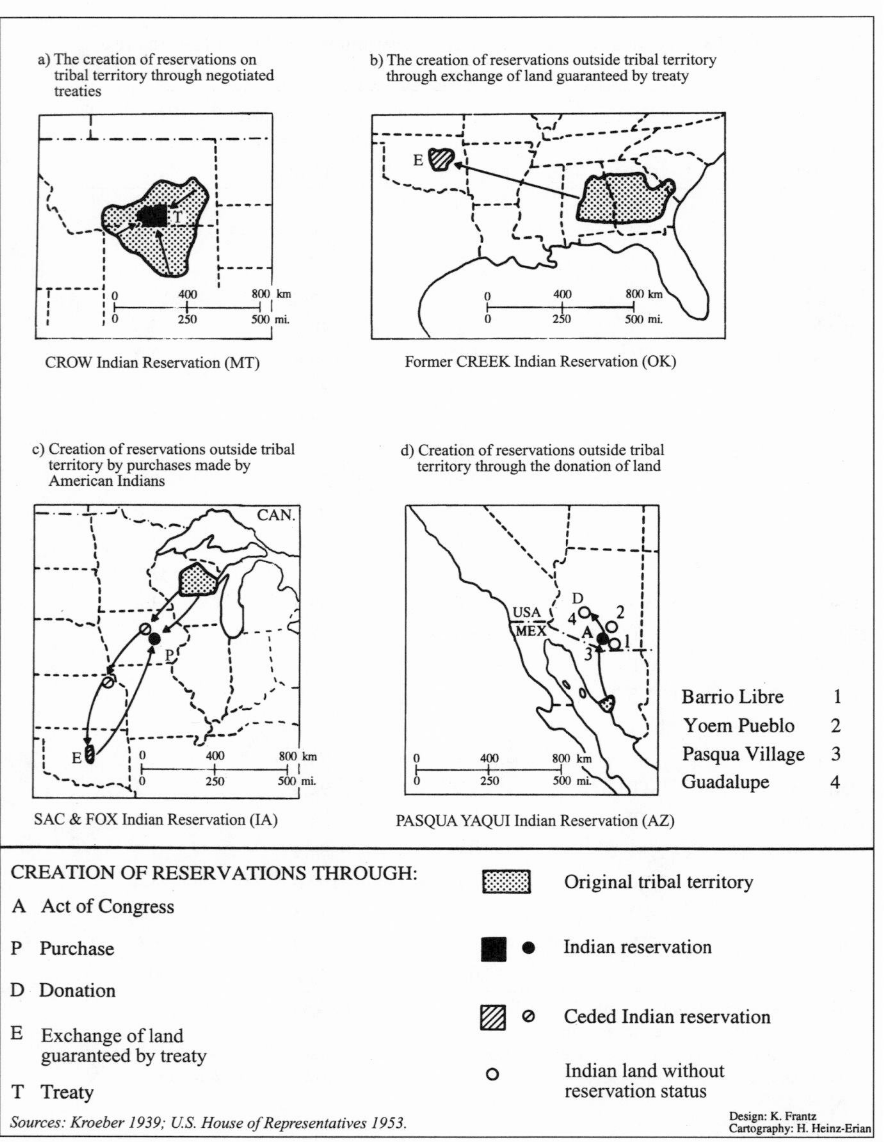

Fig. 3.5: Different ways of creating an Indian reservation in the United States

Sauk and Fox tribes took the initiative and gradually bought various parcels of land in the vicinity of their previously dissolved reservation in Iowa, which had been closed in 1836 (see fig. 3.5c). Subsequently, in 1867, this newly formed Indian community in Iowa, located between the original territory of the Sauk and Fox tribes and the reservation they were given in

Oklahoma, was granted reservation status (see Royce 1900, 778–779, 842–843, and 918).

In some cases the origins of reservations can be traced back to *gifts.* A case in point is that of the Yaqui Indians (see fig. 3.5d), whose ancestors fled from Mexico to southern Arizona at the beginning of the twentieth century to escape repression by the military there. Parcels of land near Tucson and Phoenix were placed at their disposal free of charge, chiefly from private sources through the Presbyterian Church. For a long time the federal government did not consider itself responsible for these Indians, who were not natives of the United States, yet in 1978 one of these four parcels was granted reservation status (see fig. 2.3), and its residents thus became recognized by the federal government and eligible for tribal benefits.

The great majority of the reservations founded after 1871 and their later expansions, reductions, or closures, were the products of *executive orders or acts of Congress.*[6] Compared to the treaty reservations, which in the final analysis amounted to land exchange deals, the reservations which originated through executive orders or acts of Congress had the character of a loan which the president or Congress could try to cancel at any time. This type of reservation is founded, however, on a much less secure legal basis, because the reservation could be curtailed or annihilated as needed without justification.

The present-day boundaries of most reservations were, however, formed over time through a combination of different legal procedures; to reconstruct each tribe's territorial development therefore requires the use of archival material (figs. 3.6 and 7.1). The Navajo Indian Reservation (see fig. 3.1) is a good example because in its present appearance this reservation represents a conglomerate of almost every legal procedure used to create reservations. An original treaty, executive orders from eight presidents, and four acts of Congress were used to create the present boundaries of this reservation. In adding these lands, the Hopis, another tribe that had its home in this area, were overlooked, gradually encircled, and some of their possessory rights curtailed. The ensuing legal controversy, and the large body of literature it produced (see Goodman and Thompson 1975), came to an end in 1974 through the Navajo-Hopi Land Settlement Act, yet disputes and injustices continued for both Navajos and Hopis. The original boundaries of the Hopi Indian Reservation were worked out on a drawing board to conform to the U.S. Public Land Survey and illustrate a phenomenon that is characteristic of many reservations. Many reservation boundaries used straight lines, which frequently failed to take into consideration the natural

6. From the mid-nineteenth century on, reservations were increasingly established by executive order. The president, however, was deprived of this right in 1919, and henceforth only Congress was authorized to establish reservations or to make boundary changes.

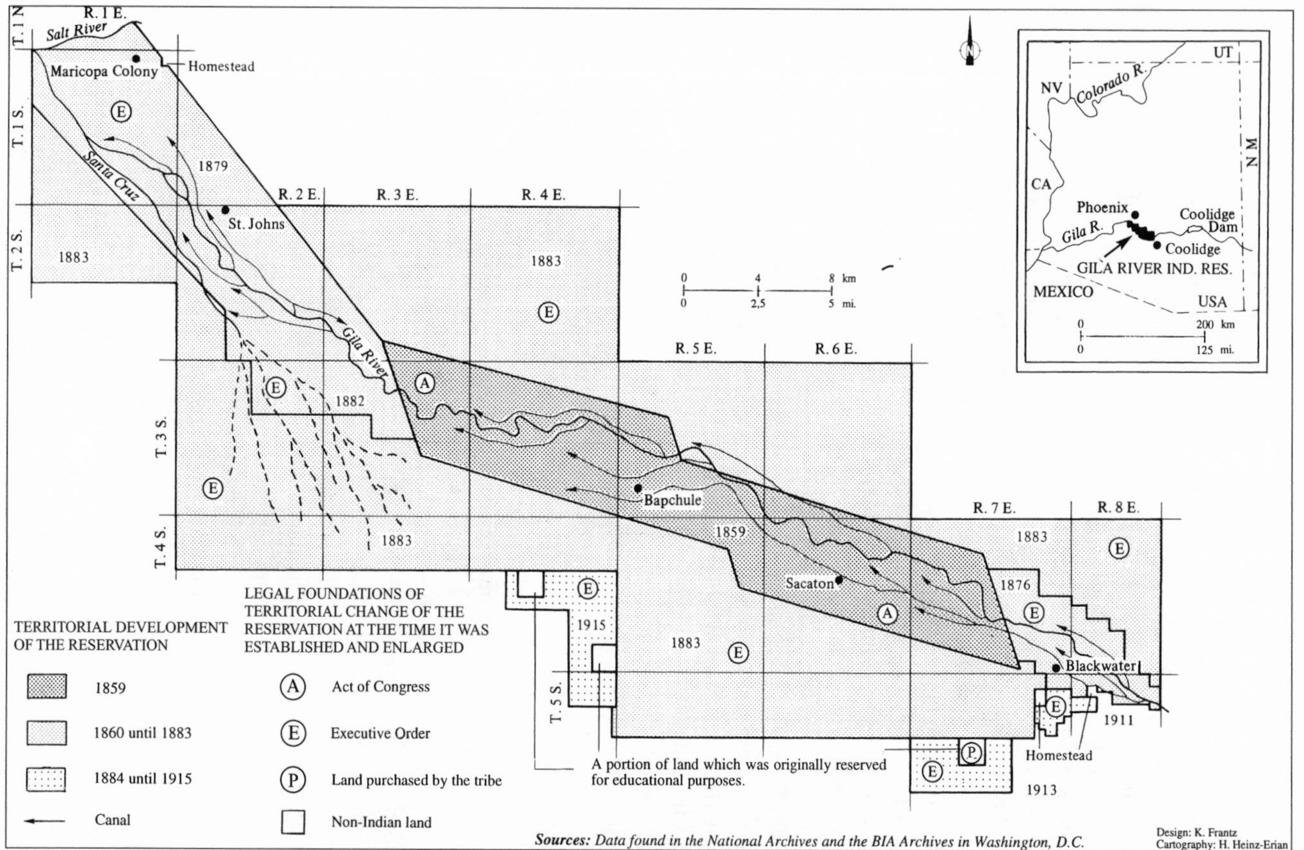

Fig. 3.6: Changes of boundary and non-Indian ownership on the Gila River Indian Reservation in Arizona

boundaries of the landscape and often resulted in the breakup of traditional tribal territory (see Hofmeister 1980), a fact which calls to mind the history of colonial Africa.

THE STATUS OF LANDOWNERSHIP ON RESERVATIONS

Landownership on today's American Indian reservations takes many different forms, and determines to a considerable extent the economic well-being of a tribe and its potentials for development. Many Americans have the impression that all Indian land is held in common, and therefore Indians have been branded by some Americans as Communists.[7] In fact most reservations are comprised of some combination of tribal land held in trust, trust land allotted to individual Indians by the federal government, trust land assigned to individual Indians by the tribe, land owned by individual Indian families (patent-in-fee), and land owned by individual non-Indians (patent-in-fee) acquired from Indian families. The complexity of these issues produces different interests which cannot easily be reduced to a common denominator. This situation often makes it very difficult not only for American Indians, but also for the BIA, which is responsible for protecting, administering, and managing the Indian country; carrying out useful measures for improvement; and initiating and maintaining the utilization of natural resources.

Most of the indigenous peoples knew only tribal or collective ownership of land before their contact with the United States, except for some tribes in the Northwest (e.g., the Makah Indians) and the Southeast (e.g., the Creek Indians) who did not follow this pattern. This, to some extent, makes it understandable why some white Americans had the idea that American Indians were Communists. Although the individual members of the tribe, family, or clan had certain parcels of land for their own use, they did not derive the right of ownership from this fact. In many tribal societies this right of use could be passed down within the family, while the tribe maintained title to the land.

When Indian land was taken over by white settlers, the tribes were compelled to subordinate the title to their land to the U.S. government, which took over the role of trustee. The federal government holds this trust responsibility, but in dealings with the individual states, the tribes could still maintain their limited sovereignty as regards land title. An analysis of the distribution of ownership on all reservations (see table 3.1) shows that in 1984 approximately 80% of all reservation land was under the control of

7. This was the view of James Watt, secretary of the interior under President Reagan, who, in this capacity, was responsible for the BIA.

the tribes. This statement applies particularly to tribes in the West, above all in the Southwest, whose territory was often spared the process of allotment. In the Southwest there are Indian reservations, such as the Papago, Hopi, and Fort Apache, which are almost entirely under tribal ownership. On the other hand, in most tribes in the Midwest and the Northwest, including tribes in Montana and Idaho, the allotment policy led to a great reduction of tribal lands.

There are great advantages for a tribe when the reservation controls its land base under trust status so that the land is then at the disposal of all members and is protected against seizure from outside. This puts an end to the gradual loss of Indian land and makes it easier to administer and manage the lands, for it is necessary only to consult the tribal council with review by the BIA in order to make a decision.

A large proportion of tribal territory, as already mentioned, is held in trust through treaty, executive orders, or acts of Congress, which means that it is tribal or tribally controlled land for which the federal government is ultimately responsible. In addition, there are generally small, scattered plots of land that once belonged to the tribe, were taken by the government, and then were bought back during the IRA era and later integrated into their respective reservations. Tribes themselves have also purchased land during the last two or three decades, only some of which has achieved reservation status (see fig. 3.7).

Today it is still customary for a tribe to assign land to its members, to a family or clan, to an Indian livestock association or agricultural cooperative, or for some designated use upon request. The rights of use for this so-called *assigned land* can be granted by the tribe, either for a definite period of time or without restriction. In each case, however, the tribe always reserves the right to cancel the arrangement when it is in the general interest of the tribe to do so. This right of use is a legal claim, but does not amount to full title to the land. It can expire when the person concerned leaves the reservation for some length of time, or it can be extended or transferred to the children. Today reservation Indians apply for assigned land when they set up their own households or need land for agriculture or ranching. Because of the rapid increase in population, however, most tribal lands are not large enough to provide all tribal members with enough land.

Less than 1% of the total reservation land belongs to the federal government (see table 3.1). This is where the BIA agency offices, the BIA schools, and the IHS hospitals and health centers are located, along with often segregated and fenced-in housing tracts for BIA and IHS employees. These lands also include irrigation projects of considerable size. Some of these facilities are returned to the tribes by the federal government when they are no longer needed.

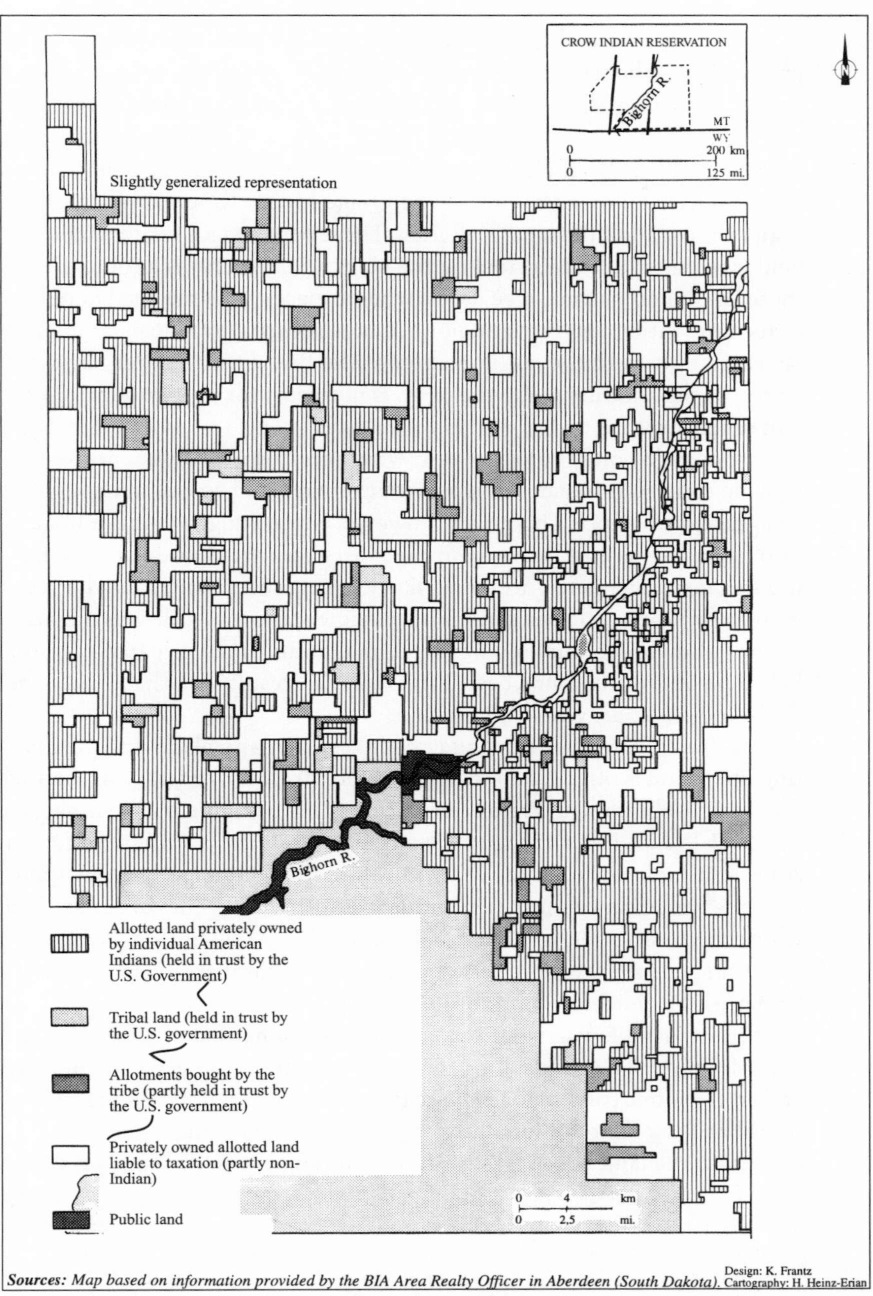

Sources: *Map based on information provided by the BIA Area Realty Officer in Aberdeen (South Dakota).*

Fig. 3.7: Present-day distribution of landed property on the Crow Indian Reservation in Montana

In addition to tribal and federal governmental land, *allotted land* comprises 19% of the reservation land (see table 3.1) which is for the most part in the hands of small or very small landowners. Today, on some of the large reservations, such as Pine Ridge in South Dakota, Crow in Montana (see fig. 3.6), and Fort Berthold in North Dakota, 60 to 85% of the land is allotted to individual owners, and on the Osage Indian Reservation in Oklahoma almost all of the land is allotted. These percentages relate to allotted land which is still held in trust by the United States, and as a consequence these landholdings cannot be used for all purposes. This means that the allottee can neither sell the land nor use it as security for a loan, and it can be leased only with the authorization of the BIA. On the other hand, the owner of the allotment, like the tribe, is not liable for property tax or certain other taxes.[8]

The declared intent of the GAA, which created the allotments, was to provide American Indian families with as much land as was necessary for an agrarian existence. This policy, however, which sought to make Indians into farmers, was effective only in the most exceptional cases, and the failure of most Indians to succeed as farmers can be explained by the fragmentation of the allotted land. This fragmentation, which was a consequence of dividing landholdings into smaller and smaller parcels for the heirs, apparently had not been foreseen. The result was that fewer and fewer Indians could live off their land.

This *fractionated heirship land* can be seen in many places and the splitting up of land is often taken to the extreme. However, contrary to some of the specialist literature (Wehmeier 1975, 144), the fragmentation is not the result of Indian inheritance traditions. The problem is rather that most American Indians do not make a will before they die, and the GAA stipulates that inheritance settlements must conform to the laws of the state in which the reservation is located (see Canby 1981, 124; Sutton 1975, 136). In most states the law of inheritance stipulates that if there is no will, the land must be divided among the heirs.

Within just a few generations this regulation has led to an utterly absurd situation. The result is a general confusion in matters of ownership and an economic basis so inadequate that it is virtually impossible for heirs to utilize their share of land properly, with the result that a considerable portion of the land lies in waste. Such conditions are to be found on many

8. The widespread belief that reservation Indians do not have to pay taxes is only partly correct. It is true that they are not obliged to pay taxes for agricultural or forest products, or for mineral resources developed on their reservations. This represents certain remnants of their original tribal sovereignty. Like any other U.S. citizen, however, American Indians must pay federal income tax. As long as they receive their salaries on the reservation, they are exempted from state income tax. The same exemption is also true for state sales tax on condition that the Indians make their purchases on the reservation.

reservations, and this has often led not only to the paralysis of agriculture, but of the whole economy. With every new generation this problem of land fragmentation increases, as in a geometrical series, so that on some reservations the mosaic of distribution of landholdings resembles an intricate patchwork, whose pattern makes no sense either to the Indian or to the outsider. In spite of a few attempts on the part of the BIA and a few tribal governments to carry out a land consolidation policy by restricting the splitting up of inheritances and by merging or purchasing small parcels of land, there are as yet no signs of a radical land reform.

For many American Indian landowners it is only by leasing their land that they can obtain even a modest income, as the ever-decreasing size of landholdings as well as the lack of capital and know-how are usually obstacles to individual farming. Moreover, many Indians by tradition have no real interest in agriculture. The possibility of leasing was at first not foreseen in the GAA, so an amendment for this purpose was enacted a few years later (Cohen 1982, 153). Until recently the leasing lay almost exclusively in the hands of the BIA, which, however, often grossly neglected its trust responsibilities when advantageous conditions for interested white parties had to be negotiated. It was only with the policy of self-determination, and the cooperation of helpful lawyers, that some tribes were finally able to revise old leases to their advantage and draw up new leases from which they would derive more benefit. The old leases, which were often concluded for twenty-five years, in some cases even for ninety-nine, brought hardly any return to the lessors and did little to prevent a ruthless exploitation of natural resources.

In 1984 altogether 13.6% of the reservation land of the United States was under lease (see table 3.1), of which 97% was for agricultural purposes (see fig. 9.3).[9] However, the statistics do not show what proportion of the respective tribal territory or of the individually owned landholdings are under lease, nor do they show the percentage of non-Indians and Indians among the leaseholders. My own investigations show that a large proportion of the land under lease is occupied by non-Indian leaseholders. Further, it is a striking fact that in the very states where the fragmentation of land has been carried to the greatest extreme the amount of leasing is far above the national average. In some states of the Midwest, as well as in Oklahoma, Montana, and Idaho, it amounts to between 40 and 80% (see table 3.1).

Apart from the conditions of ownership and leasing already discussed,

9. With regard to the latter 97%, one must, however, bear in mind that the same parcel of land can be leased simultaneously for agriculture, forestry, and mining. There are a number of reservations with pastureland which have also sold prospecting rights, and forests that are grazed are also quite common on reservation land.

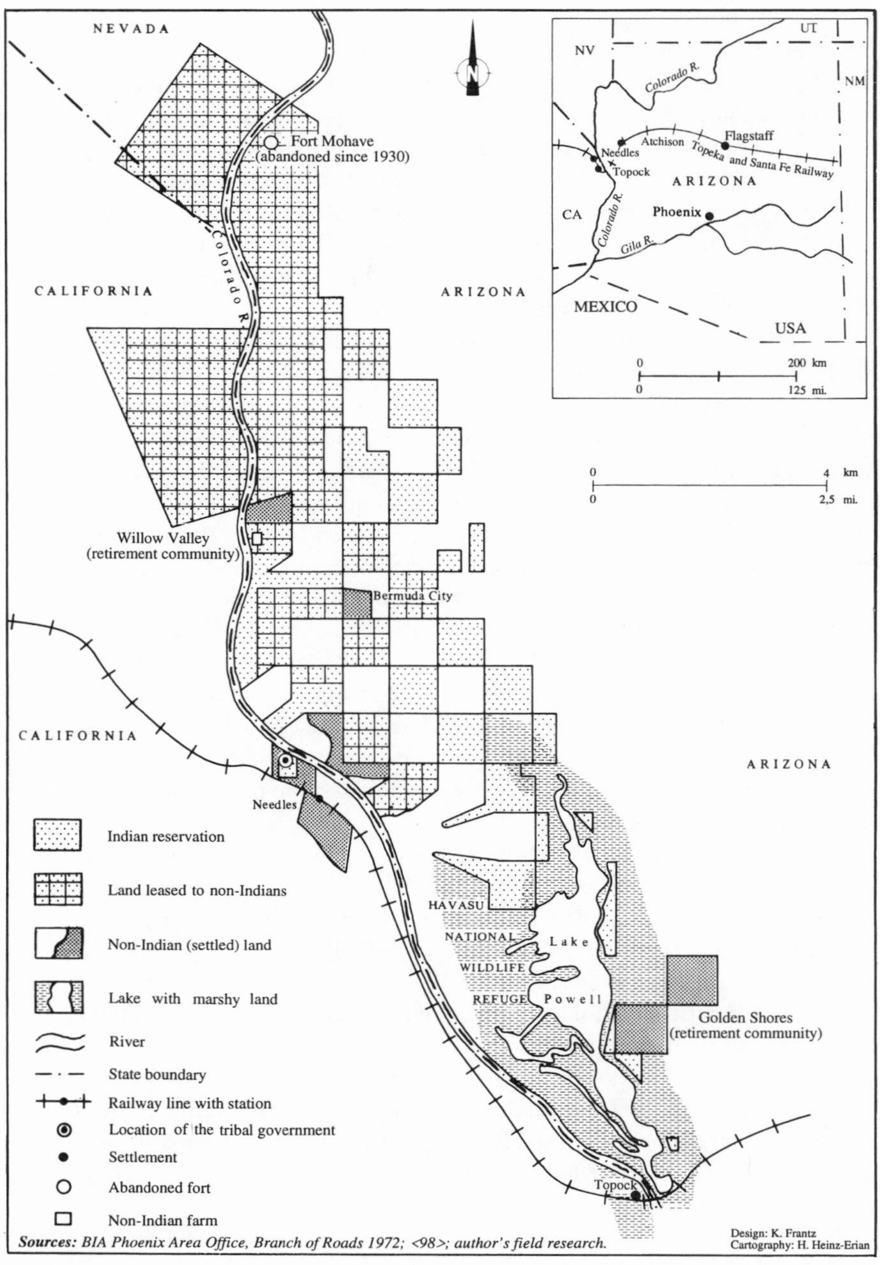

Fig. 3.8: The checkerboard pattern of the Fort Mohave Indian Reservation (Arizona, California, Nevada), 1986

it is often overlooked that on many reservations individual, non-Indian landownership is to be found within the reservation boundary, although this is usually not shown on the large-scale maps, not even on the official BIA maps. Surprisingly, the Indian agency has never carried out a survey to find out more about this situation, which is, no doubt, disadvantageous to the reservations concerned. One must therefore rely on one's own observations in order to understand this problem. It can be seen that on some reservations the distribution of landed property between Indians and non-Indians often resembles a chunk of Swiss cheese, for the territory that was once reserved entirely for the Indians is now interspersed with "white" patches and thereby, to some extent, escapes the control of the tribe.

For example, on the Crow Indian Reservation in Montana one-quarter of the reservation area is in non-Indian hands. Moreover, around 65% of this area is under lease to non-Indian farmers (see fig. 3.7). On some reservations of the Midwest and Northwest, such as the Sisseton Indian Reservation in South Dakota and the Port Madison Indian Reservation in Washington, more than 50% of the land is owned by non-Indians.

Non-Indian landownership on reservations can have a number of different causes. In some cases these are homesteads which were made over to white families or individuals before the reservation was created. This is the case in the southeastern parts of the Gila River Indian Reservation in Arizona where non-Indian landownership has caused the jagged boundary (see fig. 3.6). Moreover, in the southern part of the same reservation there are two sections, each 640 acres, which in the surveying of Arizona were reserved for school purposes, a practice also customary in other states of the union. Today these areas are either wasteland or in the possession of non-Indians.

On some western reservations the distribution of landholdings between non-Indians and Indians follows a checkerboard pattern that resulted from the Railroad Act of 1866. These include the Indian Pueblos in New Mexico and the Navajo and Fort Mohave Indian Reservations as well as the Hualapai Indian Reservation in Arizona of the late nineteenth century (see fig. 3.8). On the basis of this act, railroad companies in the western part of the United States acquired large territories of land, as every other section was allotted to them, like on a chess board, within broad strips on both sides of the railroad line. As a result of this generous policy of land distribution, some Indian tribes lost as much as half their territory shortly before or even while their reservation was being established. Sometimes a railroad station and the surrounding land remained non-Indian territory in the middle of a reservation (see Frantz 1990a, 34). Such a railway station could later develop into a larger settlement for non-Indians, as, for example, in the case of Parker, Arizona, on the Colorado River Indian Reservation (figs. 3.9, 3.10).

Parker is a settlement that spread out around a railway station at the close of the nineteenth century. The station was first built because it was easy to bridge the Colorado River at this point, and because there was water available for the steam engines. In 1980, Parker was a settlement with an area of barely 1,000 acres and around 2,100 residents. Today, population is declining and the economy is stagnant. Moreover, the settlement is gradually being taken over by the Colorado River tribe through the purchase of land as well as successful litigation. It is a striking fact that the grid of the land survey on the reservation, as in most parts of the United States, is aligned with east-west baselines and north-south meridians, whereas the arrangement of plots of land in Parker conforms to the railway line, with the consequence that the surveying lines of the non-Indian enclave Parker come together with those of the surrounding reservation at an acute angle. This is an anomaly of the system of surveying that is also to be found on other reservations.

Nevertheless, the largest percentage of non-Indian landholdings on reservation land has resulted from the General Allotment Act. Because of this legislation some reservations were swamped with non-Indian landowners within two generations. The extent of this non-Indian landownership reached its peak in the years before the world depression and the dust bowl of the 1930s. As a result of these years of crisis many non-Indian farmers left the reservations. Their land lay fallow, and some tribes were even able to buy some of this land back again at a later stage (see fig. 3.7).

CHARACTERISTIC PATTERNS OF TERRITORIAL DEVELOPMENT OF THE INDIAN RESERVATIONS AND THE REMAINS OF TRIBAL SOVEREIGNTY

Ideal-typical observations concerning a multilayered, complex subject matter often carry the risk of generalizations that are wrong or, at least, misleading. This applies not only to generalizations about large cities of a specific cultural realm, but also, and more particularly, to the study of the territorial development of the U.S. Indian reservations. The course of this development depended on such factors as the date when a particular reservation was established, its geographical location, and its economic potentials, as well as the specific cultural values and socioeconomic features of the tribe inhabiting the reservation (see Frantz 1990a, 31–43). Most phases of this development, the manifold traces of which cannot all be found on any one reservation, have already been discussed in detail in the earlier chapters. They will, therefore, simply be summarized in this

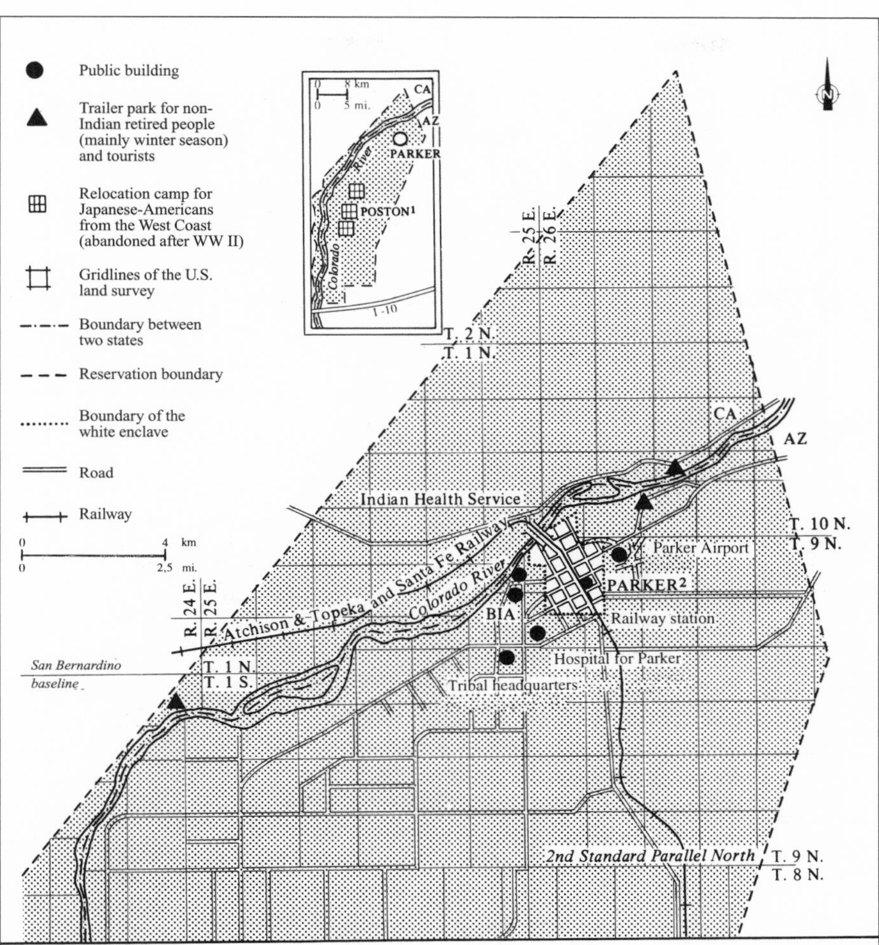

1) Between 1942 and 1945 around 200,000 Americans of Japanese descent—of whom 70,000 did not yet have American citizenship—were relocated from the West Coast states to the interior of the country, where they were settled in various camps. Around 20,000 of these deportees were interned in Poston.

2) While the gridlines used in U.S. land survey run parallel to the baseline and base meridian, such as were used in the Colorado River Indian Reservation and in many other U.S. regions, the surveying system of Parker follows the railway line.

Design: K. Frantz
Cartography: H. Heinz-Erian

Fig. 3.9: The northern area of the Colorado River Indian Reservation with the non-Indian settlement, Parker, Arizona, 1985

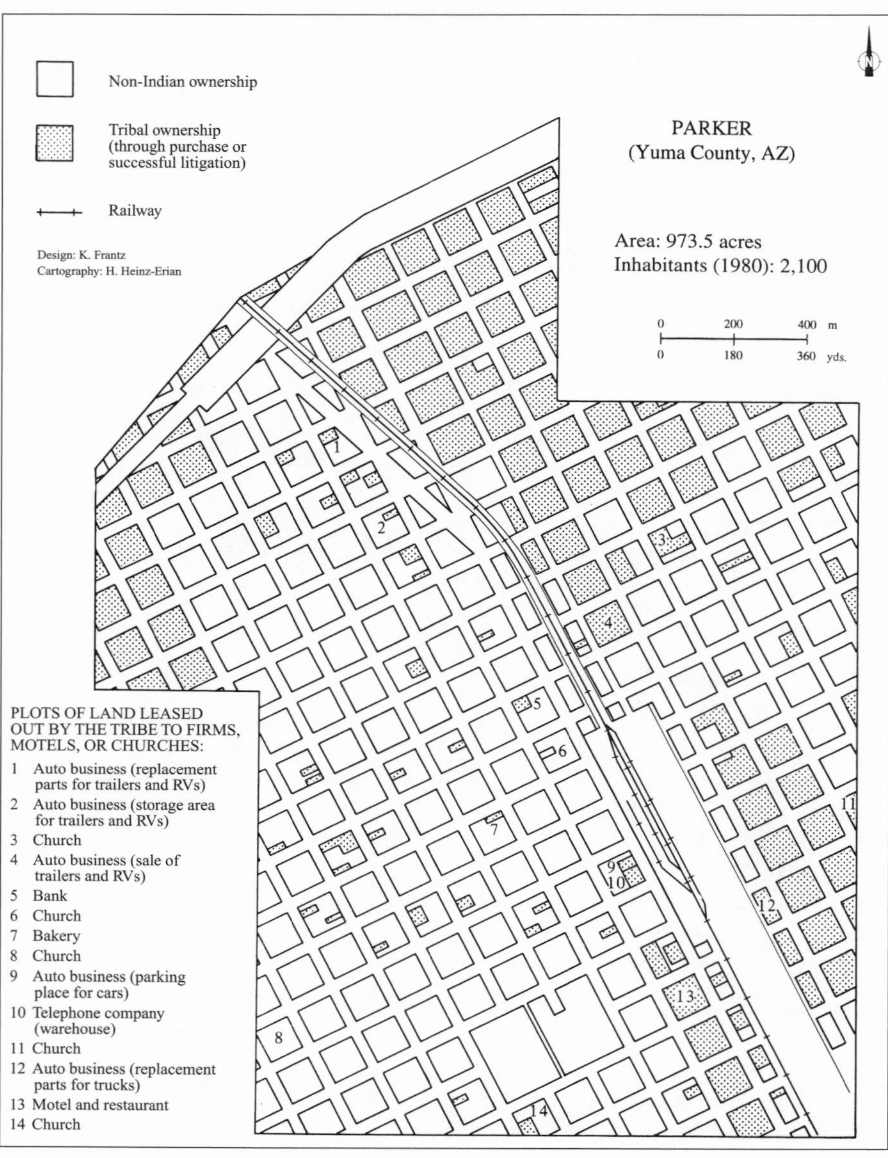

Non-Indian ownership
Tribal ownership (through purchase or successful litigation)
Railway
Design: K. Frantz
Cartography: H. Heinz-Erian
PARKER
(Yuma County, AZ)
Area: 973.5 acres
Inhabitants (1980): 2,100
0 200 400 m
0 180 360 yds.
PLOTS OF LAND LEASED OUT BY THE TRIBE TO FIRMS, MOTELS, OR CHURCHES:
1 Auto business (replacement parts for trailers and RVs)
2 Auto business (storage area for trailers and RVs)
3 Church
4 Auto business (sale of trailers and RVs)
5 Bank
6 Church
7 Bakery
8 Church
9 Auto business (parking place for cars)
10 Telephone company (warehouse)
11 Church
12 Auto business (replacement parts for trucks)
13 Motel and restaurant
14 Church

Fig. 3.10: Property and leasing arrangements in Parker, Arizona, 1986

concluding discussion and represented graphically in a pair of spatiotemporal diagrams.

Tribal Territory prior to White Settlement

Before the arrival of Europeans most American tribes, and, in particular, the nonsedentary Indians of the Midwest and West, laid claim to territories that were considerably larger than their present-day reservations. Their utilization of these territories usually showed well-defined spatial sequences, involving central and peripheral areas (see fig. 3.11a). The great majority of the settlements and sacred places of the tribe and, in many cases, also the ancestral graves were to be found in what could be termed a *core area* (see Meinig 1965, 213). Depending on tribe and region, this was also the zone where much of the farming was done. The boundaries of this tribal core area were, for the most part, clearly marked in the eyes of its inhabitants, and proved relatively stable over long periods of time despite occasional attacks by neighboring tribes.

The core area of tribal territory was usually surrounded by a *domain,* an area which provided the natural resources indispensable to the tribe's livelihood and traditional economy. In this second zone or domain the younger men of the tribe hunted during certain seasons of the year, while the women and children sought to extend the sometimes meager diet of the tribe by gathering various plants. Also, extensive dry farming methods were applied from time to time, as, for example, by many tribes in the mountainous regions of the Southwest. The boundaries of this domain area were far from clearly defined and often contested by neighboring tribes. The domains of two tribes often overlapped, but were accepted by both sides.

The area beyond this zone could be referred to as the *tribal sphere* (Meinig 1965, 216). This could be a very extensive region with fluid boundaries which might reach for many miles into neighboring tribal territory. Here trade missions and raiding expeditions often occurred.

The Creation of the Reservation

The creation of the United States and the gradual penetration of white settlers led to a fundamental transformation of American Indian territory. Absolutely straight boundary lines were imposed, which followed parallels and meridians within and between the individual states, as well as between the United States and Mexico or Canada. These boundaries ignored existing tribal territory and caused land losses which, when they occurred outside the United States, could no longer be reclaimed through the ICC (compare figs. 3.11a and 3.12e). Furthermore, in the interior of the country, the U.S. army curtailed the size of Indian territory through

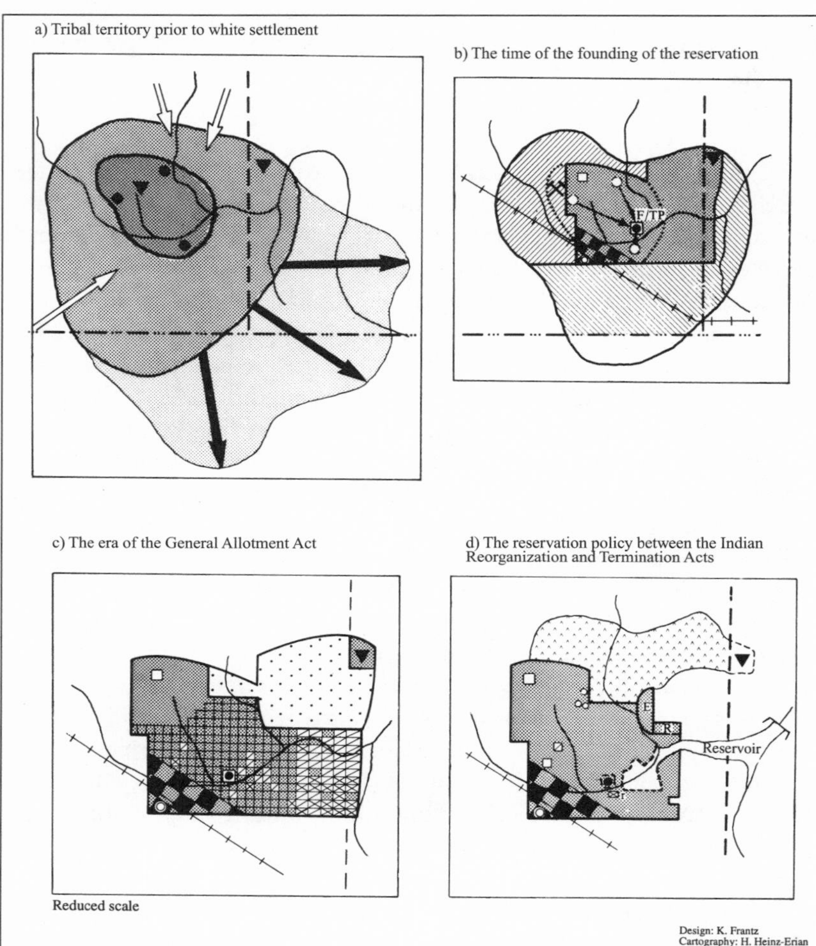

Fig. 3.11: Spatiotemporal diagram of the phases in the territorial development of an Indian reservation (part 1) *(continued)*

forced land cessions, often carried out in stages. The consequence was that well-established indigenous settlements and economic structures which previously had formed an ecological unit with the surrounding natural environment were broken up. What remained of the reservations was generally only a fragment of the traditional tribal lands (see fig. 3.11b).

Even the tribes' core areas were subject to drastic reductions in size when this was in the economic interest of the whites. Often this happened even before the establishment of a reservation and independently of the official policy of land cession. Thus, for example, numerous plots of land were made available to settlers and railroad companies through the

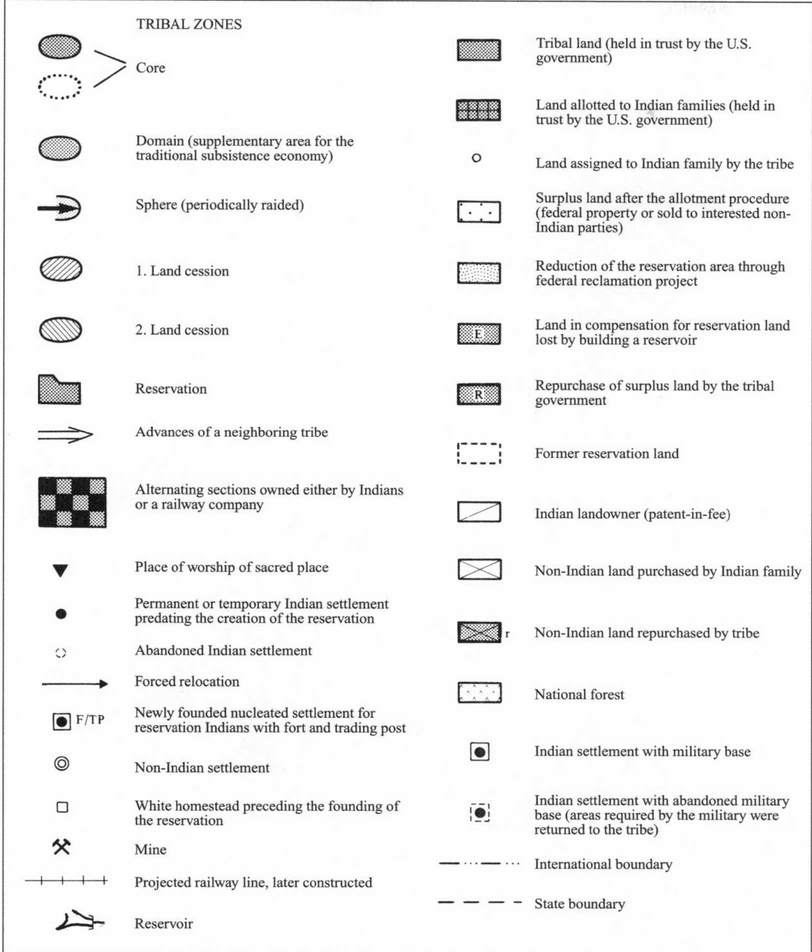

(Fig. 3.11 continued)

Homestead and Railroad Acts. These areas, as white enclaves in the midst of reservation land, are today a thorn in the flesh of the tribal governments. Particularly complicated situations could occur when an enclave developed into a small Anglo town over the course of time. This has happened, for example, on the Colorado River Indian Reservation in Arizona.

In the early years of the reservations the authorities often set out to destroy the network of Indian settlements, for it was military and Indian agency policy to concentrate all the tribal members in the newly established nucleated settlements, as this would make it easier to keep American Indians under control.

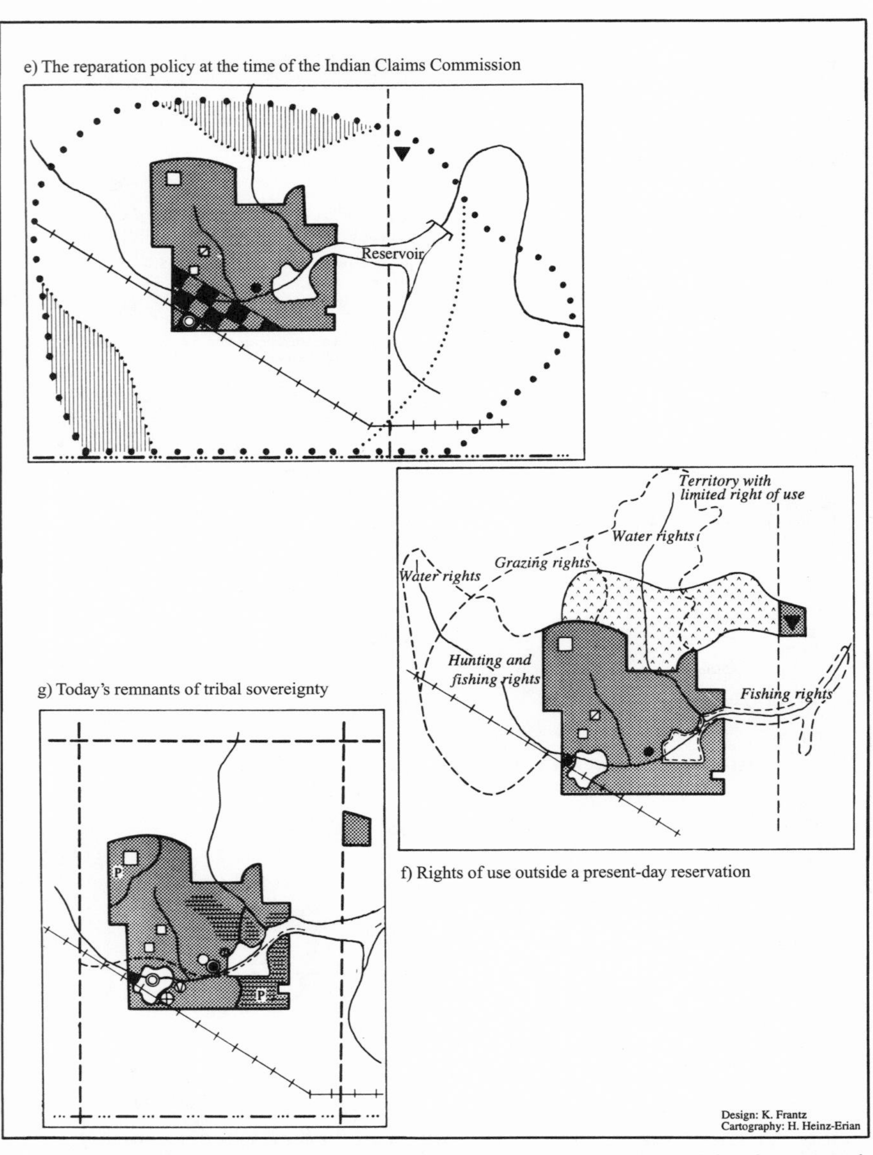

Fig. 3.12: Spatiotemporal diagram of the phases in the territorial development of an Indian reservation (part 2) *(continued)*

Indian land claims before the founding of the Indian Claims Commission (ICC)
Indian land claims recognized by the ICC (financial compensation instead of restitution of land)
Sacred place
Reservation with tax sovereignty, limited authority in administration and jurisdiction, and its own regulations for hunting and fishing
P Leased land
Partial liability to taxation
Area of fractionated heirship land
National forest
Area with overlapping territorial claims
Area outside the reservation with rights defined by treaty
Reservoir
Alternating sections (owned either by Indians or a railway company)
Non-Indian urban area (extended by exchange of land)
Indian settlement
Indian settlement with tribal government. Indian agency (BIA), tribal law-court, hospital (IHS) etc.
Trailer settlement for non-Indian technical personnel
Non-Indian settlement
Industrial park (taxed by Indians)
Gambling hall (taxed by Indians)
Smoke shop (owned by Indian family)
International boundary
State boundary
County boundary
Railway

(Fig. 3.12 continued)

The Era of the General Allotment Act

Not long after the reservations were created many tribes realized that even their rather small reservations were not secure from seizure by whites. The GAA gave free rein to the land-hungry whites and positively encouraged settlers to appropriate land at the expense of American Indians. The means to this end was to allot tribal land to individual Indian families. Thus many American Indians became unrestricted landowners for a short time before they sold their land to non-Indians (see fig. 3.11c). Even the so-called surplus land, which was at first administered by the federal government, passed into hands of whites to a considerable extent. What remained was largely federal land, and as such, particularly in arid and semiarid regions, was used frequently for the conservation of forest and water reserves. As a result of this permissive policy with Indian territory, some once large reservations survived only in the form of a few scattered fragments.

Reservation Policy at the Time of the Indian Reorganization and Termination Acts

After half a century of systematically shrinking Indian territory, there followed a brief period of land consolidation in the 1930s, for the first time in the history of Anglo-Indian relations. Even so, large dam projects further reduced reservation land (see fig. 3.11d). Now, for the first time, the tribes were given the opportunity to buy back previously lost land from the government or from private landowners. In addition, on many reservations military and BIA facilities that were no longer needed were returned to the Indians without charge, although this increased their territory only very slightly. At the time of these attempts to reinforce the collective landholdings of American Indians, the Indian agency was still pursuing its policy of land redistribution by asking the tribes to assign land to their individual members.

This short period of a liberal Indian policy came to an end with the *reparation policy* of the Indian Claims Commission. This was a commission, founded in 1946, to which the Indian tribes could bring their land claims, when they believed land had been taken from them illegally. Generally the ICC with its staff of historians and anthropologists saw the reconstruction of tribal territory much more narrowly than the tribes and their consultants (see fig. 3.12e). In addition, the reparation commission had to consider competing territorial claims of neighboring tribes. The consequence was that the financial reimbursements paid by the government were far less than the Indians had expected.

While the ICC was still engaged with reparation payments for illegal deprivation of land, the U.S. government with its termination policy once again started to reduce the Indian land base. Through negotiations, however, some of these lost reservation areas have been recently retrieved (fig. 3.12e, f).

The Remains of the Former Tribal Sovereignty

During the last two decades many Indian tribes have been concerned not only about the defense of their land claims, but also increasingly about the reassertion of their residual sovereignty and traditional rights of use, which frequently extend beyond reservation land. In the past, although these rights of use were guaranteed by treaty, the agreements were all too frequently disregarded (see fig. 3.12f). Particularly serious conflicts have arisen with some states in the Northwest and Southwest in which fishing and water rights are of prime importance for tribal economies. In these conflicts of interest the U.S. government is in a difficult position. On the one hand, as trustee of the Indian reservations, it is bound by treaty to protect the different tribes from state or private seizures; yet, on the other hand, it must support the states involved, for whom the partly "extraterritorial" reservations, legally speaking, are an economic obstacle.

During the past few years a primary concern on many reservations has been to overcome the problems resulting from land fragmentation, which in some cases was extreme. This fragmentation was a consequence of the

Railroad Act, as well as land inheritance. On some reservations on which railroad companies owned alternate sections of land, the bewildering pattern of ownership, which so easily produced conflict, could be cleared up through land exchanges (see fig. 3.12g). Overall, however, the increasing trend toward fractionated heirship land remains an unsolved problem to this day.

In conclusion, tribal sovereignty remains a fundamental question still to be clarified, one closely bound up with the territorial status of the Indian reservations (see fig. 3.12g). Although all reservations are located within states, they are largely independent entities with regard to politics, administration, and liability to taxation. In matters of legal procedure they each go their own way. All this shows that down to the present day the Indian reservations have been able to preserve aspects of their former sovereignty, even if they are compelled time and again to fight for this special status in court.

Generally speaking, reservations are not integrated into the political systems of individual states and their counties. Their members elect their own tribal council and a leader whom they designate as chairman, governor, or president. Moreover, they have no political representatives on the federal level. The tribal administrations can therefore make many regulations, with which even nontribal members must comply, provided that these regulations have been previously approved by the BIA.

In the field of taxation tribal sovereignty proves to be particularly striking. Even if the allegation that reservations are not liable to taxation is true only with qualifications (see Feest 1976, 118), it is nevertheless correct to say that they enjoy a considerable degree of tax exemption. The tribe and its members pay taxes neither to the states in which they are located nor to county authorities. This regulation applies even to Indian landowners with unlimited right of disposal. This means that reservation Indians are exempt from property and sales taxes as well as the other taxes levied by the state. This special status, of course, results in a considerable loss of income for states that have a large share of reservation land, and it is therefore understandable that such states are generally reluctant to give financial support to their Indian reservations.

The tribe and its members are also exempt from federal taxes. This applies to income derived from landed property and natural resources. Thus, for example, income from mining, agriculture, and forestry and goods processed from this primary sector are exempt from taxation. This applies also to income from leases. On the other hand, a federal tax is usually levied on income from a privately or tribally operated motel or store on the reservation (see Canby 1981, 182). The regulations mentioned here regarding federal tax exemption are only in part applicable to Indian landowners with

Fig. 3.13: Bingo hall at the Fort McDowell Indian Reservation, near Phoenix, Arizona

unlimited right of disposal. The white landowner on reservation land is, however, liable to taxation (see fig. 3.12g).

Taxing authority also means that the tribe itself has the right to levy taxes, though for the most part, this right is not exercised on tribal members. However, non-Indian businesspeople on the reservations, who in many respects enjoy the advantages of a tax haven, must often pay moderate tribal taxes. In the past few years some reservations, on the strength of their geographical locations, have shown considerable imagination in exploiting their right to levy taxes. For example, the Navajo tribe taxes a coal-fired thermal power station for its air pollution despite initial protests from the white owners. Bingo operations and other gambling businesses, such as the casino on the Fort McDowell Indian Reservation near Phoenix, Arizona (see fig. 3.13), have for some time now been springing up on reservations close to densely populated areas all over the country, in spite of the vehement opposition of the states in which they are located. This gambling is also taxed by the tribe.[10] On the other hand, there is no

10. According to the BIA, gambling operations have arisen on 85 of the 264 reservations since 1981. Twenty-five reservations even show profits of more than a million dollars a year. For the treasury of the respective states this, of course, involves a great loss of internal revenue. In the past few years many law courts have sought to prevent the rapid spread of gaming. However, charities in most states of the Union are allowed to run orga-

tax on the so-called smoke shops, which for the most part, belong to Indians (see fig. 10.9). They are located near reservation boundaries in order to benefit their non-Indian customers, who can thereby avoid paying state taxes on tobacco.

Finally, a word about the administration of justice, which is closely linked to the territorial status of Indian country, a privilege peculiar to the tribes. This privilege varies from state to state: in the wake of the termination policy, Public Law 280 (1953) took effect in some states, and there the administration of justice was mostly taken over by state courts (see Cohen 1982, 175–176; Ranquist 1975, 701–704; Sutton 1975, 147–154; Washburn 1971, 86–87; and Witt 1968, 65). In California, Oregon, Montana, Minnesota, Wisconsin, and Alaska, tribes have thus lost this remnant of their original sovereignty. In other states, including Arizona, Colorado, New Mexico, Utah, Wyoming, the Dakotas, Mississippi, and North Carolina, Indian legal jurisdiction is, in contrast, still intact, based on the special relations between the federal government and reservation Indians as regulated at the time of the controversy between the Cherokees and Georgia. Although serious crimes such as robbery and murder fall under the jurisdiction of the federal courts, the tribal courts' jurisdiction over minor offenses is a symbol of these reservations' special status.

---

nized games of chance, with the exception of roulette and the so-called one-armed-bandit slot machines. These states have, therefore, been powerless to do much against this new trend in Indian country.

4 The American Indian Population

Being Indian is paying 15 dollars a piece for eagle feathers today when you don't have enough food for tomorrow's meals. Being Indian is to be the best you can possibly be at what you do, but not to openly compete with your fellow man to your own aggrandizement and glorification and his shame and humiliation. Being Indian is having at least one alcoholic relative put the touch on you once a day. Being Indian is having at least a dozen missionaries from twelve different faiths trying to save your heathen soul every year. Being Indian is missing work at least two days a month because so many of your friends and relatives are dying. Being Indian is living on borrowed time after your forty-fourth birthday. Being Indian is feeling Grey Wolf, Thunder Chief and Smoke Walker are more beautiful names than Smith, Brown or Johnson. Being an Indian is forever!

The Winnebago Reuben Snake, Jr., 1972

COUNTING THE UNCOUNTABLE? PROBLEMS IN DEMOGRAPHIC ANALYSIS FOR AMERICAN INDIANS

In a country in which great attention has long been paid to population statistics and their evaluation, it is surprising to note that the number of its indigenous population has always been largely unknown. The results of the various censuses are, it is true, classified with all possible precision, yet a careful examination shows that at least with regard to the American Indian these findings are, even today, only estimates. This is equally true of the data from the U.S. Bureau of the Census (see Frantz 1990b; Meister 1978) and the data from the Bureau of Indian Affairs (BIA), the Indian Health Service (IHS), and even of the individual tribal administrations.

The data on American Indians are imprecise, and in some cases, extremely variable from one data source to another owing to a number of factors. First, since the beginning of the U.S. census, there has never been general agreement on the question of who is an Indian. Furthermore, the census takers who collect data in the field did not have an adequate knowledge of the Indian way of life. Not until 1980 were investigations carried out by people with sufficient geographical and linguistic knowledge to locate and enumerate tribal members accurately.[1] There is a further problem in that some authors regard the expressions "American Indian" and "Native American" as synonymous when interpreting the results of the census, and then make comparisons with earlier periods. This overlooks the fact that census authorities regard Eskimos, Aleuts, and Hawaiians, alongside American Indians, as distinct ethnic groups, a distinction which, in my opinion, is of little value. Thus, for example, in the population statistics for the United States from 1960 on, American Indians, Eskimos, and Aleuts are listed separately. If this distinction is made, it would be only logical to distinguish between Sioux, Apaches, Hopis, Papagos, and so forth instead of lumping them all together as "American Indians," as was done until 1980.

The methods by which the Census Bureau has collected data have changed over time, and must be studied to understand the development of the American Indian population.

Until the census of 1950 (inclusive) it was up to the individual census takers to decide which race a person belonged to. These investigators received instructions from census authorities and from each state before starting, which threw open the door to wrong conclusions. The governor of Virginia, as recently as 1924, decreed that in order to count as an American Indian outside the two tiny state reservations still remaining at that time, it was necessary to have at least one-sixteenth of Indian blood and no African ancestry (Forbes 1981, 407). This meant that someone who was seven-eighths American Indian would be listed as "colored" or "half-breed" if they were one-eighth African American. A similar situation existed in North Carolina, and this is the only possible explanation for the low Indian population numbers there until 1960 (see table 4.1). Many of the American Indians removed to Oklahoma were partly of African ancestry, and therefore not counted. Because of these factors the American Indian population figures were too low and not reliable in most states.

With the census of 1960 the authorities adopted the principle of *ethnic self-enumeration*. That meant that it was left to each individual to identify which ethnic group they felt they belonged to. This was reflected in increased American Indian populations in the census in all states (see table 4.1), even though many Indians until 1970 were afraid to declare themselves as such, probably because of latent racial discrimination.

1. Census takers, who were not familiar with the region, did not count many Navajos because their homes were in areas remote from main roads. On the other hand, some Navajos were counted twice, either because they had two homes on the reservation or because they were on a visit to another family at the time when the census was taken (see Frantz 1990b, 67; König 1980, 34–35).

Table 4.1. American Indian population: Number and percentage of selected state populations, 1890–1980

	1890		1910		1930		1950		1960		1970		1980	
	No.	%	No.	%	No.	%	No.	%	No.	%	No.	%	No.	%
Arizona	29,981	34.0	29,201	14.3	43,726	10.0	65,804	8.7	83,387	6.4	94,310	5.3	152,498	5.6
California	16,624	1.4	16,371	0.7	19,212	0.3	19,886	0.2	39,014	0.2	88,263	0.4	198,275	0.8
Colorado	1,092	0.3	1,482	0.2	1,395	0.1	1,567	0.1	4,288	0.2	8,002	0.4	17,734	0.5
Idaho	4,223	4.8	3,488	1.1	3,638	0.8	3,791	0.6	5,231	0.8	6,646	0.9	10,418	1.1
Kansas	1,682	0.1	2,444	0.1	2,454	0.1	2,381	0.1	5,069	0.2	8,261	0.4	15,256	0.6
Michigan	5,625	0.3	7,519	0.3	7,080	0.1	6,941	0.1	9,701	0.1	16,012	0.2	39,734	0.4
Minnesota	10,096	0.8	9,053	0.4	11,077	0.4	12,507	0.4	15,496	0.4	22,322	0.6	34,831	0.9
Missouri	2,036	0.2	1,253	0.1	1,458	0.1	2,493	0.1	3,119	0.1	3,791	0.2	6,131	0.2
Montana	11,206	7.8	10,745	2.6	14,798	2.8	16,709	2.8	21,181	3.1	26,385	3.8	37,598	4.8
N. Carolina	1,516	0.1	7,851	0.4	16,579	0.5	3,730	0.1	38,129	0.8	44,195	0.9	64,536	1.1
N. Dakota	8,174	4.3	6,486	1.1	8,387	1.2	10,756	1.7	11,736	1.9	13,565	2.2	20,120	3.1
Nebraska	6,431	0.6	3,502	0.3	3,256	0.2	3,952	0.3	5,545	0.4	6,671	0.4	9,145	0.6
New Mexico	15,044	9.4	20,573	6.3	28,941	9.8	41,886	6.1	56,255	5.9	71,582	7.0	107,338	8.2
Nevada	5,156	10.9	5,240	6.4	4,871	5.3	4,973	3.1	6,681	2.3	7,476	1.5	13,306	1.7
New York	6,044	0.1	6,046	0.1	6,973	0.1	10,427	0.1	16,491	0.1	25,560	0.1	38,967	0.2
Oklahoma	64,456	24.9	74,825	4.5	92,725	3.9	53,602	2.4	64,689	2.8	96,803	3.8	169,292	5.6
Oregon	4,971	1.6	5,090	0.8	4,776	0.5	5,806	0.4	8,026	0.5	13,210	0.6	26,591	1.0
S. Dakota	19,854	5.7	19,137	3.3	21,833	3.2	23,318	3.6	25,794	3.8	31,043	4.7	44,948	6.5
Utah	3,456	1.6	3,123	0.8	2,869	0.6	4,188	0.6	6,961	0.8	10,551	1.0	19,158	1.3
Washington	11,181	3.1	10,997	1.0	11,253	0.7	13,796	0.6	21,076	0.7	30,824	0.9	58,186	1.4
Wisconsin	9,930	0.6	10,142	0.4	11,548	0.4	12,191	0.4	14,297	0.4	18,776	0.4	29,320	0.6
Wyoming	1,844	2.9	1,486	1.0	1,845	0.8	3,237	1.1	4,020	1.2	4,717	1.4	7,057	1.5
Total	240,622	1.2	256,054	0.8	317,841	0.7	323,942	0.5	466,186	0.6	648,965	0.8	1,120,439	1.1
U.S.	248,253	0.39	276,927	0.30	343,352	0.27	357,499	0.24	523,591	0.29	792,730	0.39	1,366,676	0.60

Sources: U.S. Department of Commerce 1910, Indian Population, 11; U.S. Department of Commerce 1930, 5 and 8; U.S. Department of Commerce 1950, Nonwhite Population by Race, 3B-60-61; U.S. Department of Commerce 1960, Characteristics of the Population, vol. 1, table 56; U.S. Department of Commerce 1970, 1 and 18–26; U.S. Department of Commerce 1980, Population (Supp.), American Indian, 14–26.

The BIA, on the other hand, as the responsible agency for Indian affairs in the Department of the Interior, has data with much lower population figures (see table 4.2). This is hardly surprising, as it has no interest in having a large number of American Indians on its records, because each additional Indian puts a new burden on the BIA's limited budget. In order to be recognized as an American Indian by the BIA and thus to have this legal status and title, one had to either live on or near a reservation and be an enrolled member of a federally recognized tribe or be able to prove at least one-quarter Indian ancestry.

As a consequence the following paradoxical situation has arisen. Today thousands of people who can give no evidence of Indian ancestry are officially recognized as American Indians, while, on the other hand, tens of thousands American Indians, some of them full-blooded, are deprived of this status. Among the latter are many urban Indian groups who regard themselves as Indian tribes; these groups, generally located east of the Mississippi or in the west coast states, are not federally recognized, however (see fig. 4.1). On the other hand, whites who enjoy the legal status of American Indians can be found on a number of reservations rich in natural resources, as Feest (1976) has so pointedly described in his example of the Osage tribe in Oklahoma. At the time when the tribal register was composed, so-called "instant Indians" sprang up overnight on the Osage Indian Reservation. With the help of corrupt BIA officials, they became tribal members. It was later proved that at least 151 of the 2,229 Osage Indians, and more likely considerably more than that, turned out to be white. Since 1916 these few white people, together with their descendants, have received more than $32 million in royalties from petroleum found on tribal lands, and they own tribal land which in the mid-1970s had a value of more than $10 million (Feest 1976, 130–132).

Although the BIA generally requires that all "status Indians" have a certain quantum of Indian blood, today the tribes decide this matter, though different tribes have very different criteria for determining whether someone may be a tribal member. The Ute Mountain Indian Reservation in Colorado, for example, requires that anyone born after 1958 must be at least five-eighths Ute Indian to be accepted as a member of the tribe. For the White Mountain Apaches a quantum of one-quarter blood is enough, and for the Wyandottes of Oklahoma a two hundred and fifty-sixth part of Indian ancestry is sufficient. The Pine Ridge Sioux of South Dakota require that one be born on the reservation and that one's father or mother be a member of the tribe. For the Ottawas in Oklahoma one is a tribal member if one has an ancestor as far back as 1853 who was registered as a member. On the Fort Berthold Indian Reservation in North Dakota any Indian ancestry is qualification enough as long as membership is endorsed by seven

Table 4.2. Population trends of American Indians, 1820–1980

	American Indians on Reservations	Reservation Indians of total U.S. Indian population (%)	BIA pop. (status Indians)	Bureau of the Census	Population change (%)	American Indians of total U.S. pop. (%)
1820	—	—	471,036	—	—	4.89
1850	—	—	388,229	—	-17.6[a]	1.67
1860	—	—	254,300	—	-34.5[a]	0.81
1870	287,640	91.7	298,000	313,712	23.4	0.79
1880	240,136	74.5	256,127	322,534	2.8	0.64
1890	230,437	92.8	246,834	248,253	-23.0	0.39
1900	247,522	91.5[a]	270,544	237,196	9.0[a]	0.36
1910	256,320	84.1[a]	304,950	276,927	12.7[a]	0.33
1920	236,539	70.3[a]	336,337	244,437	10.3[a]	0.32
1930	185,377	54.0	340,917	343,352	2.0	0.27
1940	224,731	62.1[a]	361,816	345,252	6.1[a]	0.26
1950	244,906	68.5	315,389	357,499	-1.2	0.24
1960	285,600	54.5	344,951	523,591	46.5	0.29
1970	332,582	42.0	453,227	792,730	51.4	0.39
1980	339,836	24.9	755,201	1,366,676	72.4	0.60

[a]Based on BIA statistics.

Sources: ARCIA 1880, 1930, 1940; U.S. Department of Commerce 1890, Indians Taxed . . . ; U.S. Department of Commerce 1910, 1930, 1940, 1950, 1960, 1970, 1980; U.S. Department of the Interior 1955, 1960.

tribal council members. In California, it is sufficient simply to establish a two hundred and fifty-sixth quantum of Indian blood in order to be eligible for an equal share of funds from the Indian Claims Commission, but one must have at least one-quarter Indian ancestry to benefit from the scholarship program of the BIA or the services of the IHS (Forbes 1981, 409).

Finally, a word about the population statistics of the tribal governments. Many tribes have a right to financial compensation from the federal government, for example, in connection with the ICC. When funds are appropriated, it is in the tribe's interest to have fewer members, so that individual grants will be correspondingly larger. In times when payments are made in accordance with the number of its members, however, the tribe seeks to establish more liberal criteria of tribal membership. These different viewpoints sometimes lead to very considerable changes in the population statistics of a tribe and should therefore be taken into consideration when tribal enrollment registers are used as fundamental data. The same caution is called for with regard to data of the U.S. Bureau of the Census, the BIA, and the IHS.

THE HETEROGENEITY OF AMERICAN INDIANS

The word "Indian" as a collective name to designate all the different peoples who were indigenous inhabitants of the territory of the United States originated in a mistake made when America was discovered. Even today the "First Americans," as they are sometimes called, must still live with this erroneous term. In the pre-Columbian era there were no "Indians" in the European sense of the word on the American continent. This can also be seen from the fact that in the more than a hundred American Indian languages which are still spoken in the United States today there is no uniform collective name for the indigenous people of America. If so-called Indians are asked what name they give themselves, they will usually reply that they are Diné (Navajos), Pima, Pais, or Apaches, for their primary feeling is that they belong to their own tribe, although the terms "Diné," "Pima," and "Pai" in their respective languages mean "the people." The various Apache tribes, on the other hand, had no collective designation for themselves. The Spanish word Apache, translated into English and German, is probably derived from "'a·paču," which, in the language of the neighboring Zuni Pueblo, means something like "enemy" (Ortiz 1983, 385).

The great number of languages previously mentioned, as well as the diversity in way of life, settlement patterns and types of dwellings, religions, customs, and political structures prove that the various tribes of the United States generally had very little in common with one another. This is still true today, though to a lesser extent. In many cases, what the tribes do have in common first developed from contact with white settlers and often resulted from the U.S. policies toward them.

From what has been said it follows that one cannot regard American Indians as one homogeneous people, and that caution is called for when making generalizations about them. General statements may have some validity with regard to some of the Indians living in large cities, however, who have been, and still are, particularly exposed to the pressure to assimilate that comes from mainstream America. It is not surprising, therefore, that pan-Indian movements and organizations should have developed in the melting pot of large metropolitan areas such as Chicago and Los Angeles. In these cities many non-Indian racial and ethnic groups and Indians from perhaps a hundred different tribes can be found (Schulze-Thulin 1976, 263–264; Wax 1971, 39).

Based upon the BIA figures in 1980, there were 291 Indian tribes or communities <4> that were federally recognized, not counting Alaska (U.S. Department of the Interior 1984, 24), yet these figures do not mean very much

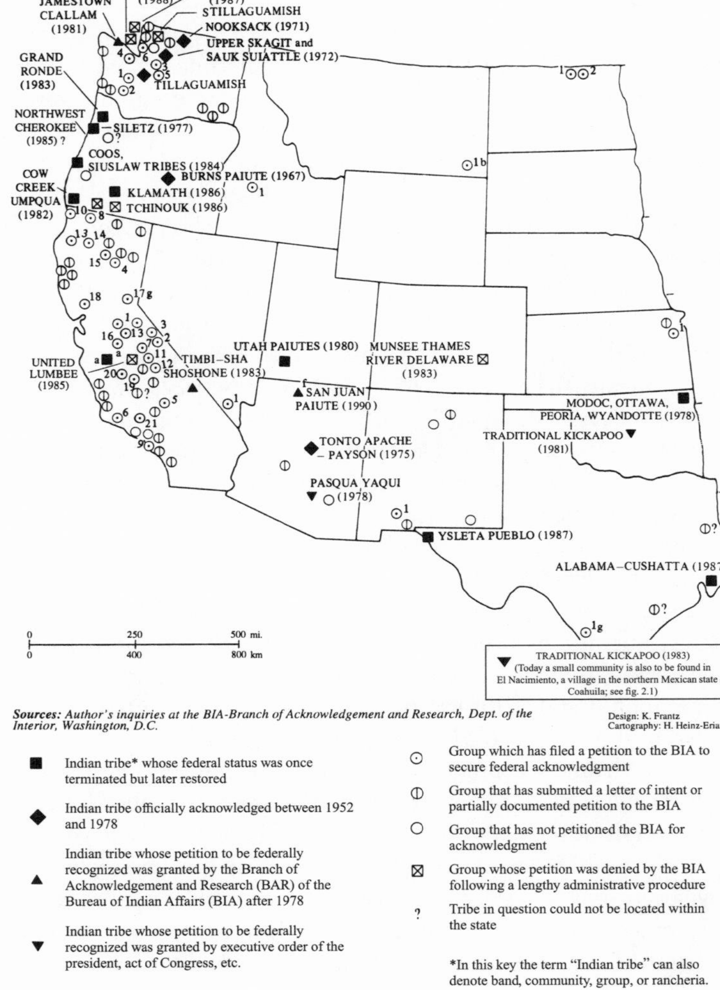

SNOHOMISH (1988)
SAMISH (1987)
JAMESTOWN CLALLAM (1981)
STILLAGUAMISH
NOOKSACK (1971)
UPPER SKAGIT and SAUK SUIATTLE (1972)
GRAND RONDE (1983)
TILLAGUAMISH
NORTHWEST CHEROKEE (1985) ?
SILETZ (1977)
COOS, SIUSLAW TRIBES (1984)
BURNS PAIUTE (1967)
COW CREEK UMPQUA (1982)
KLAMATH (1986)
TCHINOUK (1986)
UNITED LUMBEE (1985)
TIMBI–SHA SHOSHONE (1983)
UTAH PAIUTES (1980)
MUNSEE THAMES RIVER DELAWARE (1983)
SAN JUAN PAIUTE (1990)
TONTO APACHE – PAYSON (1975)
PASQUA YAQUI (1978)
YSLETA PUEBLO (1987)
MODOC, OTTAWA, PEORIA, WYANDOTTE (1978)
TRADITIONAL KICKAPOO (1981)
ALABAMA–CUSHATTA (1987)
0 250 500 mi.
0 400 800 km
TRADITIONAL KICKAPOO (1983)
(Today a small community is also to be found in El Nacimiento, a village in the northern Mexican state of Coahuila; see fig. 2.1)
Sources: Author's inquiries at the BIA-Branch of Acknowledgement and Research, Dept. of the Interior, Washington, D.C.
Design: K. Frantz
Cartography: H. Heinz-Erian
Indian tribe* whose federal status was once terminated but later restored
Indian tribe officially acknowledged between 1952 and 1978
Indian tribe whose petition to be federally recognized was granted by the Branch of Acknowledgement and Research (BAR) of the Bureau of Indian Affairs (BIA) after 1978
Indian tribe whose petition to be federally recognized was granted by executive order of the president, act of Congress, etc.
Group which has filed a petition to the BIA to secure federal acknowledgment
Group that has submitted a letter of intent or partially documented petition to the BIA
Group that has not petitioned the BIA for acknowledgment
Group whose petition was denied by the BIA following a lengthy administrative procedure
? Tribe in question could not be located within the state
*In this key the term "Indian tribe" can also denote band, community, group, or rancheria.

Fig. 4.1: Indian tribes that have gained or regained federal acknowledgment since 1952 and federally unacknowledged groups, 1989

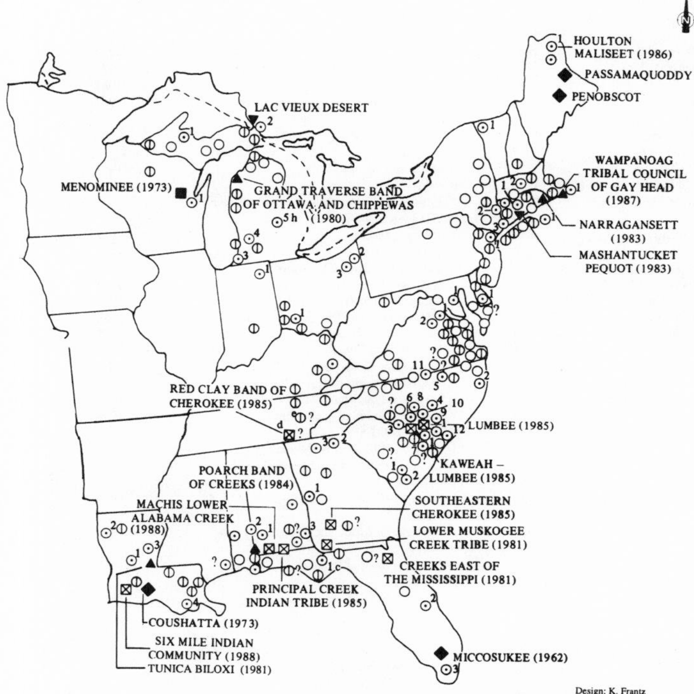

a. On the following California rancherias federal status was terminated but later restored by decree: Robinson (1977), Hopland (1978), Big Lagoon (1979), Upper Lake (1979), Table Bluff (1981), Big Sandy, Table Mountain, Big Valley, Blue Lake, Buena Vista, Chicken Ranch, Cloverdale, Elk Valley, Greenville, Mooretown, North Fork, Picayune, Pinoleville, Potter Valley, Quartz Valley, Redding, Redwood, Rohnerville, Smith River (all 1983).

b. The Little Shell Chippewa are dispersed in various locations of Montana. Their main location is at Lame Deer.

c. The Florida tribe of Eastern Creek is to be found both in Tallahassee and in Bruce.

d. The Red Clay Band, resident in Ooltewah (Tennessee), belongs to the South Eastern Cherokee Confederacy in Leesburg (Georgia).

e. Splinter groups of the Melungeon Indians are also to be found in West Virginia and in various locations along the southeast coast.

f. The San Juan Paiute live in the vicinity of Tuba City on the Navajo Indian Reservation. The Navajo have protested against their federal acknowledgment because of the territorial claims in question.

g. The Texas Band of Traditional Kickapoo was acknowledged by the BIA outside the normal administrative procedures.

Sources: The data for this map were very kindly provided by the staff of the BAR at the headquarters of the BIA in Washington, DC. I am particularly indebted to Bud Shapard <4> and R. E. Johnson <49>. The present map, revised and enlarged, is based on a previously published map by the author (see Frantz 1990b, 72).

(Continued)

GROUPS WHO HAVE FILED PETITIONS WITH THE BRANCH OF ACKNOWLEDGEMENT AND RESEARCH WITHIN THE BIA TO BECOME FEDERALLY ACKNOWLEDGED INDIAN TRIBES, 1988

	Date of application
ALABAMA	
1 Cherokees of Jackson County	1981
2 **Mowa Band of Choctaw ("Creoles")**	1983
3 Cherokees of Southeast Alabama	1988
CALIFORNIA	
1 Ione Band of Miwok Indians	1916
2 Mono Lake Indian Community	1976
3 Antelope Valley Indian Community	1976
4 Maidu Tribe of Greenville Rancheria	1977
5 Kern Valley Indian Community	1979
6 Coastal Band of Chumash Indians	1982
7 American Indian Council of Mariposa County	1982
8 Shasta Nation	1982
9 Juaneno Band of Mission Indians	1982
10 Tolowa Nation	1983
11 North Fork Band of Mono Indians	1983
12 Dunlap Band of Mono Indians	1984
13 Nor-El-Muk Band of Wintu Indians	1984
14 San Luis Rey Band of Mission Indians	1984
15 Wintu Indians of Central Valley	1984
16 Wintoon Indians	1984
17 Chukchansi Yokotch Tribe, Coarsegold	1985
18 Yokayo Tribe	1987
19 Wukuchumni Council	1988
20 Choinumni Council	1988
21 Coastanoan Band of Carmel Mission Indians	1988
CONNECTICUT	
1 Eastern Pequot Indians	1978
2 Schaghticoke	1981
3 Golden Hill Paugussett Tribe	1982
DELAWARE	
1 **Nanticoke Indian Association**	1978
FLORIDA	
1 Florida Tribe of Eastern Creeks	1978
2 Tuscola United Cherokee Tribe	1979
3 Seminole Nation	1983
GEORGIA	
1 Cherokee Indians of Georgia	1977
2 Georgia Tribe of Eastern Cherokees	1979
3 Cane Break Band of Eastern Cherokees	1979
IDAHO	
1 Delawares of Idaho	1979
INDIANA	
1 Miami Nation	1980
KANSAS	
1 Delaware-Muncie	1978
LOUISIANA	
1 Clifton-Choctaw	1978
2 Choctaw-Apache Com. of Ebarb	1978
3 Jena Band of Choctaws	1979
4 United Houma Nation	1979
MASSACHUSETTS	
1 Mashpee Wampanoag	1975
2 Nipmuc Tribal Council	1980
MARYLAND	
1 Piscataway-Conoy Confederacy	1978
MAINE	
1 Aroostook Band of Micmacs	1985
MICHIGAN	
1 Huron Potawatomi Band	1996
2 **Bahwetig Ojibwas and Mackinac Tribe**	1979
3 Potawatomi of Indiana and Michigan	1981
4 Potawatomi Indian Nation	1981
5 Burt Lake Band of Ottawa and Chippewa, Brutus	1985
MINNESOTA	
1 Warroad Chippewa	1979
MISSOURI	
1 Northern Cherokee Tribe	1985
MONTANA	
1 Little Shell Tribe of Chippewa of Montana	1978
NORTH CAROLINA	
1 Hatteras Tuscarora	1978
2 Coree	1978
3 Cherokee Indians of Robeson	1979
4 Hattadare Indian Nation	1979
5 Haliwa-Saponi Tribe	1979
6 Lumbee Regional Develop.Association	1980
7 Tuscarora Tribe	1981
8 Coharie Intra Tribal Council	1981
9 Waccamaw Siouan Develop Assoc.	1983
10 Cherokee Indians of Hoke County	1983
11 Cherokee-Powhattan Indian Assoc.	1984
12 Tuscrora Nation of N.Carolina	1985
NORTH DAKOTA	
1 Little Shell Band of N.Dakota	1975
2 Christian Pembina Chippewa	1984
NEW JERSEY	
1 Ramapongh Mountain Indians	1979
NEW MEXICO	
1 San Juan de Guadalupe Tiwa	1971
NEVADA	
1 Pahrump Band of Paiutes	1987
NEW YORK	
1 Shinnecock Tribe	1978
OHIO	
1 Shawnee Nation United Remnant Band	1979
2 N. Eastern U.S. Miami Inter-Tribal Council	1979
3 **Allegheny Nation**	1979
SOUTH CAROLINA	
1 Four Hole Indian Organization	1976
2 Santee Tribe (White Oak Ind. Community)	1979
VIRGINIA	
1 Unitad Rappahannock Tribe	1979
2 The Upper Mattaponi Ind. Tribal Assoc.	1979
VERMONT	
1 St. Francis/Sokoki Band of Abenakis	1980
WASHINGTON	
1 Steilacoom Tribe	1973
2 Cowlitz Tribe	1975
3 Snoqualmie Indian Tribe	1976
4 Duwamish Indian Tribe	1977
5 Chinook Indian Tribe	1979
6 Snoqualmie Tribe of Whidbey Island	1988
WISCONSIN	
1 Brotherton Indians of Wisconsin	1980

(Fig. 4.1, continued)

when closely examined. The development of reservation policy with its forced removals and relocations meant that some tribes were broken up and divided among different reservations, while other times, different tribes were put together on one reservation. All these population groupings, often artificially created, now constitute individual tribes in the eyes of the BIA, albeit only in a political, not necessarily cultural, sense of the term.

Today the Sioux are dispersed among various Indian reservations in Nebraska, Minnesota, Montana, North Dakota, and South Dakota, for example,[2] while most of the Cherokees are divided between North Carolina and Oklahoma. The Mohaves and the Chemehuevis live together in a very narrow strip of land on the Colorado River Indian Reservation, and they are compelled to share their reservation with dozens of Hopi and Navajo families (McIntire 1969). These Hopis and Navajos, who had long been neighboring enemies (see fig. 3.1), left their reservations in the late 1940s as they did not have enough arable land there, and the BIA suggested that they move to the Colorado River Indian Reservation in Arizona. Through this policy "new" tribes have developed on some reservations in the course of time, such as the Brotherton and Stockbridge Indians, originally from New Jersey and New England, who are now joined together in one tribe in Michigan (Spicer 1981, 58). There are more frequent cases in which the feeling of solidarity of the individual tribe has been sustained down to the present day despite hundreds of miles of separation. There have also been cases where, even prior to contact with white settlers, certain tribes have allowed other tribes to settle on their territory. The Hopis in Arizona, for example, granted asylum to the Tewas, as did the Pimas to the Maricopas. In each case the inhabitants of the reservation were officially declared to be one tribe, although in reality two distinct tribes existed on one reservation.

The BIA is aware of an additional 230 groups claiming to be of Indian ancestry, with an estimated total membership of between 180,000 and 250,000 people, who are not officially recognized by the U.S. government (see fig. 4.1). They are at a serious disadvantage compared to the federally recognized tribes. Without the special status of the latter, they are denied social and economic benefits as well as certain sovereignty rights. Since the BIA founded the Branch of Acknowledgement and Research (BAR)[3] in 1978 to determine whether a particular group should be granted Indian status or

2. The thirteen reservations in question are the following: Cheyenne River, Crow Creek, Lower Brule, Pine Ridge, Rosebud, Yankton (all South Dakota); Sisseton, Standing Rock (North Dakota and South Dakota); Santee (Nebraska); Shakopee Sioux, Upper Sioux (Minnesota); Fort Peck (Montana). Assiniboine and Yanktonai Sioux live alongside one another on the Fort Peck Indian Reservation.

3. In 1985, at the time of my inquiries, the BAR had a tremendous amount of work to do with only six permanent and two temporary employees, including two sociologists, an anthropologist, a historian, and a genealogist.

not, 180 groups have submitted letters of intent to petition for acknowledgment. The petitioner must submit a detailed account to the BAR and comprehensive documentation, which can be time-consuming and costly to prepare and which must contain the records which prove that the tribe has existed continuously since its first contact with white settlers. After administrative procedures that lasted for about two years so far only nine petitioners have become fully acknowledged tribes, fifteen groups were turned down for recognition, and eighty-six more tribes are still awaiting active consideration. Official contact has been established with an additional ninety-one groups who plan to petition for acknowledgment; for an additional thirty-seven groups this contact has not yet been established (fig. 4.1). Apart from the cases mentioned here, fifty-eight other Indian communities have either gained or regained federal recognition since 1952,[4] some of them prior to the founding of the BAR. This recognition was granted to them either directly by Congress or through the federal courts.

Anthropologists and historians have calculated the number of Indian tribes in a different manner and come up with somewhat more than 170 tribes (see Spicer 1980, 13) in the United States today.[5] This is about 30 fewer than at the time when America was discovered. Although at least a quarter of the tribes from that time have disappeared as a result of European diseases, warfare, or integration into other tribes (Spicer 1981, 58), new tribes have also evolved.

The discrepancy between the officially recognized number of tribes and the number having a scientific basis is partly to be explained by the fact that the word "tribe" is itself rather vague and can be interpreted in different ways (see Spicer 1980, 16; Sutton 1975, 192; Weatherhead 1980, 11), and partly by the fact that the Indian peoples in the course of history were designated by white people in different ways. During the seventeenth and eighteenth centuries it was still customary to regard Indian tribes as "nations," a designation which was replaced by "tribe" and sometimes "band" in the course of the nineteenth century.[6] The process of degrading the In-

4. Among these, the status of 34 Indian communities had been revoked during the termination era.

5. Among these are the Lumbees of North and South Carolina, who are not federally recognized and have no reservation. Today with an estimated 40,000 members, they would be the fourth largest Indian tribe in the United States, after the Navajos, the Sioux, and the Cherokees, if they were recognized. Although BIA documents state that they have clearly perceptible white and Negroid racial characteristics, their ancestry is a subject of controversy among ethnologists and historians. It is a fact that land was given to them in 1732, but research has not been able to establish whether they ever had an Indian language of their own. (Unpublished documents and information from the Branch of Acknowledgement and Research of the BIA in Washington, DC.)

6. Contrary to the BIA's policy, the Canadian government designates all Indian tribes within Canada as "bands."

dians of North America can thus be seen, even in the language. Combined with the revival of Indian self-confidence during the past twenty years, many tribes have again applied the term "nation" to their reservations, a designation which, though not recognized by the federal government, is nevertheless tolerated.

It is not easy to give a commonly accepted definition of the term "tribe." Somewhat simplified, it means a number of families, clans, or groups who speak a common language, have similar institutions, customs, and traditions, show a more or less pronounced group consciousness, and perhaps also have a common ancestry. Yet tribes have commonly accepted members of other tribes, indeed sometimes even a whole tribe, into their own (see Adam and Trimborn 1958, 176–177). Contrary to this definition of a tribe, the BIA, like many non-Indian Americans, generally regards the population of a reservation as a "tribe" as a matter of convenience.

What has been said so far shows that American Indians are a minority, and this minority in turn is broken up into many other minorities. Apart from that, neither the population of a reservation nor a tribe is necessarily a homogeneous ethnic group.

It has already been pointed out that on many reservations non-Indians own a considerable portion of land, and it is equally true that on a good many reservations the population includes a considerable percentage of non-Indians. The population statistics from the U.S. Bureau of the Census show that on several reservations American Indians even constitute a minority in relation to whites. Particularly striking cases include the Agua Caliente Indian Reservation in California, the Isabella Indian Reservation in Michigan, and the Puyallup Indian Reservation in Washington, where the proportion of Indians ranges from only 0.5% to 3.4%. American Indians constitute less than one-fifth of the total population on nine of the twenty-five most populous reservations of the United States (see table 4.3).

There are various reasons for this situation. First, there is a correlation between a high percentage of white landownership and a correspondingly high percentage of white residents. At the time of the IRA a considerable number of non-Indians were successful in having their names included in the required tribal enrollment register, as was the case for the Osage Indian Reservation in Oklahoma, a reservation rich in petroleum. Long-term leases have led, in other cases, to the development of white colonies on some reservations which today are located in the vicinity of large cities. This is the case in the metropolitan areas of southern California, but also in Phoenix, Tucson, Albuquerque, Santa Fe, and Seattle, where suburban growth has already encroached upon reservation land (Frantz 1996b). Furthermore, every reservation employs a certain number of non-Indian technical personnel who work for the tribe, the BIA, or other federal agencies.

Table 4.3. The twenty-five largest Indian reservations in the United States by population, 1980

Reservation	State	Total population (including non-Indians)	Number of American Indians	Percentage of American Indians
Navajo	AZ, NM, UT	110,443	105,031	95.1
Osage	OK	39,327	4,759	12.1
Yakima	WA	25,363	4,971	19.6
Puyallup	WA	25,188	857	3.4
Isabella	MI	23,373	515	2.2
Wind River	WY	23,166	4,170	18.0
Flathead	MT	19,628	3,769	19.2
Nez Perce	ID	17,806	1,460	8.2
Uintah and Ouray	UT	16,909	2,046	12.1
Agua Caliente	CA	13,743	69	0.5
Sisseton	ND, SD	13,586	2,704	19.9
Oneida	WI	13,389	1,821	13.6
Pine Ridge	SD	13,229	11,946	90.3
Fort Peck	MT	9,921	4,276	43.1
White Earth	MN	9,486	2,552	26.9
Standing Rock	ND, SD	8,816	4,796	54.4
Leech Lake	MN	8,441	2,760	32.7
Colorado River	AZ, CA	7,873	1,968	25.0
Fort Apache	AZ	7,774	6,880	88.5
Allegany	AZ	7,681	922	12.0
Gila River	AZ	7,380	7,070	95.8
Rosebud	SD	7,328	5,687	77.6
Papago	AZ	7,203	6,958	96.6
Colville	WA	7,047	3,502	49.7
Hopi	AZ	6,896	6,593	95.6
Total population		450,996	198,081	43.9

Source: U.S. Department of Commerce 1980, Population (Supp.), American Indian, 16–20.

This segment of the population, constantly fluctuating, is usually housed in BIA or IHS residential quarters separated from the local reservation Indians (see fig. 3.12g).

On almost all reservations, however, in addition to non-Indians, there are Indian residents who are not members of the tribe where they live. These are American Indians of widely varying ancestry who, usually for professional or personal reasons, live on the reservation. As no official statistics regarding these people are available, individual research on each tribe is required. My own study of the Hualapai Indian Reservation in Arizona (see Frantz 1990b, 67–69) in January 1986 identified 727 Hualapais, 7 whites, and 111 Indians from seventeen other

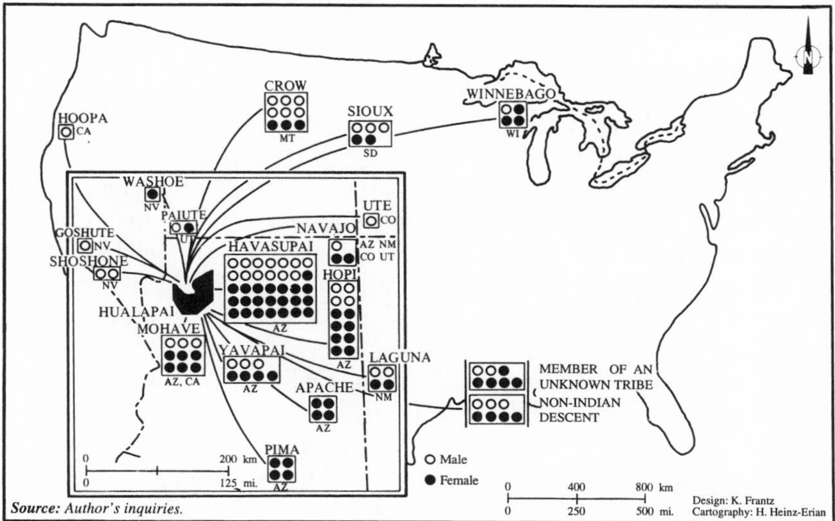

Source: Author's inquiries.

Fig. 4.2: Number and descent of residents of the Hualapai Indian Reservation in Arizona who are not members of the tribe

tribes, though it was not possible to establish the ancestry of some of these Indians (see fig. 4.2). This group of non-tribal member Indians consisted largely of Indians from neighboring tribes and comprised over 13% of the total population.

The heterogeneity of the people living on reservations is not simply a result of the historically documented merging of tribes or parts of tribes. It is in the nature of things that many American Indians do not take their marriage partners exclusively from their own tribes but practice some form of exogamy. This has always been the case, but it was not until the time of the IRA and the legally prescribed enrollment registers of tribal membership[7] that this led to complications for the people concerned, and, in particular, for their descendents. Everyone whose name was put on the official roll at that time was considered a tribal member, yet the question of specific ancestry was often ignored in those days. Every Indian reservation was also asked to include in its constitution a passage which stated the extent of direct descent a future member had to possess from the respective tribe for tribal membership.

7. There were already registers of this kind when the reservations were first established. The military had often assigned letters to the band leaders or to the individuals who were given this role, and the number 1 was added to these letters. Thus the leader of the North Fork Band on the White Mountain Apache Indian Reservation in Arizona was simply designated as A1. All married male members of the band were respectively designated as A2, A3, A4, etc.. Among the Apaches it was sufficient if this label, intended to prevent mistakes or cheating when they were rationed food, clothing, and other items, was visible on a necklace, but it was actually tattooed on many Navajos.

The Hualapai tribal constitution, for example, first drawn up in 1938, required that a tribal member must be of at least one-half Hualapai descent, a proportion which thirteen years later was reduced to one-quarter. This rigid regulation has the consequence that, at present, after the third generation, the children of a Hualapai who, like their parents and grandparents before them, took non-Hualapais as marriage partners, will no longer have the prescribed percentage of Hualapai blood and, therefore, their children cannot be included in the tribal role. It remains to be seen whether today's Hualapais, along with most of the other American Indian tribes, will be able to obtain the BIA's approval for a new constitutional change in regulations, or whether in the future the official tribal population will decline.

The enrollment register of the Hualapais shows the extent to which Indian marriages have contributed to the tribe's mixed ancestral composition within two generations.[8] When the total percentage of Indian blood of tribal members under twenty-eight years of age is added up and averaged, only a little more than 50% within this age group is of Hualapai descent (see fig. 4.3). The ancestry of tribal members includes relatives among white Americans, Mexicans, African Americans, Chinese, and twenty-two other American Indian tribes. This example of the Hualapais is certainly not an extreme case, and because their reservation is isolated and poor in natural resources, their degree of racial purity is probably above average. Nevertheless, even in the population of this tribe there has been a considerable degree of heterogeneity.

THE DEVELOPMENT AND DISTRIBUTION OF THE AMERICAN INDIAN POPULATION

Contrary to the idea that was once widespread even among anthropologists and historians (see Williams 1978, 2) the discoverers of North America did not find a virgin, largely unpopulated continent upon their arrival. Indeed, more recent estimates assume that in the pre-Columbian era as many as five to nearly ten million people lived within the present boundaries of the United States (see Denevan 1976, 291; Dobyns 1966, 415; 1976a, 97; 1976b; Jacobs 1974). Earlier literature suggested only somewhat more than a million inhabitants (see Kroeber 1939, 143; Mooney 1928, 33).

Whether there had previously been one million or ten million American Indians, the fact remains that between the colonial period and the nineteenth century the pre-Columbian population decreased drastically as land was acquired by non-Indian settlers. The main reasons for the decrease

8. Only in exceptional cases are outside observers permitted to examine the register of blood ratios. For access to these lists, which include all Hualapais born since 1958, I am indebted to the tribal chairman and the BIA agency superintendent in Valentine <14>.

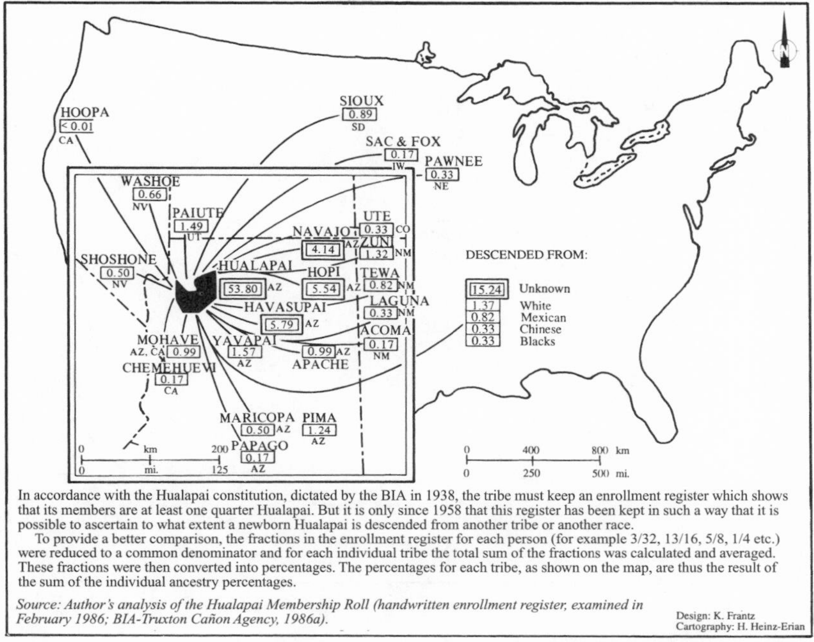

Fig. 4.3: Descent of the Hualapai Indians of Arizona up to the age of twenty-eight calculated on the basis of the membership roll of the tribe, 1986

were warfare with the military and the settlers; the forced removal of Indians, often to regions with totally different ecological settings; and the influence of undernourishment, alcohol, and, above all, diseases that were brought over from Europe. Although around 70,000 American Indians from the Southeast were removed to the newly established Indian Territory in Oklahoma, 20,000 did not survive to see their new destination. The Indians of the Ohio Valley suffered a similar fate (see Spicer 1981, 59), while in California, during the era of the gold rush, a succession of massacres by whites caused the number of Indians to shrink from an estimated 100,000 in 1850 to less than 20,000 in 1906 (see Heizer and Almquist 1971, 23–64). Smallpox epidemics killed over half the Pawnee tribal population in the southern prairies (see McNickle 1973, 4) and reduced the Mandan Indians, located more to the north, from a population of around 1,600 to a mere 31 survivors in 1837 (see Wax 1971, 32). The list of these terrible losses could go on and on.

Although these figures must be treated with caution, one can see that despite the high birthrate, the American Indian population, estimated at 600,000 for the United States in 1800, was reduced to less than 250,000 by 1890 (see table 4.2 and fig. 4.4). During these 90 years the proportion of

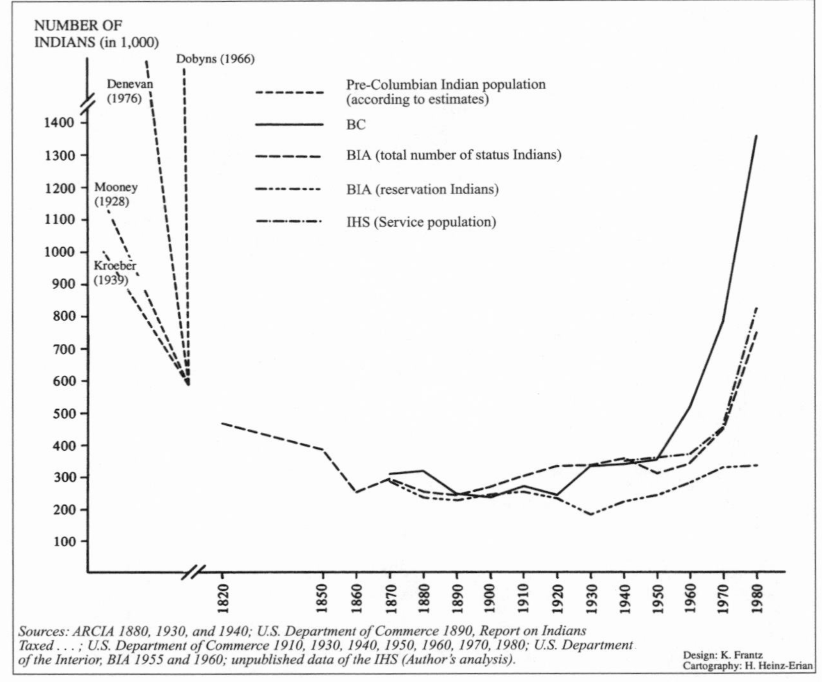

Sources: ARCIA 1880, 1930, and 1940; U.S. Department of Commerce 1890, Report on Indians Taxed . . . ; U.S. Department of Commerce 1910, 1930, 1940, 1950, 1960, 1970, 1980; U.S. Department of the Interior, BIA 1955 and 1960; unpublished data of the IHS (Author's analysis).

Fig. 4.4: The development of the American Indian population since the pre-Columbian era

American Indians compared to the total U.S. population declined from 11.3% to barely 0.4% (table 4.2).

Of the previously mentioned quarter of a million American Indians, around 133,000 lived on the 133 reservations existing at that time and another 56,000 in New York State and what was then the Indian Territory. These 189,000 American Indians were not taxed. The remaining 61,000 Indians were more or less assimilated and lived outside the official Indian country (see U.S. Department of Commerce 1890, 24–28).[9]

From 1890 to 1920 the Indian population stagnated, while its proportion of the total population declined to just a little more than 0.3%. The appearance of Zane Grey's novel *The Vanishing American* in 1925, in which he saw little chance for the survival of American Indians as an independent racial and ethnic group, is understandable in the light of these figures.

After a continued period of population stagnation between 1930 and 1950, the general improvement of the public health service, the expansion

9. Incidentally, 1890 was the first year in which the U.S. Bureau of the Census assembled precise statistics about American Indians. Up until that time the Census Bureau had been satisfied with rough estimates and data provided by the BIA.

of the sanitary infrastructure, and the development of public housing caused a clear reversal of the trend, and led to an almost fourfold increase of the Indian population between 1950 and 1980, according to census statistics. This recorded increase is the result not simply of a natural population growth, but also of changed methods for collecting statistics. Nevertheless, the figures of the last census appear to correspond fairly closely to the facts, so that today, according to the census authorities, for every American Indian there are 166 neo-Americans, including, among others, 138 whites, 19 African Americans, and 3 Asians.

The distribution of the American Indian population at the beginning of the nineteenth century was very different from what it is today. Formerly the majority of Indians were settled in the East, where there were a number of regions with a relatively high population density. Such concentrations were to be found in Massachusetts and Connecticut as well as the coastal areas of Virginia and North and South Carolina, but also in the tribal territory of the Choctaws, Creeks, and Cherokees (see Spicer 1981, 59). In the West, too, there were several regions of dense population, such as along the Rio Grande, in the lower Gila and Colorado River valleys, and in Washington, Oregon, and California, although generally speaking these regions in the West had lower total populations. On the other hand, the large arid or semiarid regions were very sparsely settled at that time because of the predominantly nonsedentary way of life of the people living there.

At the close of the nineteenth century, however, the pattern of the American Indian population distribution had already changed quite radically (see table 4.1), and this pattern has not changed much since then, except for the Lumbees of North Carolina. Today there are no longer any federally recognized Indian communities in Georgia, Tennessee, or Kentucky, states that once had so many Indians. Today 73% of all American Indians and 92% of all reservation Indians live west of the Mississippi. In numbers California, Oklahoma, Arizona, and New Mexico each have more than 100,000 Indians, followed by Washington and South Dakota. A total of 46% of all U.S. American Indians live in these four states.

While California and Oklahoma have small proportions of Indians living on-reservation, far under the national average of around 25% (see table 4.4 and figs. 4.5, 4.6), in Arizona, New Mexico, South Dakota, and also Montana, reservation Indians have always been in the majority. Here the American Indians, constituting from 4.8% to 8.2% of the population of these states, are still a minority that cannot be ignored, even if, with the exception of South Dakota, they are a shrinking minority relative to the total population (see table 4.1). Eight of the ten reservations with the largest number of American Indians (see table 4.3) are in Arizona or South Dakota.

Table 4.4. American Indian population on- or off- reservation by state, 1980

State	American Indian population	% of state/ regional population	American Indians on reservations	% reservation Ind. of state/ regional Amer. Ind. pop.
NORTHEAST	77,430	0.2	7,997	10.3
Connecticut	4,431	0.1	27	0.6
Maine	4,057	0.4	1,235	30.4
Massachusetts	7,483	0.1	1	
New Hampshire	1,297	0.1	—	—
New Jersey	8,176	0.1	—	—
New York	38,967	0.2	6,734	17.3
Pennsylvania	9,179	0.1	—	—
Rhode Island	2,872	0.3	—	—
Vermont	968	0.2	—	—
MIDWEST	246,365	0.4	64,677	26.3
Illinois	15,846	0.1	—	—
Indiana	7,682	0.1	—	—
Iowa	5,369	0.2	492	9.2
Kansas	15,256	0.6	715	4.7
Michigan	39,734	0.4	1,607	4.0
Minnesota	34,831	0.9	9,901	28.4
Missouri	12,129	0.2	—	—
Nebraska	9,145	0.6	2,846	31.1
North Dakota	20,120	3.1	11,287	56.1
Ohio	11,985	0.1	—	—
South Dakota	44,948	6.5	28,468	63.3
Wisconsin	29,320	0.6	9,361	31.9
SOUTH	370,198	0.5	15,597	4.2
Alabama	7,502	0.2	—	—
Arkansas	9,364	0.2	—	—
Delaware	1,307	0.2	—	—
District of Columbia	996	0.2	—	—
Florida	19,134	0.2	1,303	6.8
Georgia	7,442	0.1	30	0.4
Kentucky	3,518	0.1	—	—
Louisiana	11,969	0.3	210	1.8
Maryland	7,823	0.2	—	—
Mississippi	6,131	0.2	2,756	45.0
North Carolina	64,536	1.1	4,844	7.5
Oklahoma	169,292	5.6	4,749	2.8
South Carolina	5,665	0.2	728	12.9
Tennessee	5,013	0.1	—	—
Texas	39,740	0.3	859	2.2
Virginia	9,211	0.2	118	1.3
West Virginia	1,555	0.1	—	—
WEST	672,683	1.6	251,565	37.4
Alaska	21,869	5.4	942	4.3
Arizona	152,498	5.6	113,763	74.6
California	198,275	0.8	9,265	4.7
Colorado	17,734	0.6	1,966	11.1
Hawaii	2,655	0.3	—	—
Idaho	10,418	1.1	4,771	45.8

(Continued)

(Table 4.4 continued)

State	American Indian population	% of state/ regional population	American Indians on reservations	% reservation Ind. of state/ regional Amer. Ind. pop.
Montana	37,598	4.8	24,043	63.9
Nevada	13,306	1.7	4,400	33.1
New Mexico	107,338	8.2	61,876	57.6
Oregon	26,591	1.0	3,072	11.6
Utah	19,158	1.3	6,868	35.8
Washington	58,186	1.4	16,440	28.3
Wyoming	7,057	1.5	4,159	58.9
ALL OF U.S.	1,366,676	0.6	339,836	24.9

Source: U.S. Department of Commerce 1980, Population (Supplementary Report), 14–26.

Following what has been said about the development and geographical distribution of the American Indian population of the United States, let us now consider the size of the respective reservations in relation to the number of their Indian and non-Indian inhabitants (see fig. 4.7). Considering only the Indian population, 159 (54.6%) of the 291 reservations had fewer than 250 inhabitants. Only 7 reservations had more than 5,000 Indian residents (see table 4.3), and these reservations contained almost half of all reservation Indians. If the non-Indian residents are also taken into account, 25 reservations (12.7%) had more than 5,000 inhabitants.

Finally a word about Indians who live in cities outside the reservations. For the United States as a whole the move to cities and the attendant transformation from a predominantly rural to a predominantly urban society took place in the course of the second half of the nineteenth century, so that by the time of the First World War more than half the total U.S. population lived in urban areas (see Frantz 1987, 76–78). It was not until much later, however, that this change took place among American Indians. As recently as 1940 most American Indians still lived in rural areas, 62% on Indian reservations. This percentage dropped to 54.5% in 1960. In only a few states, notably New York, California, and Oklahoma, was the proportion of urban Indians over 10% of the total Indian population of the state (see table 4.5 and fig. 4.8).

Today almost 54% of all American Indians (44.6% in 1970) live in cities, although in the states with the largest Indian populations the percentages vary greatly. It is only in California, New York, Minnesota, and Washington that the Indian minority populations are primarily urban. A great many urban Indians are residents of large metropolitan areas (see fig. 4.9), and in metropolitan Los Angeles alone there were 95,000 Indians in 1980, though there are also large numbers of urban Indians in San Francisco, Tulsa, and other metropolitan areas (see table 4.6 and fig. 4.10).

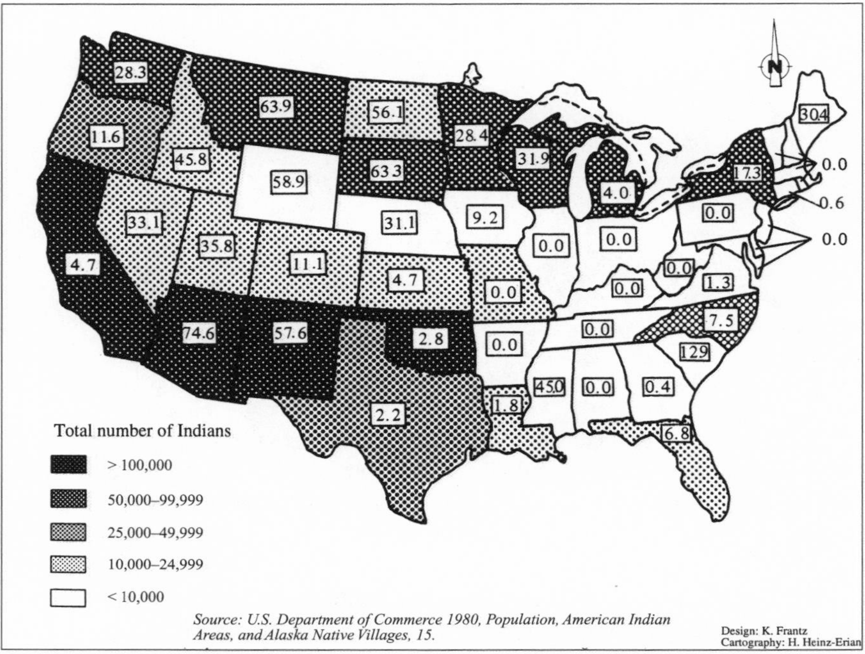

Fig. 4.5: Total number of American Indians and the percentage of reservation Indians, by state, 1980

During the next few decades it can be predicted that the number of American Indians living in large cities will continue to increase at the expense of the reservation population, although in absolute numbers reservation populations will continue to increase, especially on economically successful reservations.

Two reservations in Arizona can serve as examples of the difference among reservations in the proportion of their tribal members who live in urban areas and their geographical distribution within the United States. According to my data, on the Hualapai Indian Reservation (see fig. 4.11), where it is hard to find work, 484 (39.5%) of the 1,227 tribal members lived off the reservation in 1986. Large cities were the place of residence for 164 (13%) tribal members, most of whom lived in the metropolitan areas of Phoenix (47%), San Francisco (20%), and Los Angeles (16%). For the White Mountain Apaches (see fig. 4.12), who can find work on their reservation more easily than the Hualapais, only 347 (4.7%) of its members chose to live off the reservation. There were a total of 199 (2.7%) who lived in large cities, with most of the urban Apaches in the three cities (Phoenix (52%), Los Angeles (13%), and San Francisco (11%).

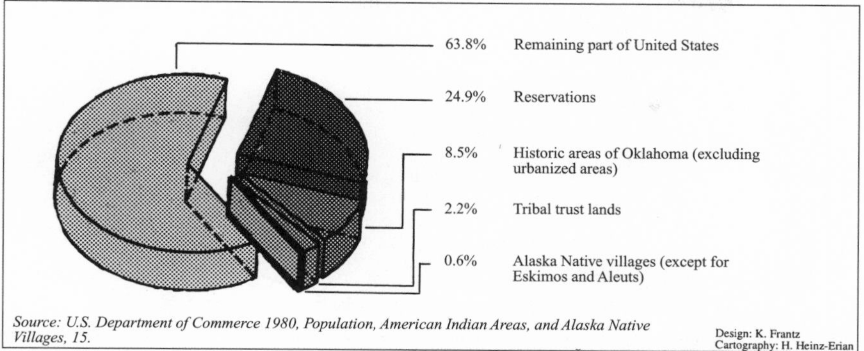

Fig. 4.6: Distribution of American Indians on and off identified American Indian areas and Alaska Native villages, 1980

On the basis of the extensive literature concerning urban Indians[10] it is possible to make a few general statements about this population. Most urban Indians are relatively young and better educated than reservation Indians and have a better command of English. The great majority have left the reservation to move directly to large cities without making any intermediate stops, that is, there is a direct and selective migration pattern. These migrants, once they have arrived in a large city, can be divided into three groups.

The largest group has settled in run-down neighborhoods, the so-called zones of blight near the center of the city, which have long been occupied by the lowest income groups. It is often only regarding location that these Indians have made the jump to "white" America, because they have not been able to establish themselves socially, economically, or culturally in mainstream American society. Many from this group go back and forth time and again between city and reservation in the course of their lives, and even though they have always maintained contact with the predominantly traditional way of life back home, they have usually resigned themselves to lives between two different kinds of society, even after a final return to the reservation.

10. In comparison with the abundant literature about other racial and ethnic groups in American cities not much has been written about urban American Indians, in particular on the part of geographers. The following is a brief list of studies on this subject, grouped according to topic and region. General studies: Jorgensen 1971; Neils 1971; Waddell and Watson 1971; American Indian Policy Review Commission 1976; Stanley and Thomas 1978; Sorkin 1978; Higgins 1982; Lazewski 1982. Studies of particular cities: Los Angeles: Price 1968; Weibel 1976, 1977; San Francisco: Ablon 1964. Large cities in Arizona: Chaudhuri 1974. Denver: Graves 1972; Weppner 1972. Chicago: Lazewski 1976. Syracuse: Flad 1972. Studies about urban residents from particular reservations: Navajos: Hodge 1980; Papagos: Hackenberg and Wilson 1972.

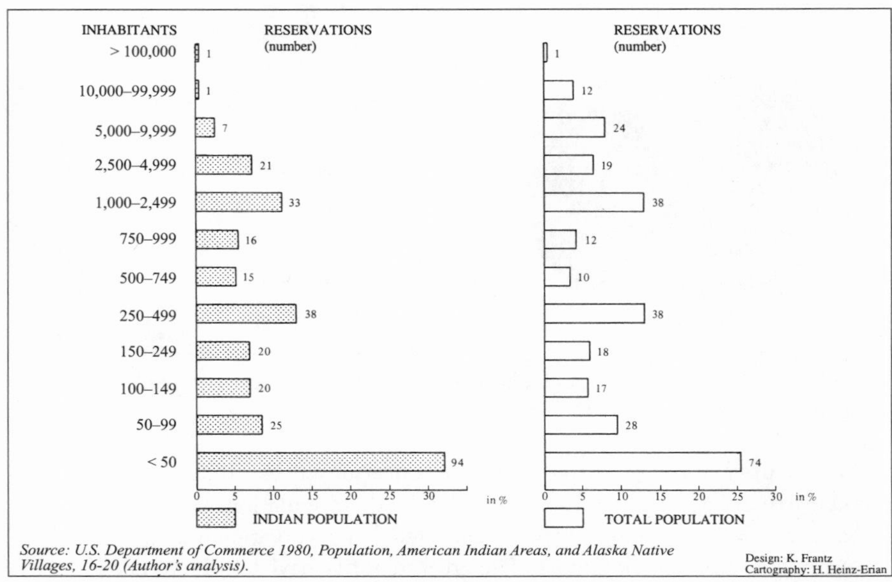

Fig. 4.7: U.S. reservations with their American Indian and total populations ranked according to size, 1980

Table 4.5. American Indian urban population in selected states, 1940 and 1980

State	1940 American Indian urban population	1940 % of urban Indians	1980 American Indian urban population	1980 % of urban Indians
Arizona	876	1.6	47,996	31.5
California	4,078	21.8	161,192	81.3
Minnesota	875	7.0	20,316	58.3
Montana	465	2.8	9,748	26.2
New Mexico	647	1.9	31,316	29.5
New York	2,482	28.7	27,035	69.3
North Carolina	139	0.6	14,261	22.1
North Dakota	237	2.3	4,120	22.5
Oklahoma	7,454	11.8	83,936	49.6
South Dakota	650	2.8	11,816	26.3
Washington	934	8.2	32,843	56.4
Wisconsin	1,075	8.8	13,625	47.2

Note: The Bureau of the Census classifies any American Indian living in a settlement with more than 2,500 inhabitants as urban.

Source: U.S. Department of Commerce, 1980, General Population Characteristics, 125 and 128.

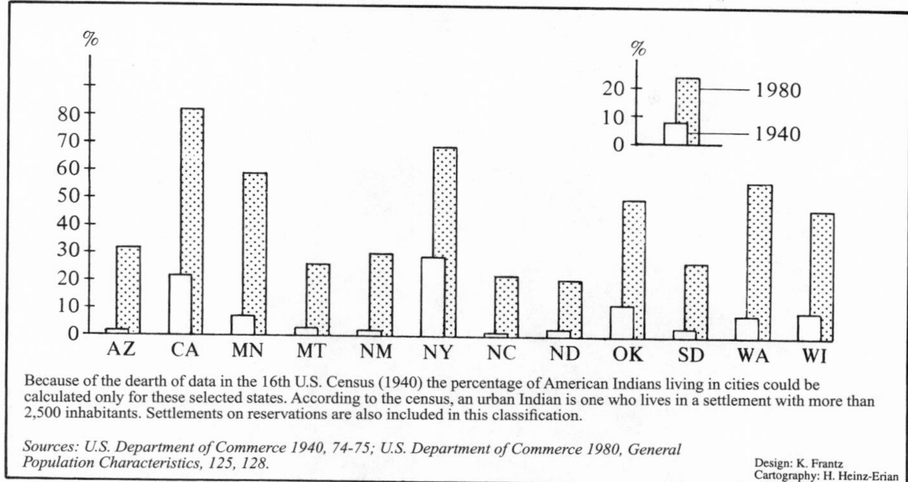

Because of the dearth of data in the 16th U.S. Census (1940) the percentage of American Indians living in cities could be calculated only for these selected states. According to the census, an urban Indian is one who lives in a settlement with more than 2,500 inhabitants. Settlements on reservations are also included in this classification.

Sources: U.S. Department of Commerce 1940, 74-75; U.S. Department of Commerce 1980, General Population Characteristics, 125, 128.

Fig. 4.8: The percentage of American Indians living in urban areas in selected states, 1940 and 1980

The second group of migrants, generally Indians with a larger income, live for the most part in working-class districts bordering the blighted area. As skilled and unskilled workers, these urban Indians find permanent employment which corresponds to their qualifications and which would generally not be available on the reservation, so they become dependent on an urban environment for employment. With their move to a large city many from this group cut the umbilical cord that connected them with the reservation and the extended family as ever-ready providers, and make the transition to the isolated nuclear family of the city.

The third group consists of a small number of highly educated and well-paid Indians whose ethnic origins have also faded but who nevertheless still feel that they are Indians, and they are therefore frequently active in Indian organizations within the city. Some of them later return to the reservation, sometimes as the future generation of leaders.

THE DEMOGRAPHIC STRUCTURE OF AMERICAN INDIANS

The development and structure of the American Indian population since 1950 is similar in some respects to the situation in the Third World, particularly as regards the so-called population explosion (see Bähr 1983, 249–260; Hauser 1974, 58–145; Hofmeister 1978, 331). Between 1955 and 1965, the birthrate of American Indians lay between 36 and 42 births per thousand, figures comparable with those of the developing countries. Between 1970 and 1981 it leveled off at around 27 births per thousand. Moreover, this figure is

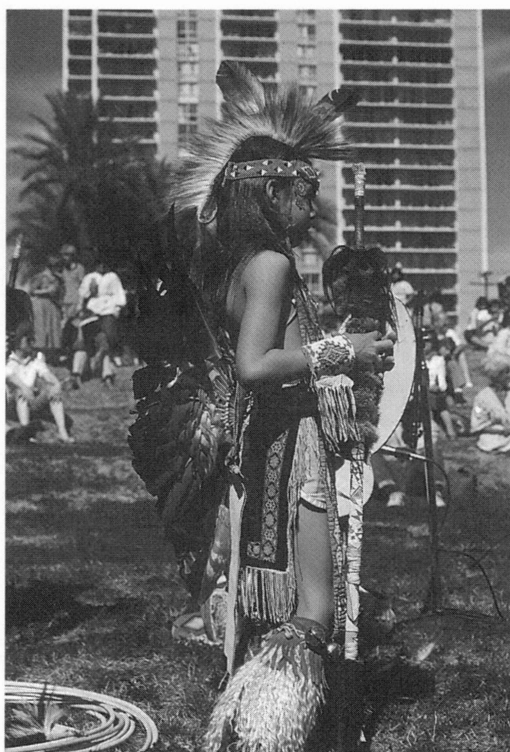

Fig. 4.9: Powwow on Indian Day at Heard Museum, Phoenix, Arizona, 1986

almost 74% above the birthrate of the total U.S. population (see table 4.7and fig. 4.13a). On the other hand, the death rate and in particular the infant mortality rate have drastically declined (see fig. 4.13b). In the period from 1955 to 1982 the latter fell from 61 to 11.6 per thousand live births,[11] a rate considerably lower than that of African Americans and even slightly below the infant mortality rate of the total U.S. population.

This population development, based on data of the Indian Health Service,[12] can be seen in the light of Jürgen Bähr's reflections on the *demographic transition model* (1983, 250–251). After a pretransformative phase of the early 1950s and an early transformative phase of the early 1970s, the American Indian population has today reached the point of an inter-

11. In neighboring Mexico, for example, the birthrate was 45 births per thousand in 1960, which decreased 33.9 per thousand in 1983 (Macmillan & Co. Ltd. 1964, 1229; and Macmillan & Co. Ltd. and Walter de Gruyter & Co. 1984, 847; United Nations 1985, 290 and 397). The infant mortality rate in Mexico decreased from 75.5 per thousand in 1960 to 53.0 per thousand twenty three years later.

12. The service population of the Indian Health Service include not only reservation Indians but also members of some Indian groups that are not federally recognized and urban Indians who still have close ties with their reservations.

Table 4.6. Standard Metropolian Statistical Areas with more than 50,000 American Indians, 1980

Greater Los Angeles, CA	94,706	Tucson, AZ	15,230
Los Angeles-Long Beach, CA	52,809	Detroit, MI	14,267
Riverside-San Bernardino-Ontario, CA	19,844	Dallas-Ft. Worth, TX	13,578
Anaheim-Santa Ana-Garden Grove, CA	16,445	Chicago, IL	11,843
Oxnard-Simi Valley-Ventura, CA	5,608	Denver-Boulder, CO	10,599
Greater San Francisco, CA	42,038	Fort Smith, AR, OK	9,957
S.F.-Oakland, CA	19,675	Portland, OR, WA	9,143
Sacramento, CA	12,561	Houston, TX	8,282
San Jose, CA	9,802	Washington, DC, MA, VA	7,464
Tulsa, OK	38,547	Buffalo, NY	7,162
Oklahoma City, OK	25,858	Milwaukee, WI	6,970
Phoenix, AZ	23,635	Bakersfield, CA	6,822
Greater Seattle, WA	22,355	Salt Lake City, UT	6,792
Seattle-Everett, WA	16,991	Yakima, WA	6,669
Tacoma, WA	5,364	Philadelphia, PA, NJ	6,223
Albuquerque, NM	20,417	Kansas City, MO, KS	6,007
Minneapolis-St. Paul, MN	17,034	Fresno, CA	5,971
San Diego, CA	16,187		
New York City, NY, NJ	15,587	Total	469,343

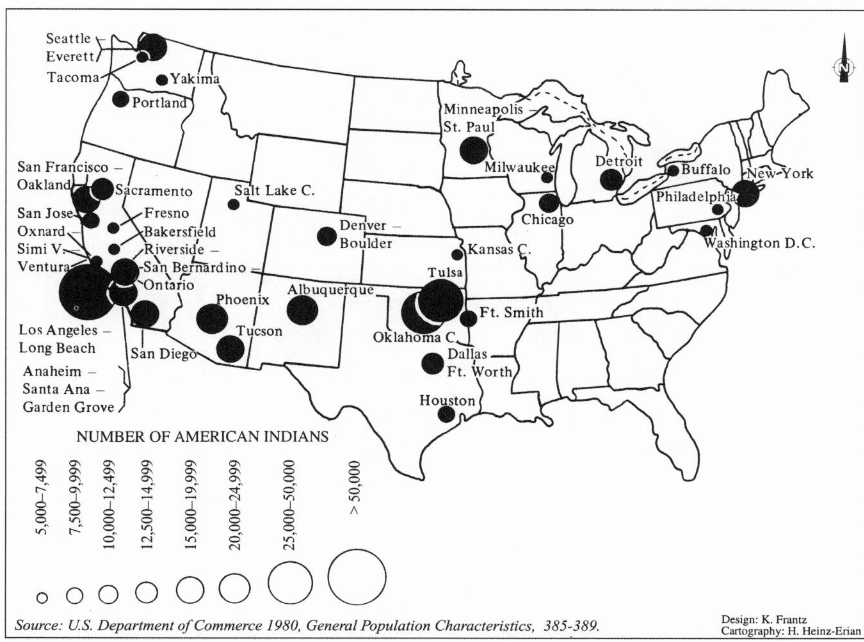

Fig. 4.10: Standard Metropolitan Statistical Areas (SMSAs) with more than 5,000 American Indians, 1980

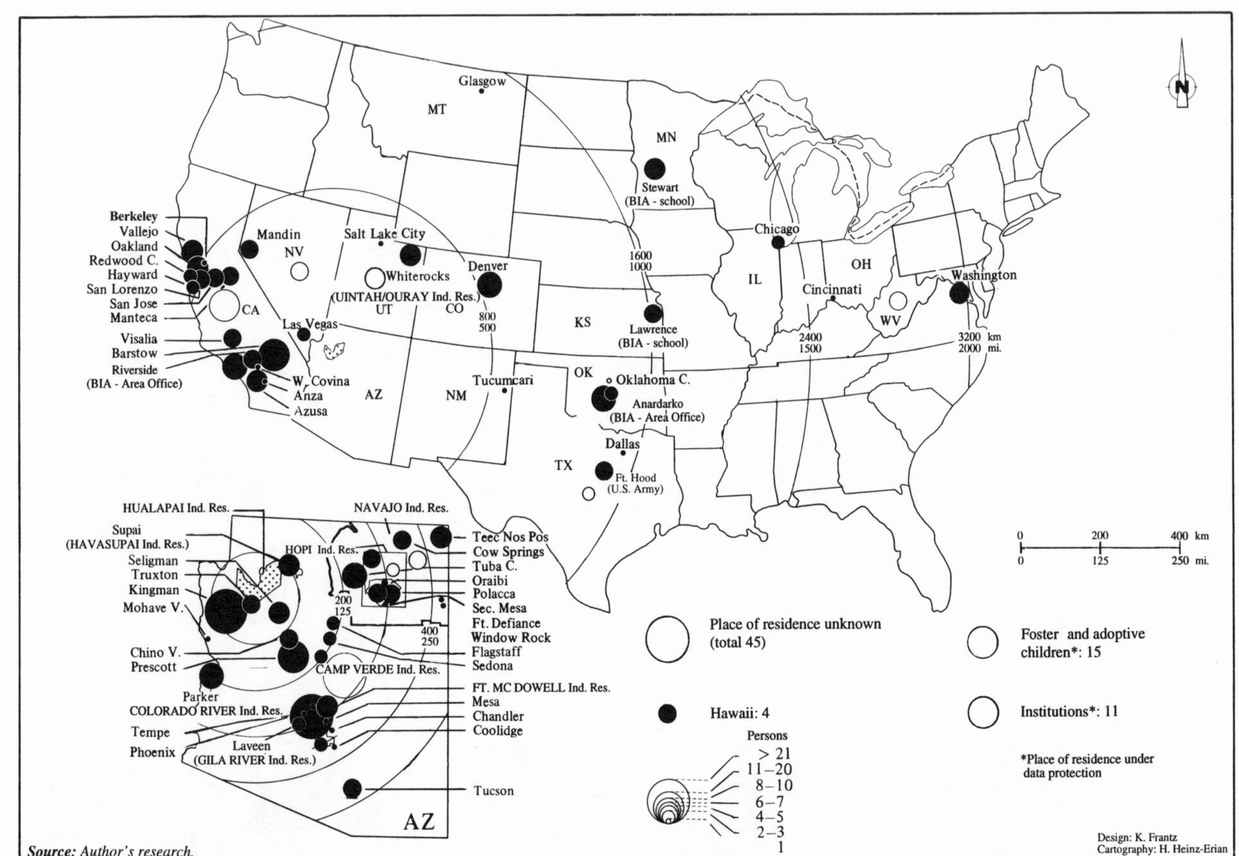

Fig. 4.11: The number and distribution of members of the Hualapai tribe in Arizona residing outside the reservation, 1986

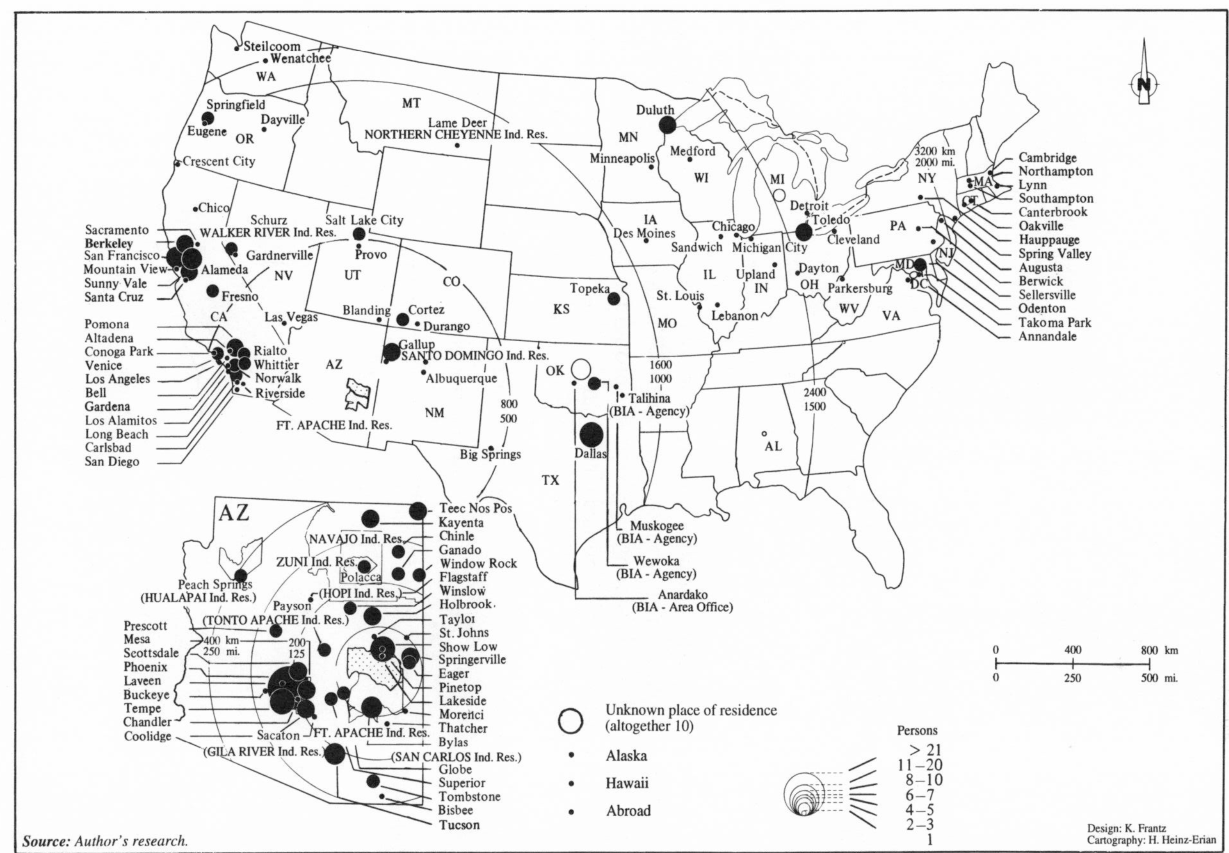

Fig. 4.12: The number and distribution of members of the White Mountain Apache tribe in Arizona residing outside the reservation, 1985

Table 4.7. Minority birthrate and infant mortality rate for the United States, per thousand, 1955–1981

	Birthrate			Infant mortality rate			
	American Indians	Total U.S. pop. (non-white)	Total U.S. pop.	American Indians	African American	Total U.S. pop. (non-white)	Total U.S. pop.
1955	36.5	33.1	24.6	60.9	43.1	42.8	26.4
1960	41.7	32.1	23.7	45.5	44.3	43.2	26.0
1965	36.0	27.6	19.4	40.2	41.7	40.3	24.7
1970	32.0	25.1	18.4	24.3	32.6	30.9	20.0
1975	26.7	21.0	14.6	18.2	26.2	24.2	16.1
1980	26.7	22.5	15.9	13.4	21.4	19.1	12.6
1981	27.5	22.0	15.8	11.6	20.0	17.8	11.9

Sources: U.S. Department of Health, Education, and Welfare, IHS 1984, 18 and 20; U.S. Department of Health, Education, and Welfare, NCHS 1984, Natality Statistics; unpublished data of the IHS Vital Events Staff.

mediate transformative phase in which the birthrate is now declining, while at the same time the mortality rate continues to decline.

This decline of the mortality rate was brought about by external factors and was attended by a marked change in the causes of death, as seen from unpublished data of the IHS (see table 4.8 and figs. 4.13, 4.14). While it has

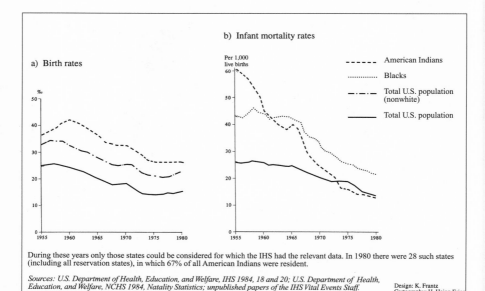

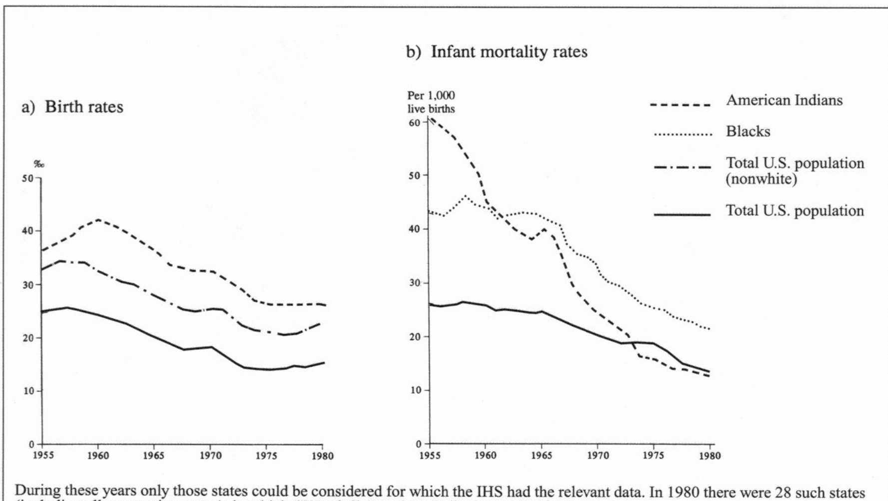

Fig. 4.13: Birth and infant mortality rates in different population groups of the United States (per thousand), 1955–1980

Table 4.8. Leading causes of mortality for American Indians (age-adjusted rates, per 100,000 inhabitants), 1955–1982

	Tuberculosis			Alcohol		
	American Indian and Alaskan Natives	Total U.S. pop. (nonwhite)	Total U.S. pop.	American Indian and Alaskan Natives	Total U.S. pop. (nonwhite)	Total U.S. pop.
1955	57.9	24.1	8.4	*	*	*
1960	32.3	15.1	5.4	*	*	*
1965	27.3	10.9	3.6	*	*	*
1970	11.4	6.8	2.2	56.2	*	8.1
1975	8.6	4.0	1.2	62.2	*	8.6
1980	3.6	2.4	0.6	41.3	*	7.5
1982	2.0	2.0	0.6	35.8	*	6.4
	All accidents			Accidents by motor vehicle		
1955	184.0	71.1	54.3	97.6	28.1	24.6
1960	186.1	67.3	49.9	91.9	24.4	22.5
1965	186.7	70.8	53.4	91.9	26.6	29.2
1970	181.8	72.8	53.7	98.5	30.9	27.4
1975	143.6	56.9	44.8	78.5	22.5	21.3
1980	107.3	49.5	42.3	61.3	20.3	22.9
1982	93.3	40.8	36.5	49.1	16.8	19.3
	Homicide			Suicide		
1955	*	*	*	*	*	*
1960	19.5	25.8	5.3	16.8	5.4	10.6
1965	19.7	29.8	6.3	12.9	6.1	11.4
1970	23.8	41.3	9.1	17.9	6.5	11.8
1975	21.9	41.1	10.5	21.1	7.5	12.6
1980	18.1	35.0	10.8	14.1	6.7	11.4
1982	14.6	30.0	9.7	13.4	6.4	11.6

*No data available

Sources: U.S. Department of Health, Education, and Welfare 1978; unpublished Chart Series of the IHS; personal interviews with <9> and <10> from the IHS; and with <11>, <12>, and <13> of the NCHS.

been possible to successfully treat puerperal fever, tuberculosis, and gastrointestinal diseases and lower the infant mortality rate owing to progress in combating infections, there has also been a relative increase in deaths linked not only to the conditions of modern civilization such as heart disease, diabetes, chronic liver disease, and cirrhosis, but also to accidents, homicide, and suicide.

The American Indian population is still somewhat distinctive in their leading causes of death, differing from both the total U.S. population and

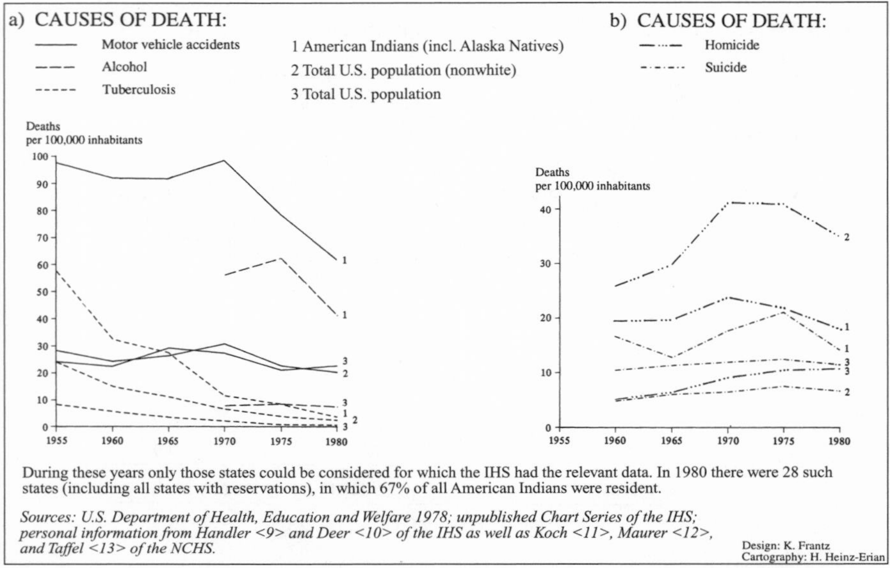

Fig. 4.14: Age-adjusted mortality rates of selected population groups of the United States, by cause of death (per 100,000 inhabitants), 1955–1982

the total nonwhite population. The age-adjusted mortality rate from tuberculosis was 58 deaths per hundred thousand population in 1955, almost seven times as high as the rate of the total American population at the time, while the present rate of 2 per hundred thousand population indicates the great progress in medicine that has since been made. The age-adjusted alcoholism death rate for American Indians was seven times the national average in 1970 (5.5 to 1 today), and from accidents almost 3.5 times the national average (2.5 to 1 today). The homicide and suicide death rates for American Indians are also much higher than for the total U.S. population, but the homicide rate among the entire nonwhite population of the United States is more than twice as high.

Unfortunately there are no comparable data regarding the leading causes of death of reservation Indians throughout the country. The inquiries that I carried out on two reservations, however, would appear to indicate that the figures for the various leading causes of death on the reservations are probably much higher. For example interviews with hospital, emergency, and social service personnel on the Fort Apache Indian Reservation in Arizona <92, 112–114> have shown that from 1982 to 1984 the annual number of deaths caused by alcoholism (liver disease and cirrhosis) was 2.5 times higher than that of the total American Indian population and 14.1 times higher than that of the total U.S. population (fig. 4.15). The

Fig. 4.15: Anti-alcohol campaign among the White Mountain Apache, Arizona, 1985

Table 4.9. Average life expectancy of American Indians (including Alaska Natives) and whites in the United States, 1940–1980

	American Indians			Whites		
	Total	Male	Female	Total	Male	Female
1940	51.6	51.3	51.9	64.9	62.8	67.3
1950	60.0	58.1	62.2	69.0	66.3	72.0
1960	61.7	60.0	65.7	70.7	67.6	74.2
1970	65.1	60.7	71.2	71.6	67.9	75.5
1980	71.1	67.1	75.1	74.4	70.7	78.1

Note: Based on only those states in which the IHS collected data. In 1980 there were twenty-eight such states (including all states with reservations), in which 67% of all American Indians in the United States lived.

Sources: Unpublished data of the Vital Events Staff of the IHS (Rockville, MD).

corresponding numbers for diabetes were 2.7 times that of the total American Indian population and 3.6 times that of the total U.S. population; for automobile accidents, 3.4 and 8.7;[13] for homicide 4.0 and 6.1;[14] and for suicide 3.2 and 3.9. Data collected for a one-year period on the Gila River Indian Reservation in Arizona resulted in similar figures <117>, although they are not significant, since data could be acquired only for a short period.

This decline in the mortality rate of American Indians has been clearly reflected during the past few decades in the rise in average Indian life expectancy (see table 4.9 and fig. 4.16). In 1940 the Indian life expectancy at birth was little more than fifty-one years, the difference between men and women being very small. At this time the white population on the average lived fourteen years longer. Twenty years later the Indian life expectancy had risen to sixty-two years compared to seventy-one years for the white population. In 1980 the average life expectancy of seventy-one years for American Indians was only about three years fewer than the corresponding figure for the white population. It is striking that between 1940 and 1980 the average life expectancy of male Indians, like that of the white population, increased at a decidedly lower rate than for Indian women.

Another characteristic of the American Indian population is its youthfulness, although, since the close of the 1960s, this has been less pronounced than previously. In 1930 some 51% of all American Indians were under 20 years of age (52% in 1910) compared to 44% of the white population (see fig. 4.17).

13. Most of these auto accident fatalities were caused by excessive consumption of alcohol.

14. During the period from 1982 to 1984 almost 26 times as many murders were committed on the Fort Apache Indian Reservation as in Austria <116>. There is no generally accepted explanation as to why there is such a high incidence of homicide among the Apaches and some other tribes. Excessive consumption of alcohol, the generally poor economic situation, the widespread unemployment, the pressure to assimilate, and broken family relationships are among the factors which might contribute to these outbursts of aggression, which usually take place within the family <115>.

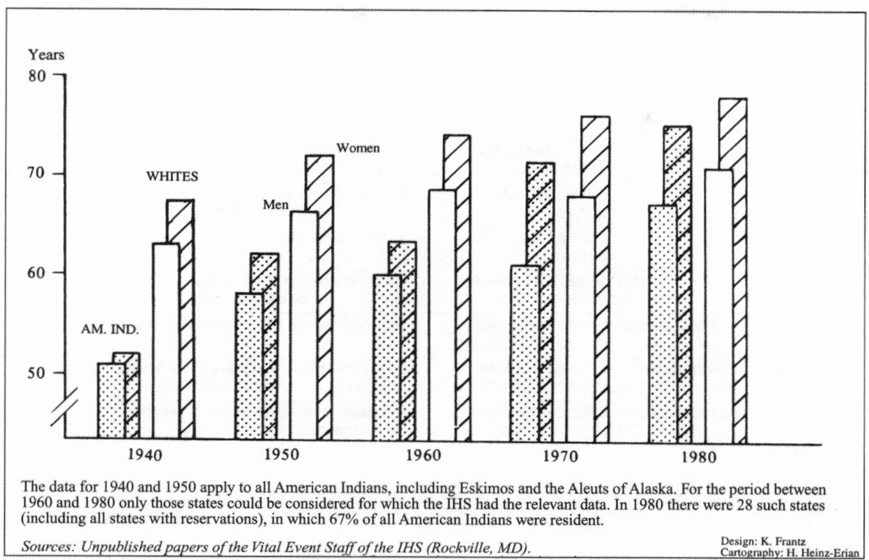

Years
80
70
60
50
WHITES
Women
Men
AM. IND.
1940
1950
1960
1970
1980
The data for 1940 and 1950 apply to all American Indians, including Eskimos and the Aleuts of Alaska. For the period between 1960 and 1980 only those states could be considered for which the IHS had the relevant data. In 1980 there were 28 such states (including all states with reservations), in which 67% of all American Indians were resident.
Sources: Unpublished papers of the Vital Event Staff of the IHS (Rockville, MD).
Design: K. Frantz
Cartography: H. Heinz-Erian

Fig. 4.16: Average life expectancy of American Indians and whites in the United States, 1940–1980

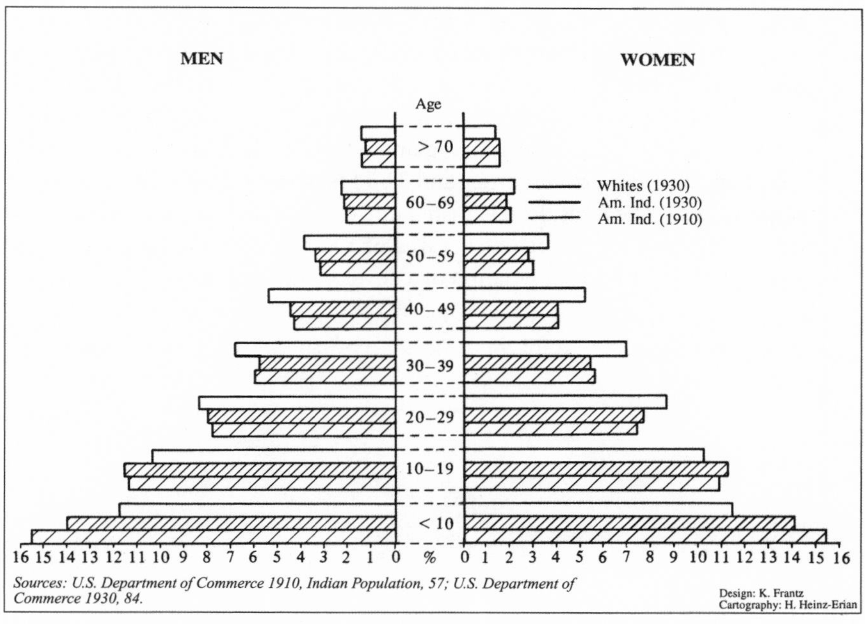

MEN
WOMEN
Age
> 70
60–69
50–59
40–49
30–39
20–29
10–19
< 10
Whites (1930)
Am. Ind. (1930)
Am. Ind. (1910)
16 15 14 13 12 11 10 9 8 7 6 5 4 3 2 1 0 % 0 1 2 3 4 5 6 7 8 9 10 11 12 13 14 15 16
Sources: U.S. Department of Commerce 1910, Indian Population, 57; U.S. Department of Commerce 1930, 84.
Design: K. Frantz
Cartography: H. Heinz-Erian

Fig. 4.17: Percentage of American Indians and whites in the United States by age and sex, 1910–1930

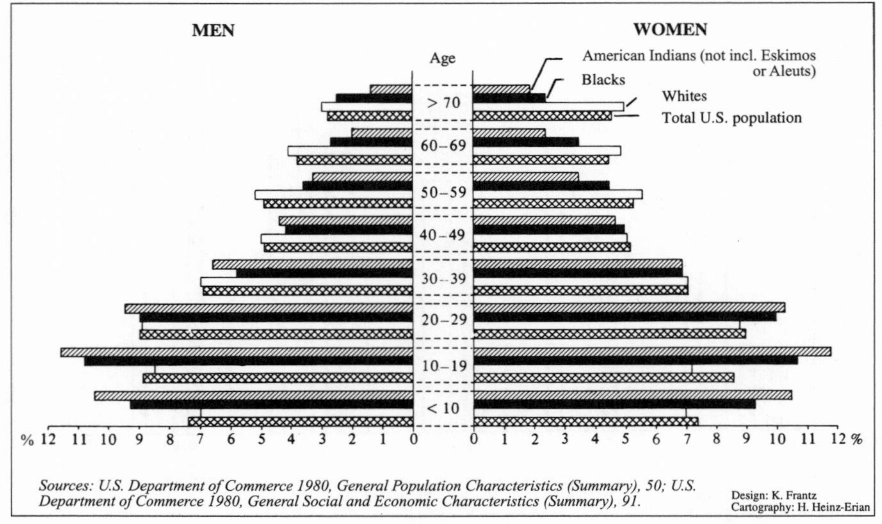

Fig. 4.18: Percentage of specific racial groups in the United States by age and sex, 1980

Only 7% (in 1910 a little more than 7%) had reached the age of 60 as compared with 8.5% of the white population. By 1980 the percentages for the age group under 20 declined to 44% for American Indians and 29% for whites, while the percentages for people over 60 years of age increased to 7.7% for American Indians and 17.5% for whites (see fig. 4.18). While there has been an increase in the ratio of old people to the total white population, this has not been the case with the American Indians, whose median age in 1980 was still less than 23 as compared with 25 and 31.5 for African Americans and whites respectively. If one takes these statistics into consideration, it would appear justifiable to compare the age structures of the aboriginal population of the United States with the population of a developing country.

5 The Socioeconomic Status of American Indians

A COMPARISON OF THE AMERICAN INDIAN STANDARD OF LIVING WITH OTHER RACIAL GROUPS IN THE UNITED STATES

Statistics show that American Indians have made great progress to improve their socioeconomic conditions during the past twenty-five years, not least because of substantial federal expenditures. Nevertheless, the official figures for average life expectancy, state of health, income, housing conditions, unemployment rate, and educational attainment show that American Indians are still one of the two most disadvantaged racial groups in the United States along with African Americans.

The per capita income or the average family income is often used as a measure for the standard of living of a country or a specific population group. The reliability of these two indicators seems to be questionable, however, not only for research done on the Third World (see Hauser 1974, 20–34) but also with regard to American Indians.

The low average income of American Indians does not reflect such key factors as their specific age structure, the average size of families, the widespread unemployment, or additional sources of income. Part of the reason that a smaller percentage of the American Indian population is gainfully employed compared to all other racial groups in the United States, is their relative youthfulness. This , along with the high unemployment rate, bears a direct relation to the Indians' per capita income.

In assessing family income for American Indians one must also keep in mind that the average American Indian family of approximately 3.8 persons (1980) is larger than the average white family (3.2 persons), but comparable to the average African American family (3.7 persons) or Asian American family (3.8 persons). This difference is even greater

Fig. 5.1: Navajo rug from the Two Grey Hills area, Arizona; and Papago basketry, Arizona, 1976

among reservation Indians, whose average family consists of 4.8 persons.[1]

There is also a noncash or exchange economy on reservations which does not show up in statistics, but which is most beneficial to some of the reservation residents. Unreported incomes can include cash exchanges, natural produce from hunting and gathering, blankets, baskets, and other goods, as well as additional governmental or private services and benefits, which also have a certain monetary value. Some reservation Indians receive considerable income from arts and crafts products, which are often sold privately and are therefore not included in income statistics (see figs. 5.1, 5.2, 5.3).[2]

Even today a sizeable portion of the income for many reservation Indian households results from noncash income. Many families are still partially self-sufficient and produce much of what they need for their own living, though the degree of self-sufficiency has been steadily declining. Apart

1. The largest families to be found on any Indian reservations were those in the San Felipe and Santo Domingo pueblos in New Mexico, with 6.5 members in the average family (unpublished data of the twentieth U.S. Census, 1980).

2. High-quality products of Indian arts and crafts such as rugs, pottery, silver jewelry, bead work, and basketry are sold at collector's prices because of the very limited supply, and the number of persons producing these high-quality products is declining rapidly. The Gila River and Fort Apache Indian reservations, for example, both in Arizona, were once well known for their basket weavers, but only one or two people still practiced this handicraft in the mid-1980s (Author's data, 1985–86).

Fig. 5.2: Maricopa Indian pottery, Maricopa Colony, Gila River Indian Reservation, Arizona, 1960s. Photo by Lauren Post, courtesy Malcolm Comeaux.

from that, generous gifts, often exchanged on the occasion of traditional ceremonies, still form an integral part of reservation life). The government also supports needy families with food stamps and surplus commodities from the Department of Agriculture, and various religious communities provide similar help. These public benefits often do not appear in official income statistics, nor do other services which the federal government provides to federally recognized Indians in accordance with its trust obligations, such as free health service and free education, which can be a great financial burden for non-Indians in the United States. There are, however, certain restrictions on these services.

Despite problems regarding the reliability of statistics on American Indian income, the fact remains that average Indian incomes are very low. Their average per capita income amounts to around 58% of non-Indians, and their family income is 69% of the white population. Only the African American population is poorer (see table 5.1 and fig. 5.4). Furthermore, the proportion of American Indian families who lived on an income lower than the officially recognized poverty level of $7,412 in 1980 was 23.7%, more than three times that of the white population.[3] Among American

3. The figures given here were valid for those Indians who required the services of schools, hospitals, and physicians off the reservations, without making use of the BIA or the IHS.

Fig. 5.3: Pima barrette made of glass beads, Blackwater Trading Post, Gila River Indian Reservation, Arizona, 1986.

Indians the percentage of families earning more than $50,000 per year was only one-third of that for white families in the same year (see fig. 5.5 and table 5.2).

These income disparities are even more striking when the comparison is made with reservation Indians only (see tables 5.3 and 5.4). The poverty rate of these American Indians in 1980 was 40.5%,[4] almost six times that of the white population, while a regional breakdown shows that it was precisely in the regions where the proportion of reservation Indians was highest, in the Southwest and the prairie states, that American Indian poverty was most severe. More than half of the families lived below the official poverty level on some of the large reservations in these areas, such as the Papago, Hopi, and Gila River Indian Reservations in Arizona; the Pine Ridge in South Dakota; and the Navajo in Arizona, New Mexico, and Utah.[5] Among the larger reservations the economic situation was best on the

4. In 1979 when the Census Bureau questionnaire was drawn up, the poverty level for a family of four members was $7,412, for five members $8,776.

5. On a number of smaller reservations in the Southwest the percentages for families living below the poverty level were: 72.8%, Alamo (New Mexico); 72.2%, Canoncito (New Mexico); 61.9%, Ute Mountain (Colorado); 73.6%, Santa Rosa Rancheria (California); and 67%, Santo Domingo Pueblo (New Mexico). (U.S. Department of Commerce 1980, General Social and Economic Characteristics [Summary]).

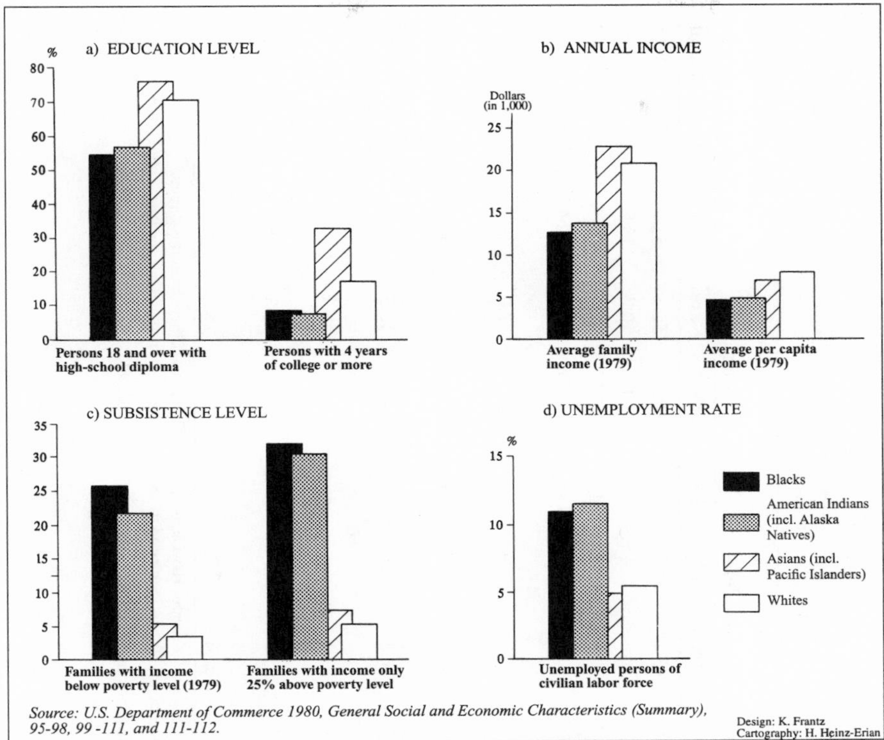

Fig. 5.4: Selected socioeconomic conditions by race in the United States, 1980

Osage Indian Reservation in Oklahoma, a reservation rich in oil, where only 16% of the families lived in poverty, and approximately 9% had an annual income of more than $35,000. Even on this reservation, however, the residents did not reach the same income level as the average for the total U.S. population.

Along with income, housing conditions are another key measure of the standard of living, although it cannot be assumed that all racial and ethnic groups make the same demands for housing. Until well into the twentieth century the overwhelming majority of reservation Indians lived in traditional dwellings of diverse appearance and materials (see figs. 5.6, 5.7, 5.8). In the prereservation era these dwellings ideally fulfilled the needs of the various tribes' lifestyles. The cone-shaped tepees of the Lakotas and the wickiups of the Apaches, both found in dispersed settlements, were perfectly suited to the mobile way of life of these nonsedentary peoples. After they were forced to settle in nucleated settlements, the tepees and wickiups, which were not meant to be lived in for too long a time, as well

Table 5.1. Socioeconomic conditions by race in the United States, 1979 or 1980

	White			Asian[a] (incl. Pacific Islanders)			African American			American Indian (incl. Alaskan Natives)		
	Total	Urban	Rural	Total	Urban	Rural	Total	Urban	Rural	Total	Urban	Rural
Labor force over 16 (%)	62.2	63.0	59.9	66.6	67.1	59.6	59.4	60.5	52.6	58.6	63.2	52.2
Unemployed (%)	5.8	5.5	6.6	4.7	4.6	6.4	11.8	12.0	10.7	13.2	11.9	15.0
Ave. per capita income, 1979 ($)	7,929	8,449	6,652	7,037	7,082	6,427	4,545	4,757	3,316	4,577	5,381	3,637
Ave. family income, 1979 ($)	24,166	25,410	21,218	26,439	26,410	26,870	15,684	16,053	13,381	16,643	17,958	14,983
Families with ann. income < poverty level, 1979 (%)	7.0	6.2	9.1	10.7	10.7	10.7	26.5	25.6	27.0	23.7	19.5	29.2
Families with ann. income 25% > poverty level, 1979 (%)	10.3	9.0	13.4	14.7	14.3	15.0	33.8	33.5	42.0	30.9	25.7	37.5
Persons over 25 with < 5 years of education (%)[b]	2.6	2.4	3.1	6.4	6.3	7.8	8.2	6.8	16.5	8.4	4.5	3.4
Persons 18-24 with high-school degree (%)[b]	78.7	80.6	72.5	80.7	81.1	74.6	66.5	67.8	58.4	60.2	64.6	53.6
Persons over 25 with high-school degree (%)[b]	68.8	71.3	62.4	74.8	75.4	66.7	51.2	54.2	33.3	55.5	62.7	46.4
Persons with > 4 years college education (%)	17.1	19.3	11.4	32.9	33.6	23.2	8.4	9.0	4.7	7.7	9.9	4.9

[a]Asian is defined by the U.S. Bureau of the Census to include Japanese, Chinese, Filipinos, Koreans, Hawaiians, natives of India, and others from this region.
[b]These figures are, in part, very different from those of the microcensus data (see table 5.10). An inquiry addressed to the census bureau did not shed any light on this matter.

Source: U.S. Department of Commerce 1980, General Social and Economic Characteristics (U.S. Summary), 95-98, 99-100, and 111-112.

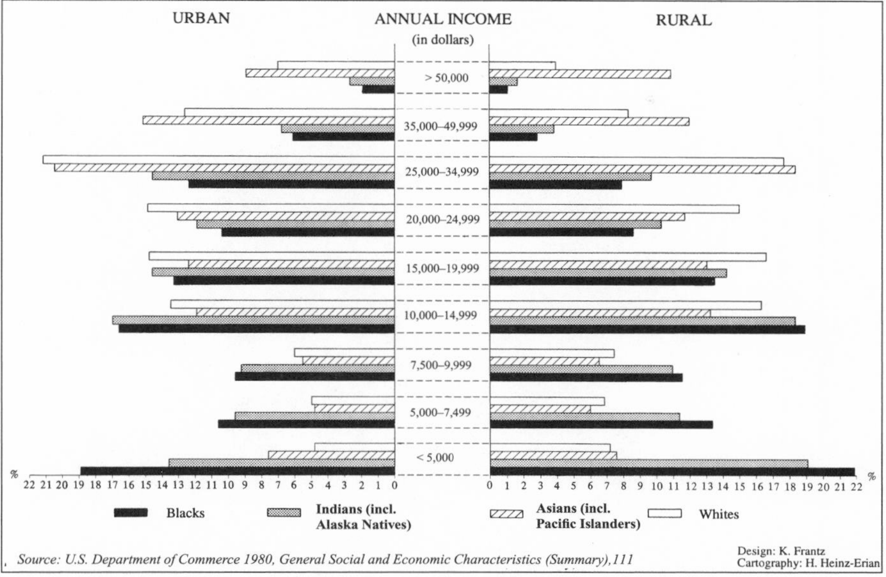

Fig. 5.5: Family income by race in the United States, 1980

as more recent makeshift huts and shacks, became a danger to their health (see fig. 5.9).

Since the Meriam Report of 1928, a report critical of federal Indian policy, the federal government knew about the deplorable living conditions on reservations. It was not until 1954, however, that the first steps were taken to arrange publicly financed housing programs on reservations, initiated at the same time as the comprehensive slum clearance projects in large U.S. cities.

The Department for Housing and Urban Development (HUD), which is responsible for public housing in large cities, was also in charge of a large majority of the new home construction on reservations (almost 70%; see table 5.5) between the mid-1960s and mid-1980s. The linkage with non-Indian housing is also apparent in the design of HUD homes for Indians. The standardized, often prefabricated, and for the most part boxlike, wooden frame houses financed and constructed by this housing program are uniform in appearance and set up in grid patterns, reminiscent of the monotony of American suburbia (see fig. 5.10). This type of design ignored not only the traditional forms and spaces of Indian housing, but also the needs of individual tribal members, and gave no attention to climatic factors. The disregard for this last factor often has led to extremely high utility costs for heating or cooling the home which the average Indian family cannot afford.

Table 5.2. Family income by race in the United States, 1980

Annual income ($)	White	African American	Asian (incl. Pacific Islanders)	Amer. Ind. (incl. Alaskan Natives)
< 5,000	5.6	19.4	7.6	16.0
5,000–7,499	5.5	11.0	4.9	10.4
7,500–9,999	6.4	9.0	5.6	10.0
10,000–14,999	14.6	16.9	11.9	17.6
15,000–19,999	15.2	13.3	12.4	14.5
20,000–24,999	15.0	10.3	13.1	11.3
25,000–34,999	20.1	11.8	20.4	12.4
35,000–49,999	11.4	5.6	15.1	5.6
< 50,000	6.2	1.8	9.0	2.2
	100.0	100.0	100.0	100.0

Source: U.S. Department of Commerce 1980, General Social and Economic Characteristics (U.S. Summary), 111–112.

These HUD subdivisions often deteriorated within a few years of their completion because the wood frame houses were usually built poorly with cheap materials, and it was expensive to keep them in good condition. Moreover, 71% of the residents rented their homes (see table 5.5), and showed little inclination to maintain their houses, apparently taking the attitude that maintenance was the responsibility of the government.

The Bureau of Indian Affairs (BIA), and the Public Housing Administration also initiated the Housing Improvement Program (HIP) in 1961.

Table 5.3. Family income and poverty rate of reservation Indians (%), 1980

Region	< $2,500	$2,500–$9,999	$10,000–$19,999	$20,000–$29,999	> $30,000	< Poverty level
East	7.1	38.1	36.0	12.4	6.4	27.5
Great Plains	11.9	37.8	31.2	12.6	6.5	41.5
N. Great Plains	12.8	38.9	30.6	11.7	6.0	43.9
S. Great Plains	7.7	32.1	34.0	16.9	9.2	26.0
Rocky Mountains	8.7	32.7	29.8	18.3	10.5	37.4
Northwest	16.2	34.5	27.7	15.2	6.5	28.1
Southwest	17.6	36.8	27.3	12.6	5.7	42.4
Navajo	18.8	37.8	24.9	12.2	6.3	49.4
Remaining Southwest	16.2	35.5	30.3	13.0	4.9	38.9
Total Amer. Ind. on reservations	14.4	36.6	29.4	13.2	6.4	40.5
Total U.S.	4.6	24.5	29.3	21.8	19.8	11.7

Source: U.S. Department of Commerce 1980, American Indians, Eskimos, and Aleuts on Identified Reservations (unpublished microfilm data); <15>.

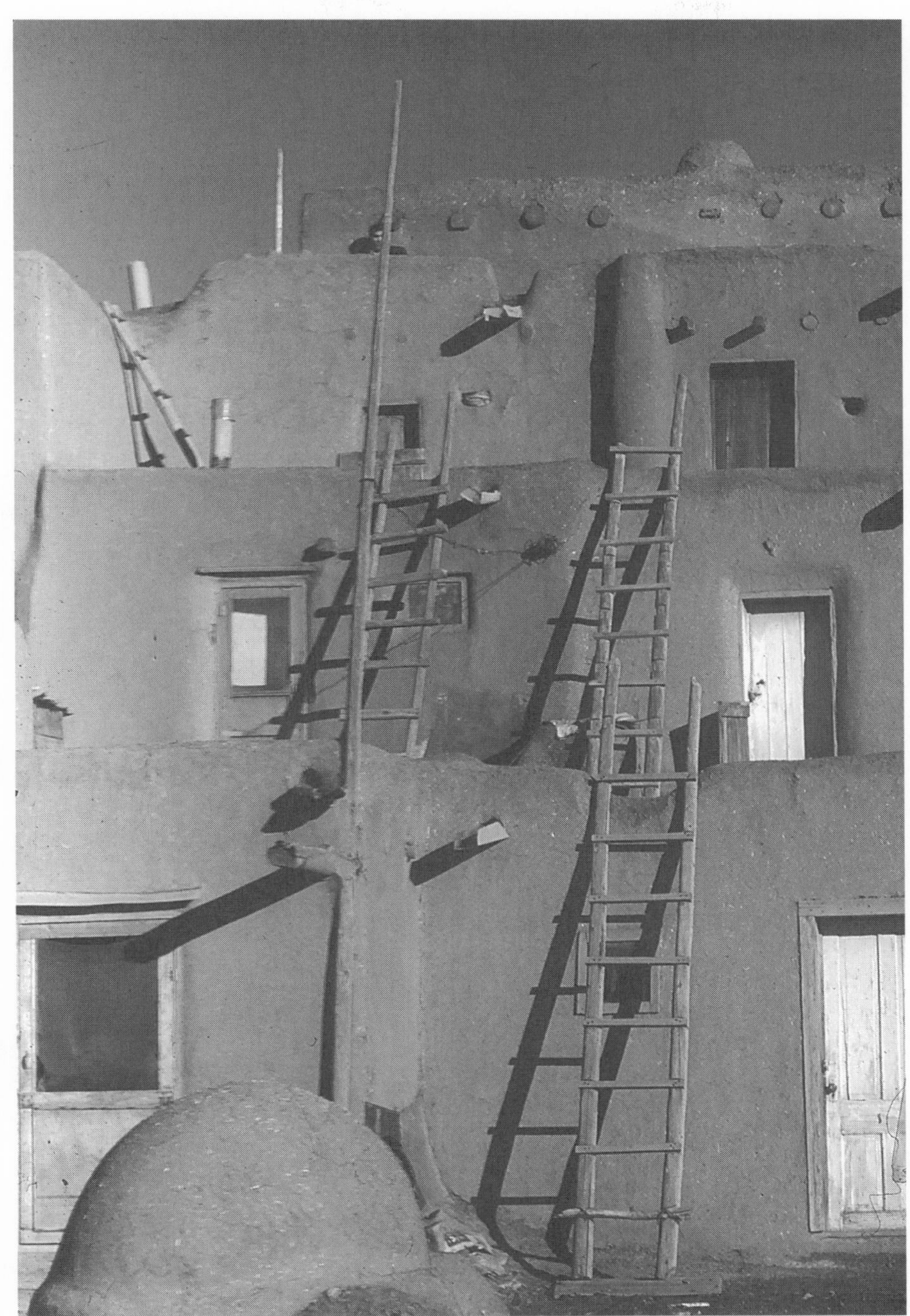

Fig. 5.6: Four-story adobe building with newly added doors, Taos Pueblo, New Mexico

Table 5.4. Selected socioeconomic characteristics for reservation Indians, 1980

Reservation	Family income ($)	Families < poverty level (%)	Persons with social welfare benefits (%)	Families with > $35,000 income (%)	Occupied housing units without: Electricity (%)	Telephone (%)	Refrigerator (%)	Running water (%)	Indoor toilets (%)
Colorado River (AZ, CA)	14,205	27.1	16.4	4.3	0.4	30.4	1.6	0.9	0.9
Crow (MT)	14,724	30.9	27.3	3.6	0.0	40.7	0.6	5.6	3.6
Eastern Cherokee (NC)	11,047	32.9	29.3	1.7	0.3	40.3	0.7	1.0	0.7
Flathead (MT)	13,009	30.0	42.2	3.9	0.6	30.2	0.6	1.2	1.0
Fort Apache (AZ)	10,689	48.2	14.4	2.7	9.9	77.4	14.8	12.6	20.0
Gila River (AZ)	9,606	51.0	17.1	1.2	6.1	72.5	6.2	11.2	12.7
Hopi (AZ)	10,163	51.2	11.1	1.9	47.1	77.9	36.3	58.1	55.6
Hualapai (AZ)	15,334	28.0	21.2	4.8	2.5	59.4	4.4	8.8	11.3
Menominee (WI)	14,146	21.4	29.7	2.7	1.1	27.0	0.9	4.2	6.0
Mescalero (NM)	10,669	43.4	17.0	0.8	1.6	60.6	0.2	0.2	0.2
Navajo (AZ, NM, UT)	11,179	50.8	14.7	3.5	45.8	79.4	46.9	50.6	49.5
Osage (OK)	17,421	15.7	10.7	8.8	< 0.1	9.8	0.4	0.3	0.2
Papago (AZ)	9,160	56.1	15.4	1.2	22.2	87.3	28.8	41.2	53.6
Pine Ridge (SD)	10,214	55.6	22.0	2.3	8.0	64.9	9.5	24.8	0.4
Red Lake (MN)	11,787	39.7	15.1	1.0	0.0	40.9	0.6	6.9	5.7
Rosebud (SD)	10,537	47.9	21.5	2.6	2.5	62.6	3.1	6.9	6.3
Salt River (AZ)	11,181	42.0	14.1	1.2	2.9	78.6	5.8	18.5	20.3
Taos (NM)	9,026	42.3	19.8	0.8	46.9	69.5	36.5	73.6	73.0
Uintah and Ouray (UT)	12,692	32.8	14.7	2.8	0.2	40.3	1.3	3.2	1.5
Warm Springs (OR)	16,227	27.3	16.9	7.4	0.2	31.9	0.6	1.3	1.7
Yakima (WA)	13,564	36.2	22.2	4.9	0.1	39.9	0.4	0.9	0.8
Total Amer. Ind. Reservations	11,867	40.5	27.1	3.1	15.9	56.4	16.6	21.0	20.8

Sources: U.S. Department of Commerce 1980, American Indians, Eskimos and Aleuts on Identified Reservations (unpublished microfilm data); <15>.

Fig. 5.7: Old, conical hogan in Monument Valley, Navajo Indian Reservation, Arizona, 1956. Photo by Lauren Post, courtesy Malcolm Comeaux.

Fig. 5.8: Box-shaped Pima "sandwich house" in St. John Mission, Gila River Indian Reservation, Arizona, 1986

Fig. 5.9: Makeshift wooden hut on the Fort McDowell Indian Reservation, Arizona, 1985

Table 5.5. New housing construction on Indian reservations and in the historic areas of Oklahoma, 1963-1981

				Owner-occupied houses		
BIA area	New housing units	HUD houses[a] (%)	Rental units (%)	HIP[a] (%)	Bank financing (%)	Other financing (%)
Aberdeen	8,560	82.0	46.4	8.8	0.2	9.0
Albuquerque	4,507	70.9	71.9	2.0	1.6	25.5
Anadarko	3,209	89.1	54.4	0.3	3.4	7.2
Billings	5,383	70.4	37.4	6.4	7.8	15.4
Eastern	2,448	67.4	72.2	13.8	1.5	17.3
Juneau	4,125	54.6	79.9	23.1	2.6	19.7
Minneapolis	3,420	54.7	79.1	29.6	4.4	8.3
Muskogee	9,884	90.4	59.3	3.6	0.3	5.7
Navajo	7,585	50.7	84.9	26.6	2.7	20.0
Phoenix	7,349	71.6	59.7	13.6	4.4	10.4
Portland	4,711	46.4	79.2	9.5	20.5	23.6
Sacramento	1,924	40.5	95.2	45.3	3.8	10.4
Total	63,105	69.3	70.9	13.0	4.0	13.7

[a]These houses were developed either by the Department of Housing and Urban Development or through the Housing Improvement Program of the BIA. Some are owner-occupied.

Sources: Unpublished statistics from the BIA Division of Housing Assistance, <17>.

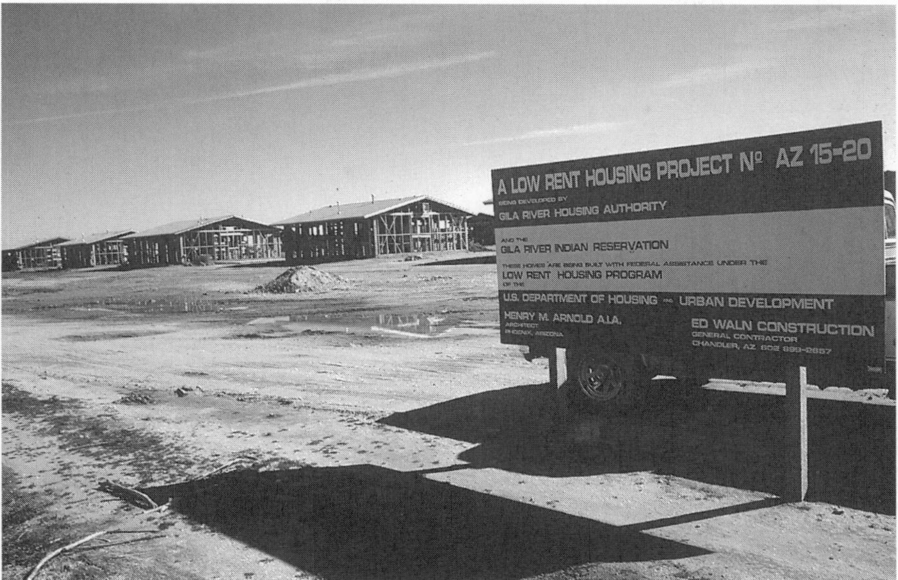

Fig. 5.10: HUD housing in Sacaton, Gila River Indian Reservation, Arizona, 1985

Under the motto "Help for Self-help" it provided Indian families on the reservations with funds for the renovation and construction of homes upon application. It was a condition of this program, however, that these families should participate in building the homes. More than 8,000 HIP homes were constructed in the period from 1963 to 1981 (see table 5.5). They were always very simple, detached, single-family homes. As a rule, there were from one to three rooms, even for a large family; the toilet and sometimes the washing facilities were outdoors, as in traditional dwellings. At a later stage of the HIP program when the houses were mainly built for the elderly, toilets and washing facilities were placed indoors.

Most people in America finance their homes with loans from a bank instead of their own resources, but this method of financing is generally not available on reservations, partly because tribal land cannot be used for collateral on loans. The fact that only 4% of reservation Indians were able to build their own homes is explained by the general lack of capital and creditworthiness owing to lack of collateral. Few residents of a reservation are able to offer a bank any security in the form of land or other property which can be repossessed outside of tribal control, with the consequence that most banks refuse to enter into such credit transactions.

There is no doubt that during the past two decades the housing conditions of reservation Indians have improved considerably thanks to public programs, and a number of recently built Indian settlements even give the

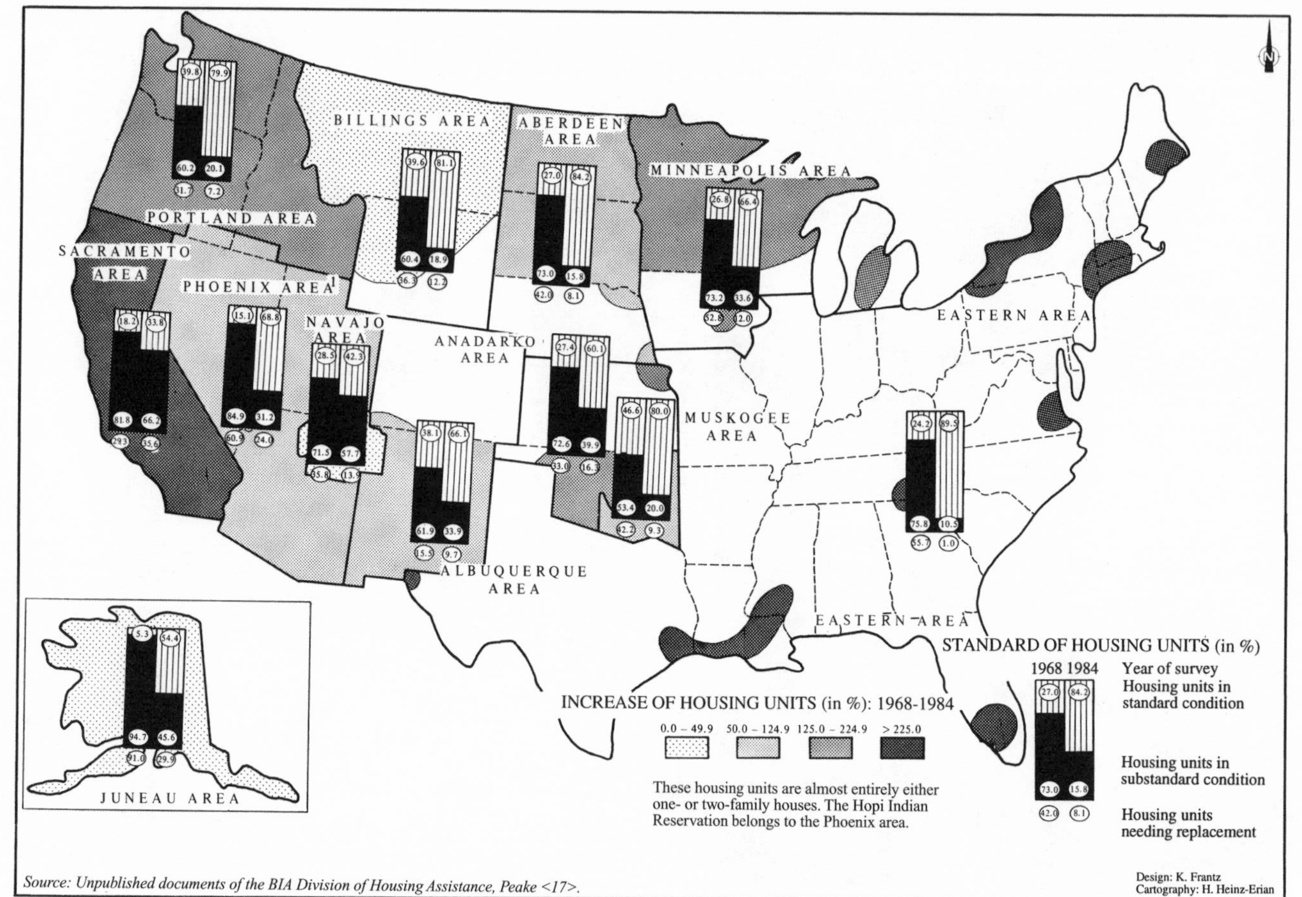

Fig. 5.11: The growth and standard of homes on Indian reservations in the United States, 1968–1984

outsider the impression of a modest prosperity, although this was not generally the outcome of the residents' work and is therefore often misleading. At the close of the 1960s President Johnson's "War on Poverty" spread to the reservations, and statistics on housing by region, provided by the BIA Division of Housing Assistance in Washington, DC, show this social progress very clearly (see fig. 5.11).

The number of Indian housing units throughout the country almost doubled from 1968 to 1984, and the percentage of substandard homes decreased by half, but was still about 36% (fig. 5.11).[6] This building activity was particularly striking in two administrative districts of the BIA, the Sacramento District and the Eastern District, where the number of housing units increased thirteenfold and fivefold respectively. While this building boom in the East resulted in the best housing conditions for reservation Indians of all BIA districts, two-thirds of all American Indian homes in California are still below the American minimum standard, and 35% of Indian housing units in the Sacramento District needed to be replaced, according to the Census Bureau. Between 1968 and 1984, however, building activity was far below the average in the Navajo and Juneau (Alaska) BIA districts. Measured by standards of American mainstream society, the average condition of Indian homes in these two regions was extremely poor.[7]

A detailed analysis of the housing statistics on Indian reservations revealed a great lack of basic facilities, with 21% of all Indian housing units in 1980 lacking running water, and an almost equally high percentage lacking an indoor toilet (see table 5.4). Around 16% of households had no electricity and were obliged to do without a refrigerator.

On some reservations in the Southwest the lack of these facilities is even more striking. From one-half to three-quarters of the residents on the Hopi, Navajo, and Taos Indian Reservations have neither running water[8] nor a toilet; barely 50% have electricity; and between 36% and 47% of the households do not have their own refrigerators. In the summer the temperature there often rises to well above 100° F, and the nearest grocery store is often many miles away. This makes everyday life for these families very difficult and can cause health risks.

The low standard of living of reservation Indians compared to the average American in 1980 is also shown in two further indicators, the lack

6. All homes classified as substandard by the BIA-HUD program and the Bureau of the Census have at least one of the following features: no running water, no indoor toilet, no electricity, inadequate heating, or generally poor condition.

7. Unpublished documents of the BIA Division of Housing Assistance, Peake <17>.

8. For around 43% of Hopi and Navajo households the water needed for daily use was more than 300 feet away.

Table 5.6. Households of federally recognized American Indians without telephone or motor vehicles, by region, 1980

Region	On-reservation Indians		Off-reservation Indians	
	Without telephone (%)	Without car (%)	Without telephone (%)	Without car (%)
East	28.0	17.8	19.9	19.7
Great Plains	48.8	19.4	19.9	11.1
N. Great Plains	53.0	20.5	8.7	14.7
S. Great Plains	27.3	13.3	20.2	11.2
Rocky Mountains	49.1	12.6	22.2	12.8
Northwest	29.7	7.7	14.9	14.1
Southwest	70.6	25.7	23.0	14.2
Navajo	79.2	25.6	—	—
Remaining Southwest	60.0	25.9	—	—
Total status Indians	56.4	21.3	19.8	15.5
Total U.S. pop.	4.0	12.9	4.0	12.9

Sources: U.S. Department of Commerce 1980, American Indians, Eskimos, and Aleuts on Identified Reservations (unpublished microfilm data); <15>.

of an automobile and of a telephone (see tables 5.4 and 5.6). Many reservations are in remote areas with no means of public transportation to cover the often great distances to reach places of work, shops, banks, and so forth. This striking lack of infrastructure leaves the local population seriously disadvantaged. In 1980 about 56% of American Indian households on reservations had no telephone, fourteen times greater than the average for the total U.S. population, and 21% had no motor vehicle, almost twice the percentage for the total American population. The situation was worst on the reservations in the Southwest, where more than a quarter of the families did not own a motor vehicle and where more than 70% had no telephone. The situation was particularly deplorable on the Papago Indian Reservation in Arizona, the second largest reservation in the United States, where 29% of households had no motor vehicle and 87% no telephone.

EMPLOYMENT AND LABOR FORCE CONDITIONS ON INDIAN RESERVATIONS

During the past three decades the number of American Indians both on and off reservations has increased tremendously. One of the principal factors in this population explosion is the high birthrate, which on most reservations has led to a marked increase in the percentage of young people. In 1980 the median age on the reservations was 19.7, almost twelve years below the white population.[9] Those under 18 constitute 46% of the entire reservation population,[10] so it is understandable that on most reservations a race has developed between population growth and the creation or maintenance of jobs, housing, and services, a race which seldom works to the advantage of the younger generation. Because of the rapid growth in population, the limited number of jobs often produces a very tense employment situation on reservations.

One characteristic of all reservations is the one-sided orientation of the labor market, where by far the most important source of employment is the public sector, which includes tribal governments, the BIA, the Indian Health Service (IHS), and other agencies concerned with Indian issues. These agencies together employed almost 60% of the Indian workforce on reservations in 1980 (see table 5.7), and nearly 27% of all employees have administrative positions (see table 5.8), compared to only 5% among white Americans.[11] Many of these agencies have greatly inflated bureaucracies so that a considerable portion of the annual budget which ought to be used for program activities and services in the struggle against poverty and underdevelopment disappears in red tape. In the mid-1980s, shortly after the peak of BIA personnel growth, this agency alone had 15,800 permanent employees nationwide, the equivalent of one BIA employee for every 21 reservation Indians. Many critics are therefore quite right to denounce the BIA for its size and inefficiency, and indeed, many of its programs are little more than occupational therapy for its employees. On the other hand, a reduction in the number of BIA employees would lead to such a drastic shrinkage of the employment base on many reservations that many Indians would lose their jobs and the basis of their livelihood.

The BIA staff today is made up almost entirely of American Indians, except for some non-Indians working in technical branches, who usually live on the reservations. Most key positions in the BIA are occupied by either "mixed-bloods" or members of tribes other than those they work for, and these people generally feel that their first obligation is to the BIA and to

9. On some Indian reservations the median age was exceptionally low: 15.8, Tigua (Texas); 17.1, Fort Totten (North Dakota); 17.4, Crow Creek (South Dakota); and 17.6, Mescalero Apache (New Mexico) (U.S. Department of Commerce 1980, General Social and Economic Characteristics [Summary]), 451–456.

10. Although the BIA has a different way to divide age groups (Indians under sixteen) for status Indians in 1985, its percentage for Indians under eighteen would most likely be lower than that of the U.S. Census. Ibid.

11. According to the U.S. Department of Commerce 1980, General Social and Economic Characteristics, (Summary) 99–106.

Table 5.7. Federal, state, and tribal government employment off and on selected reservations, 1980

Reservation	Number gainfully employed		Federal and state agencies (%)		BIA and IHS (%)		Tribal government[b] (%)		% employed by public sector	
Colorado River (AZ, CA)	650	(43.4)[a]	33.8	(49.5)[a]	20.5	(42.9)[a]	27.1	(53.4)[a]	60.9	(51.3)[a]
Crow (MT)	1,039	(45.9)	42.2	(57.5)	13.9	(49.3)	30.3	(43.8)	72.5	(51.8)
Eastern Cherokee (NC)	1,800	(46.6)	28.2	(48.5)	8.1	(45.5)	15.4	(52.5)	43.6	(49.9)
Flathead (MT)	1,216	(42.2)	28.1	(47.4)	9.0	(44.0)	25.4	(47.2)	53.5	(47.3)
Fort Apache (AZ)	1,837	(43.7)	29.2	(58.5)	14.8	(53.7)	30.2	(50.8)	59.4	(54.6)
Gila River (AZ)	1,762	(39.6)	26.8	(51.3)	12.0	(48.8)	28.3	(43.7)	55.1	(47.4)
Hopi (AZ)	1,435	(49.5)	36.0	(47.3)	23.8	(48.8)	30.5	(53.0)	69.5	(47.7)
Hualapai (AZ)	236	(45.3)	40.7	(44.8)	19.5	(28.3)	39.0	(55.4)	79.6	(50.0)
Menominee (WI)	699	(44.9)	24.2	(56.2)	1.0	(85.7)	36.6	(63.3)	60.8	(60.5)
Mescalero (NM)	548	(43.1)	44.7	(47.3)	33.9	(44.1)	32.1	(45.5)	76.8	(46.6)
Navajo (AZ, NM, UT)	22,636	(45.2)	34.4	(57.0)	17.8	(59.9)	23.6	(46.9)	58.0	(52.9)
Osage (OK)	1,755	(40.1)	21.4	(51.9)	2.9	(58.8)	8.1	(60.8)	29.6	(54.3)
Papago (AZ)	1,607	(40.9)	37.1	(51.8)	14.9	(49.8)	42.4	(32.2)	79.5	(41.3)
Pine Ridge (SD)	2,482	(48.5)	32.6	(50.0)	20.5	(56.3)	42.3	(49.8)	74.9	(50.1)
Red Lake (MN)	707	(45.7)	57.8	(52.8)	34.7	(48.6)	20.9	(46.6)	78.8	(51.2)
Rosebud (SD)	1,279	(46.1)	39.4	(55.8)	15.2	(55.9)	29.0	(43.4)	61.1	(56.6)
Salt River (AZ)	880	(47.8)	18.3	(57.8)	7.6	(65.7)	23.6	(40.4)	41.9	(48.0)
Taos (NM)	307	(41.0)	42.7	(38.2)	14.0	(46.5)	21.2	(32.3)	63.8	(36.2)
Uintah and Ouray (UT)	588	(43.4)	13.4	(43.0)	5.4	(31.3)	44.4	(44.4)	57.8	(44.1)
Warm Springs (OR)	602	(44.2)	12.3	(44.6)	6.6	(30.0)	40.0	(58.5)	52.3	(55.2)
Yakima (WA)	1,532	(43.8)	19.6	(43.7)	8.0	(35.0)	38.8	(50.3)	58.4	(48.0)
Reservation Indians total	89,617	(45.1)	30.4	(55.0)	12.5	(54.4)	28.1	(49.0)	58.5	(52.2)

[a]The figures in parentheses indicate the percentage of female employees.

[b]The employees of tribal enterprises are in some cases included. This explains why the figures in the last column do not always correspond to the total amount of the previous columns.

Table 5.8. American Indian employment type, by selected reservations, 1980

Reservation	Gainfully employed	Primary sector		Secondary sector		Tertiary sector				Total
		Ag., forestry, fisheries	Mining	Manu-facturing	Con-struction	Wholesale and retail trade	Trans-portation[a]	Public admin-istration	Other services[b]	
Colorado River (AZ, CA)	68.8	15.2	0.0	5.5	6.7	2.7	5.4	35.1	29.4	100.0
Crow (MT)	75.6	10.2	1.0	0.9	5.7	1.8	2.2	37.7	40.5	100.0
Eastern Cherokee (NC)	77.7	1.1	0.1	23.0	4.5	8.2	3.4	18.6	41.1	100.0
Flathead (MT)	74.0	14.6	0.2	12.8	5.9	9.4	3.6	23.0	30.5	100.0
Fort Apache (AZ)	81.2	7.8	0.0	17.1	9.7	7.0	1.5	21.3	35.6	100.0
Gila River (AZ)	65.0	14.8	0.3	5.6	6.2	3.5	2.1	29.0	38.5	100.0
Hopi (AZ)	72.7	3.2	0.7	12.6	6.4	8.5	2.7	26.1	39.8	100.0
Hualapai (AZ)	60.6	16.8	0.0	2.1	7.7	2.1	2.1	37.8	33.0	100.0
Menominee (WI)	72.4	2.0	0.0	28.9	4.1	4.0	2.6	24.1	34.3	100.0
Mescalero (NM)	31.6	17.2	0.0	0.6	5.7	1.7	0.0	33.3	41.5	100.0
Navajo (AZ, NM, UT)	69.4	3.9	7.5	6.7	7.4	5.9	6.7	21.1	40.8	100.0
Osage (OK)	45.7	8.5	12.5	7.7	7.7	14.2	4.4	18.7	26.3	100.0
Papago (AZ)	76.8	5.0	1.7	5.0	8.0	1.4	6.0	40.4	32.5	100.0
Pine Ridge (SD)	78.2	6.6	0.1	7.2	4.0	2.1	2.5	30.9	46.6	100.0
Red Lake (MN)	78.5	4.0	0.0	8.5	8.5	0.4	6.2	31.5	40.9	100.0
Rosebud (SD)	74.5	8.0	0.0	1.7	5.8	6.2	4.3	29.9	44.1	100.0
Salt River (AZ)	37.3	4.3	0.9	3.0	10.1	6.1	8.8	26.8	40.0	100.0
Taos (NM)	58.3	8.4	0.0	6.1	16.8	0.6	6.2	25.7	36.2	100.0

(Continued)

(Table 5.8, *continued*)

Reservation	Gainfully employed	Primary sector		Secondary sector		Tertiary sector				Total
		Ag., forestry, fisheries	Mining	Manu-facturing	Con-struction	Wholesale and retail trade	Trans-portation[a]	Public admin-istration	Other services[b]	
Uintah and Ouray (UT)	65.3	7.6	2.9	3.9	6.5	3.9	2.1	39.3	33.8	100.0
Warm Springs (OR)	79.2	8.0	0.0	22.9	3.8	2.9	1.8	30.0	30.6	100.0
Yakima (WA)	76.5	17.0	0.1	10.4	6.6	7.4	5.0	27.0	26.5	100.0
Reservation Indians (total)	66.4	6.9	3.5	9.3	7.0	5.0	3.6	26.9	37.8	100.0

[a]Transportation, communications, and other public utilities.

[b]Chiefly public services (health services, educational services, etc.) but also finance, insurance, real estate, and entertainment and recreation services.

Sources: U.S. Department of Commerce 1980, American Indians, Eskimos, and Aleuts on Identified Reservations (unpublished microfilm data); <15>.

their careers, not to individual reservations. To maintain this attitude and to prevent identification with the tribal population, BIA staff are transferred at regular intervals to work elsewhere.

When considering the amount of power and influence that top BIA officials still have over tribal policy, one cannot help comparing them with the local native administrative elite and former viceroys of the British colonies who, by a policy of "indirect rule," were able to protect the interests of their mother country and to shield it from serious difficulties.

The IHS is also a large-scale employer on reservations. In 1984 it operated 51 hospitals, 99 health centers, and 300 partly mobile medical clinics for Indians on and near the reservations <18>. It had more than ten thousand employees, of whom more than 70% were American Indians <19>. Unlike in the BIA, it was only in exceptional cases that American Indians could be found in key administrative positions or among the approximately 750 doctors in the IHS. This led to a disparity in the salaries.[12] One must take into consideration, however, that there were only about 90 medical doctors of Indian ancestry in the United States in the mid-1970s according to the Association of American Indian Physicians. Very few of these applied for vacant positions in the IHS because the salaries were relatively low compared to what they could earn as doctors elsewhere.

In the identified IHS service population, somewhat more than 960,000 in 1985, there was less than one (0.8) physician for every 1,000 American Indians compared to 1.7 physicians per 1,000 for the total American population. Moreover, many of the IHS doctors were completing their internships in return for scholarships from the National Health Service Corps to fund their medical education, and were obligated to render service for two years in the IHS or other services. Once this obligation had been fulfilled, most of these physicians left for more lucrative positions.

Between the mid-1960s and mid-1980s the tribal governments grew to be a third major public employer on reservations (see table 5.7).[13] The total number of all administrative, planning, and advisory positions taken by members of the tribe in proportion to the total number of reservation employees of the BIA shows to what extent a tribe has control over its reservation. When a tribe has a large number of jobs to offer, this is not only a measure of its independence but also an indicator that it has been able to loosen its bonds with the BIA and other federal agencies. The power to decide who will be hired also enhances the prestige of the tribal government.

12. A survey conducted by the Office of Personal Management within the Department of the Treasury showed that Indian IHS employees earned on average only 74% as much as their non-Indian coworkers in 1979 (quoted in Barsh and Diaz-Knauf 1983, 3).

13. Unfortunately, figures for people employed by tribal governments do not distinguish between employees in administration and those in tribal enterprises.

This power is then often exploited by newly elected tribal councils and chairpersons to their own advantage, who frequently fire the chief employees, whose positions are then sometimes taken by docile protégés, often members of the elected leader's family or clan, or by interchangeable white technical personnel.

This merry-go-round for jobs within tribal government is set in motion over and over again, usually at the close or in the middle of council terms, which, in most cases, last for only two years. This causes a number of unfortunate consequences for the reservations. Owing to the rapid change of managers at the intermediate and upper levels, maintaining continuity in the work of the reservation is often difficult. Furthermore, few well-qualified experts are willing to work for a reservation under these circumstances.

It is a striking fact that, as is the case in federal and state agencies on the reservations, many of the employees who take tribal staff positions are women (see table 5.7), not only at the lower skill levels, but also at the highest. This may pose a great additional burden for many women, often single mothers, who are responsible for both child rearing and management of their households.

Another feature of the tribal staff is the fact that many of the executive positions are taken by Anglo Americans, the so-called "white scouts."[14] This is partly because a limited number of people on reservations have the necessary qualifications. Even when there are specialists among the tribe, often they are not hired or they turn down the offered positions. The reasons for this are to be found in the reluctance of the tribal council, whose members often have had little formal education, to accept tribal members who have received a very good education outside the reservation. It is believed that because of this education, which in the eyes of some traditionalists has "infected" the person in question with the white man's values, the person could develop into a potential rival whom it might be difficult to fire. For their part, tribal applicants realize that any decision they make in the position will be subject to constant interference and discussion, for as members of an extended family and a clan they are bound into a network of mutual obligations and interdependence. This reservoir of know-how within each tribe's membership is therefore often lost to many reservations.

14. This expression is used in imitation of the term "Indian scout," which designated American Indians of the nineteenth century who were hired to support white troops in defeating a hostile tribe. Today some of the white experts working for the Indian tribes, especially attorneys, could well be called "white scouts," because in the course of this century they have helped many reservations gain victory over the white population in conflicts over lost or threatened tribal rights. The difference is that today the theater of war has been transferred to the law courts.

Compared with the public sector, which dominates the reservation economy, the remaining tertiary sector, particularly retail trade, so dominant in the United States, plays a relatively minor role on reservations as a source of employment. Whereas in 1980 over 21% of those gainfully employed in the United States were engaged in retail trade, the corresponding figure on Indian reservations was only 5%.[15] Other than in tertiary employment, which is dominated by public-sector jobs, the chances left to reservation Indians to find work are generally limited, although this is not the case on all reservations. The secondary sector, including construction work, employed about 16% of the labor force, which was 12% less than the national average.[16] The industrial enterprises, which—thanks to government subsidies—were put up on the reservations from the mid-1960s on, benefit both Indians and non-Indians. In 1985 in 103 businesses with more than ten employees each there was a total workforce of around 8,000 (see fig. 10.10), yet because of inadequate statistics it is not possible to say how many of these were American Indians. On some reservations, such as the Menominee, Eastern Cherokee, Warm Springs, and Fort Apache, these manufacturing businesses were an important source of employment for the local population (see table 5.8).

The primary sector in Indian country is also of varying importance on different reservations. Employing 10% of the labor force in 1980, it played a more important role on reservations than in the United States in general, where only 4% of the labor force were employed in this sector (table 5.8). Much of the reservation land used for agriculture was leased to farmers from outside and the extraction of mineral resources was almost purely in the hands of non-Indian companies. The consequence of this situation was that jobs in this sector could not be reserved for Indians alone.

Very few reservations had a sufficient number of jobs for the population in the early 1980s, so that approximately one-third of the Indians looking for work (see table 5.8) were obliged to take jobs outside the reservation. This is no great problem for Indians on reservations located near large cities, which usually need cheap labor. In order to reach these potential places of work, however, Indians must have their own motor vehicles or, failing that, at least the possibility of traveling with someone else, as there are usually no means of public transportation. Since many reservations are far away from major centers of economic activity, Indians looking for work must either resign themselves to traveling very long distances as commuters or find additional housing accommodation near their places of

15. According to U.S. Department of Commerce 1980, General Social and Economic Characteristics (Summary) 99–106.

16. Ibid.

work. Thus, many reservation Indians prefer to remain in their familiar environment and live from unemployment benefits and other sources of public assistance.

The high unemployment rate is one of the Indians' greatest social problems, not only on, but also off the reservations. In 1980 average unemployment among American Indians was 13% throughout the country, more than twice as high as that of the white population and even greater than that of the African American population (see table 5.1 and fig. 5.4). Although, unfortunately, reservation Indians are not listed separately in the U.S. Census socioeconomic data, a distinction is made between urban and rural Indians, and one can assume that the category of rural Indian correlates more or less to that of reservation Indians. According to the 1980 census, 15% of rural Indians of working age were unemployed, a percentage that does not correspond at all to the BIA or tribal government data, or to my inquiries on several reservations.

Contradictory statements regarding the extent of unemployment can largely be explained by different views as to who should be classified as unemployed. The U.S. Census ranks as unemployed only those individuals between 16 and 65 years of age who are able to work, who had no job during the week before enumeration, and who had been actively seeking employment during the preceding four weeks. This calculation excludes those who cannot work because of school attendance, child care responsibilities, health problems, or imprisonment.

The BIA estimates two unemployment rates, one of which is defined in basically the same terms as the unemployment rate of the census. Using the census concept, the BIA arrived at a much higher unemployment rate of 39% for all American Indians within the BIA service population, which is more than twice the unemployment rate of the census for reservation Indians in 1985.[17] The reason for this much higher rate might have been that the BIA personnel have a more intimate knowledge of reservation life and may have been counting unemployed Indians who were missed by the census (ED 1986, 45). On some of the larger reservations such as the Pine Ridge, Crow, Hualapai, and Rosebud, the overall unemployment rate was between 53% and 82% (see table 5.9).[18]

Many of the reservation Indians have already given up looking for work, believing that the job prospects are hopeless and their efforts would therefore be to no avail. These "discouraged unemployed" are not included in the U.S. Census, but they are justifiably counted by the BIA as "unemployed

17. The BIA has no precise data for the comparative year 1980.

18. According to the BIA, the highest percentage of unemployed, 85%, was on the Fort McDermitt Indian Reservation (around 500 Indians) in the border area between Oregon and Nevada.

Table 5.9. American Indian labor force status by selected reservations, 1985

Reservation	American Indian population[a]	% under 16	% over 65	% of unemployed seeking work	% of unemployed able to work
Crow (MT)	5,811	37.6	3.8	56	64
Eastern Cherokee (NC)	6,110	37.6	11.5	39	45
Flathead (MT)	3,225	52.0	1.1	25	27
Fort Apache (AZ)	8,311	34.4	3.6	5	15
Gila River (AZ)	10,556	48.3	4.9	26	31
Hopi (AZ)	8,952	43.6	7.5	33	48
Hualapai (AZ)	1,083	41.7	3.8	69	71
Menominee (WI)	3,582	42.1	6.6	35	39
Mescalero (NM)	2,899	43.3	2.1	41	54
Navajo (AZ, NM, UT)	166,665	35.3	5.0	40	52
Osage (OK)	6,743	33.1	10.1	8	17
Papago (AZ)	14,397	33.5	4.1	21	33
Pine Ridge (SD)	18,754	37.8	4.7	53	72
Red Lake (MN)	4,090	39.6	4.1	34	52
Rosebud (SD)	11,685	38.8	5.6	82	86
Salt River (AZ)	4,185	38.0	3.6	23	39
Taos (NM)	1,718	29.3	12.0	64	78
Uintah and Ouray (UT)	2,270	38.0	3.7	24	52
Warm Springs (OR)	1,967	38.1	3.2	34	49
Yakima (WA)	7,987	28.9	6.4	40	71
BIA service population	786,019	29.4	6.0	39	49

[a]These BIA employment figures, January 1985, are in some cases considerably higher than those of the Bureau of the Census 1980. The chief cause might be the BIA definition of "status Indian" for the data.

Source: U.S. Department of the Interor, BIA 1985, table 3, 1-22.

able to work, but not looking for work," and these numbers might give a more realistic picture of the labor market situation. Seen in this light, almost half of all reservation Indians of working age are unemployed (see table 5.9), which means that unemployment on Indian reservations is often more serious than in even the most depressed slum areas of large American cities.

If one refers to the internal calculations of some tribes, however, even the dire figures given by the BIA may dramatically understate the real unemployment problem. At the Crow Indian Reservation in Montana, for example, unemployment amounted to 64% in 1985 according to the BIA, while the tribal government ascertained the unemployment rate to be 80% for the same year <20>. There are various reasons for this difference, including the fact that an increase in the unemployment figures would automatically require an increase in the BIA's expenditures for unemployment benefits. Seasonally or temporarily employed, as well as grossly underemployed, reservation Indians who either depend on some subsistence farming, on livestock, or on occasionally producing jewelry, pots, or basketry in order to eke out an existence do not appear in the unemployment statistics of the BIA. On the other hand, in the tribe's unemployment statistics those engaged in subsistence farming or craft work at low levels are identified as unemployed. Furthermore, the tribes sometimes average the generally high level of unemployment in the winter with the lower level in the summer.

In view of the appallingly high level of unemployment, as well as widespread underemployment, it would therefore seem reasonable to set up businesses and to give precedence to those projects calling for larger numbers of employees instead of those requiring large capital assets. This principle is, in fact, followed by many tribal enterprises, even if it means tolerating overemployment and unprofitability. Most white entrepreneurs on the reservations, however, do not wish to follow this concept, and would generally prefer to make a profit rather than employ more people than might be necessary.

EDUCATIONAL ATTAINMENT AND THE EDUCATION SYSTEM OF AMERICAN INDIANS

Changes in the Education System

> Again, I am sure these helpless innocents for whom I plead would, if they had words in which to express their thoughts, appeal to us with a pathos inexpressibly touching to save them from the doom that awaits them if left to grow up with their present surroundings. They will become, by a law inexorable as gravity, just what their environment compels them to become, for mind makes itself like that it lives midst and on. If they grow up on Indian reservations removed from civilization, without advantages of any kind, surrounded by barbarians, trained from childhood to love the unlovely and to rejoice in the unclean; associating all their highest ideals of manhood and womanhood with fathers who are degraded and mothers who are debased, their ideas of human life will, of necessity, be deformed, their characters be warped, and their lives distorted. They can no more avoid this than the leopard can change his spots or the Ethiopian his skin. The only possible way in which they can be saved from the awful doom that hangs over them is for the strong arm of the Nation to reach out, take them in their infancy and place them in its fostering schools; surrounding them with an atmosphere of civilization, maturing them in all that is good, and developing them into men and women instead of allowing them to grow up as barbarians and savages.

> In the camp, they know but an alien language; in the school, they learn to understand and speak English. In the camp, they form habits of idleness; in the school, they acquire habits of industry. In the camp, they listen only to stories of war, rapine, bloodshed; in the school they become familiar with the great and good characters of history. In the camp, life is without meaning and labor without system; in the school, noble purposes are awakened, ambition aroused, and time and labor are systematized.
>
> These helpless little ones cry out to us: If you leave us here to grow up in our present surroundings, what can we hope for? Our highest conception of government will be obedience to the word of the chief; our patriotism will be bounded by the confines of the reservation; our lives will be at the mercy of the "medicineman"; our religion will be a vile mixture of superstition, legends and meaningless ceremonies; . . . Our homes will be hovels, our sweetest relations will be marred by corruption, and our natures will be imbruted by vices. We shall have no literature, no accumulated treasures of the past, no hopes for the future. . . . Our only hope is in your civilization, which we cannot adopt unless you give us your Bible, your spelling book, your plow and your ax. Grant us these and teach us how to use them, and then we shall be like you.
>
> —Extract from A Plea for the Papoose: An Address at Albany, New York, by Gen. Thomas Jefferson Morgan (1888)

A cursory glance at the history of Indian education during the past century and a half makes it clear that the development and financing of Indian schools was not intended to be a gift of the federal government to the indigenous population. For government and private philanthropic or religious institutions alike the Indian school, as an institution, was used as a tool to carry out the U.S. policy of forced acculturation. Within these schools subjects such as history, social studies, civics, geography, and English served as appropriate means for the transmission of mainstream American values as well as a way to convince American Indians of their cultural inferiority (see Churchill and Hill 1979).

Of course there were often differences of opinion between federal agencies and philanthropic or missionary societies as to how these objectives should be attained. Thus, for example, until after the Second World War, the BIA schools had orders to prohibit the use of Indian languages, whereas certain religious groups favored the use of native tongues at this time and even developed written languages for teaching in school.

Besides this policy of acculturation, the United States had a trust responsibility to build schools for reservation Indians, as stipulated by treaty in exchange for land cessions. The peace treaty with the Navajos in 1868, for example, stipulated that the government would provide a teacher for every thirty Indian children and construct schools, and in that same treaty the Navajos had to promise that in the future they would make sure that their children went to school. For a long time, however, neither party acted in accordance with these commitments.

Similar arrangements were still customary after 1871, the year of the Indian Appropriation Act, when Congress also granted all future reservations the right to free education.[19] At the beginning of the 1870s the federal government adopted—if only for a short period of time—the policy of entrusting schools entirely to particular religious groups (see fig. 2.10) which had established missions on Indian territory, and the BIA agreed to pay most of the attendant costs. By the beginning of the 1880s, however, the government was no longer willing to let the missionaries have sole responsibility for Indian education, as it was believed that these schools did not pay proper attention to their duty to civilize Indians and to integrate them into American society. As a result, most of the Protestant churches withdrew their teachers and only the Catholics kept all their schools open. By the turn of the century Catholic schools therefore came to receive the bulk (98%) of the greatly reduced federal funds set aside for contracted missionary schools (Provenzo and McCloskey 1981, 14).[20]

One of the provisions of the General Allotment Act (GAA), was an education system based on government boarding schools which was developed for reservation Indians. The basic principles of these boarding schools are clearly set forth in an extract from a speech given by Thomas Jefferson Morgan, the Commissioner for Indian Affairs of the Department of the Interior, quoted above. It is clear that the ideas expressed in this speech were determined by a doctrine of *social determinism*, an ideology that had many followers among social reformers of the time. According to Morgan and the self-styled Friends of the Indians closely associated with him, Indian children were fully conditioned by the sociocultural environment on the reservations. It was therefore their view that children of school age could be saved "from the doom that awaits them if left to grow up with their pre-

19. Only a few months after the founding of the first Navajo school in 1869 it had to be closed because of the lack of students. After this closure thirteen years elapsed before the next educational facility, a boarding school, was founded on the reservation. It was not until the mid-1930s that the education system of the Navajo Reservation showed real progress, with the construction of fifty day schools for around 3,500 students. The first high school on the reservation also dates from this time (see Woerner 1941; Johnston 1966, 49; Young 1976, 380–392).

20. The importance of mission schools for the education of reservation Indians has declined sharply during the course of the twentieth century. Shortly before the outbreak of the Second World War one out of every nine Indian schoolchildren attended a mission school. Thirty years later it was only one of every seventeen children (Levitan and Hetrick 1971, 37). Despite this decline in numbers, the mission schools still enjoy a particular standing on many reservations because of the often superior quality of their instruction compared to the BIA and contract schools. The reasons for this quality in education are the smaller number of students in each class, as well as teachers who are generally better educated, have a stronger motivation, and usually stay much longer at the same school.

sent surroundings" (Morgan 1888, quoted in: Prucha 1973, 243) only by being taken away from their parents and put into BIA boarding schools.

This educational policy was most strongly pursued around the turn of the century, when there were 17 mission schools with a total of around 1,000 students (ARCIA 1901, 29) and 113 BIA boarding schools in which more than 17,000 Indian children (ibid., 16–18) were enrolled.[21] Many of these boarding school students were removed from their reservations for years, so that even during the vacations they did not see their relatives and friends. When school was not in session, they either remained in their hostels or they were placed with white families, especially farmers, as cheap labor in return for modest wages. This policy, called the *outing system,* was initiated by the Indian Boarding School in Carlisle, and by the year 1900 it was applied to 80% of the students attending this school (ibid., 31), and was often followed by other government boarding schools.

In addition to boarding schools, the BIA operated 147 day schools by 1900. These were very small schools with an average of 30 to 40 registered students often located at isolated places on the reservations. According to the data of the educational authorities, however, of the total 5,090 students enrolled in these schools, only 69% actually attended school regularly (ARCIA 1901, 22) and even this attendance rate could often be maintained only through police enforcement. Many parents preferred to hide their children from the authorities, and there were even some who, despite penalties, openly refused to let their children go to these schools, not wishing to expose their offspring to the powerful influence of mainstream society.

Starting in 1891 there were also a few contract schools. Unlike the tribal contract schools that were founded much later, these public, nonfederal schools were located near reservations, and although they served predominantly non-Indian students, they also admitted Indian children. The BIA took over the additional costs incurred by Indian students. At the turn of the century, however, there were only 22 such schools, 13 of them in Nebraska and Oklahoma, the remainder in Michigan and Wisconsin and in a few western states (ARCIA 1901, 21). This type of school was not very significant, with fewer than 200 students actually enrolled and only 48% of them attending school regularly (ibid., 21). The public contract schools generally had a bad reputation, and according to the testimony of white friends of Indian families whose children attended the schools, there were frequent cases of extreme racial discrimination.

More than 84% of the 26,500 Indian students enrolled in 1900 went to

21. The oldest off-reservation BIA boarding schools were located in Carlisle (Pennsylvania), Chemawa (Oregon), Chilocco (Oklahoma), Genoa (Nebraska), Albuquerque (New Mexico), and Lawrence (Kansas), all founded between 1879 and 1884. With the exception of the school in Lawrence, which today still survives as Haskell Junior College, they have all closed.

BIA schools, which had at their disposal 2,200 employees, 68% of whom were white (ibid., 22 and 30). The classification of employees in key positions according to ethnic origins shows still greater disparities between Indians and whites. All school inspectors and school physicians were white, while only 4% of the boarding school principals and 14% of the teachers were Indian.[22] There were only four kindergarten teachers of Indian origin in the many BIA kindergartens.

The Meriam Report of 1928, gave an insight into the appalling educational situation of reservation Indians, and the official school policy changed during the IRA era. During this time the network of day schools and public school contracts was expanded at the cost of boarding schools, in particular of those outside the reservations, and the importance of the BIA with regard to Indian education began a gradual decline. Of the 25 off-reservation boarding schools that had once existed, only 10 remained by 1968, and 3 more were closed before the 1984–1985 school year.[23]

Unlike in the past, mostly handicapped and socially disadvantaged Indian children are to be found in these BIA boarding schools today. Many are either orphans or have lost one parent, and some come from broken homes. A considerable number of the students have a history of criminal delinquency, and in such cases these boarding schools often provide the last opportunity for rehabilitation.

Some boarding school students, however, come because their reservations are in particularly isolated regions where there are no suitable schools. The small Havasupai Indian Reservation in Arizona, west of the Grand Canyon National Park, is an example of one such reservation. Its only settlement, Supai Village, is located on the Havasu Creek, a tributary of the Colorado River. This remote settlement is 3,000 feet below the edge of the Coconino Plateau and can be reached only by a narrow road about sixty miles long, plus a walk of eight miles. A school was founded in Supai Village in 1894, but it was closed between 1955 and 1964. During this period even six-year-old children had to be sent to BIA boarding schools that were far away, generally in Phoenix or Riverside near Los Angeles. Even since the reopening of the eight-grade school in Supai Village, students age four-

22. Today the key positions in the BIA Office of Indian Education Programs are almost entirely taken by American Indians. This is not true of the teaching positions, however, owing to the lack of qualified Indians.

23. The following seven BIA boarding schools were still in operation in 1985 (The figures after the virgule indicate the grade levels provided at these schools): Chemawa (Oregon)/9–12, Wahpeton Indian Boarding School (North Dakota)/4–8, Flandreau Indian School (South Dakota)/9–12, Riverside Indian School (Anadarko, Oklahoma)/2–12, Sequoyah Indian School (Oklahoma)/9–12, Phoenix Indian High School (Arizona)/9–12, and Sherman Indian High School/9–12 (Riverside, California).

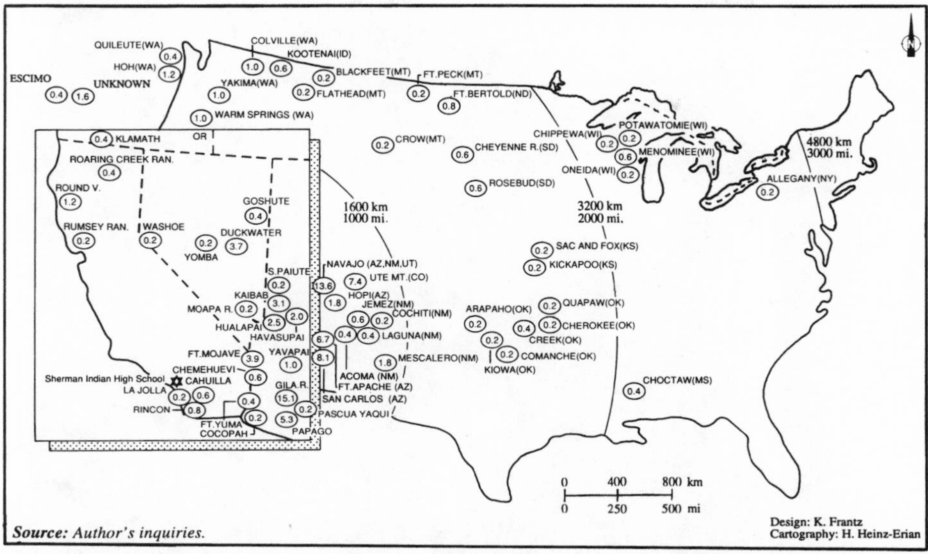

Source: Author's inquiries.

Fig. 5.12: The tribal origins of students at the Sherman Indian High School (%), Riverside, California, 1993

teen years and over who have wanted to attend high school still have to leave the reservation. Obviously this lack of adequate schools has had negative consequences on the educational attainment of these approximately 370 reservation Indians, so that according to my inquiries in 1986, only 10 had graduated from high school <23, 24>.

The diversity of ethnic origin of BIA boarding school students has changed little during the past hundred years according to school authorities. There were more than 500 students from sixty-two different reservations attending Sherman Indian School in California in 1993, for example (see fig. 5.12), and some reservations were more than 2,500 miles away from the school. In view of this ethnic diversity, it is not surprising that BIA boarding schools have always had the effect of a melting pot on the Indian population, something many American Indian people have resigned themselves to. The numerous mixed marriages (see fig. 4.3) and also some of the pan-Indian movements are doubtless the consequences of this educational policy.[24]

24. The Peyote cult, for example, which is a kind of pan-Indian religion for many of its followers from various tribes, undoubtedly spread so rapidly at the close of the nineteenth century because of the BIA boarding schools (see Feest 1976, 303); and the American Indian

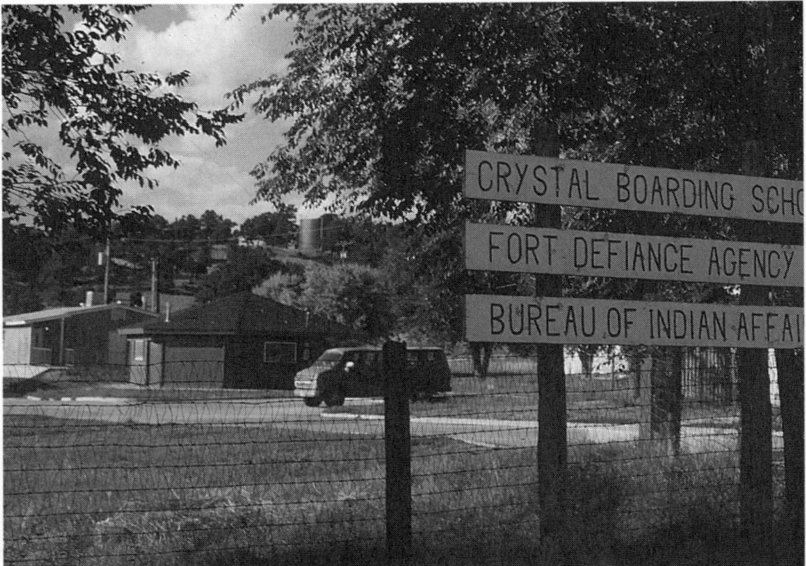

Fig. 5.13: BIA boarding school in Crystal, Navajo Indian Reservation, New Mexico, 1992

Between 1900 and 1985 there was also a decline in the number of BIA operated, on-reservation boarding schools, a decline from eighty-eight to fifty-four schools (see fig. 5.13), with all but twelve of the surviving on-reservation boarding schools to be found on the Navajo Indian Reservation. In addition, tribes under contracts and grants with the BIA operated thirteen boarding schools in 1985 (Senate 1969, xi–xiv; U.S. Department of the Interior, BIA Education Directory 1984–1985). Despite this decline of on- and off-reservation boarding schools, in 1984 there were still about 22,000 Indian students in these schools and 5,000 of these children were under nine years old. In comparison there were 35,000 students in 1968, 9,000 of whom were not yet nine years old (ibid., xii, <25>).[25]

These children have been wrenched away from their familiar sociocultural environment at a very early age, and forced to live separated from their families for years. What is more, they are subjected to strong pres-

Movement (AIM), founded in Minneapolis in 1968, had many former boarding school students among its first activists.

25. Besides these 19,000 students at BIA boarding schools there were 1,800 students at the 15 BIA dormitories in 1984. Eight of these dormitories were reserved chiefly for the Navajos (see fig. 5.17). The dormitories near the reservations were built to provide care for students attending a public school outside the boundaries of a reservation, because there were not enough public schools on the reservations.

sures to assimilate by the school and the white majority. Many of these students are inevitably affected socially and psychologically as a result.

Today's trend is toward the tribal contract schools,[26] schools on the reservations which, in the spirit of the Indian Self-Determination and Education Assistance Act, are still financed by the BIA but which are administered and directed by a tribe under Public Law 93-638, which authorizes tribes to contract for BIA programs. Since 1968 there has been a noticeable decline in the importance of the BIA day schools. Before the Second World War it had still been hoped that the day schools would develop into social and cultural centers for the many small scattered settlements on the reservations. Today barely seventy day schools remain, roughly the same number as tribally operated contract schools.

The advantages from the tribal contract schools for reservation Indians are obvious. In school affairs local tribal members are granted the right to take part in decisions regarding the membership of school boards, the selection of teachers and, above all, the composition of the curriculum. For the first time the tribes are thus given the opportunity to champion the continued existence of their own culture in the schools. As a consequence, on some reservations, such as the Navajo and Hualapai in Arizona, bilingual, bicultural school programs have been developed during the past two decades (see Shafer 1985). This has not only strengthened the self-respect of the various tribes but also improved the students' level of achievement <26>.

What worked against the BIA monopoly of Indian education more than anything else was the promotion of public schools adjacent to the reservations[27] which were willing to accept more and more Indian children from the reservations as a result of the Johnson O'Malley Act. In the mid-1980s twenty-six states participated in this program, which provides an influx of federal funds, given as reimbursement for every student of Indian origin.

Besides the government there are also private organizations, such as the Indian Student Placement Service of the Mormons, which provide financial support to make it possible for American Indians to have access to non-Indian public and private schools. The placement program has been in operation since 1954 and thus far has cared for more than 20,000 Indian children, about half of them Navajos. This program now covers only four years of schooling but in the past it took care of students for ten to twelve years <27>. Before entering school the children involved must embrace the

26. In addition to the sixty-two tribal contract schools there were thirteen "cooperative schools" in 1984 which were partly operated by a tribe under a cooperative agreement between a school district and the BIA, which financed the schools.

27. A considerable number of these public schools were also built on the reservations.

Mormon religion and survive a rigorous selection process. During the school term they are separated from their familiar environment and live with foster parents, with whose children they go to school. Owing to the long absence from their reservations they often feel uprooted and find it difficult to cope with life when they return.

The foregoing discussion has made it clear that since the Indian Self-Determination and Education Assistance Act the influence of the BIA on the school system has markedly declined. This notable decline is also reflected in the statistics. In 1968 some 33% of all federally recognized Indian children went to BIA schools, but by 1984 the percentage had dropped to 15% (see Senate 1969; U.S. Department of the Interior, BIA Education Directory 1984–1985). Three percent attended private schools in 1984 and the rest of the students went to public schools. This trend away from BIA schools and toward shifting the responsibility for Indian education to the tribes and states will continue in the future.

A description of the Pine Ridge Indian Reservation in South Dakota can again illustrate the diversity of school authorities on the reservations, and at the same time show how variable the federal policy toward Indian education has been during the past hundred years. In 1982 the Pine Ridge Indian Reservation had a total of approximately 4,200 students. There were twenty-two schools with 364 teachers, almost one-third of whom were Sioux Indians, thanks to the tribal community college that trained many teachers. Half of the schools were public schools, one-quarter were run by the BIA, and three schools were controlled either by the tribe (see fig. 5.14) or a religious community. In the Catholic mission school at the Sioux settlement of Pine Ridge the teacher-student ratio is noticeably more favorable than in other schools of comparable size. In this school there was one teacher for every seven students, whereas in the Kyle, Martin, Wanblee, and Oglala settlements and in one of the two BIA boarding schools in Pine Ridge, the teacher-student ratio ranged from 1:13 to 1:18. The favorable teacher-student ratio at the private Catholic school is bound to have a good effect on the quality of the instruction.

The Education Levels

Even today, more than two hundred years after the first attempts to integrate American Indians into the U.S. school system, the indigenous people lag behind the education levels of all other racial groups in the country in many respects (see table 5.1 and fig. 5.4). This is particularly true of reservation Indians, for whom investment in human capital resulting in higher levels of education would appear at least as helpful a key to economic development as investments in "physical capital" or the utilization and control over tribal reserves of natural resources.

Fig. 5.14: Tribal contract school in Kyle, Pine Ridge Indian Reservation, South Dakota, 1980s. Photo courtesy Peter Schwarzbauer.

During the past two decades both the federal government and a number of tribes have made major efforts to financially support educational programs. The estimated expenditures for this purpose, chiefly from funds appropriated by the Departments of the Interior and of Health, Education, and Welfare, amounted to approximately one-quarter of the total federal expenditure for American Indians in 1985 <1, 28, 29>, and were exceeded only by the costs of the health programs.

There can be no doubt that as a result of this policy the educational attainments of reservation Indians have improved considerably, yet they remain behind in relation to the total U.S. population. Only 10% of all Americans had had less than eight years of school in 1980, for example, while the corresponding figure for all American Indians was 16% (see table 5.10) and for reservation Indians it was more than 26%. In 1970 the percentage of reservation Indians with less than eight years of formal schooling was around 50%. These differences become still more striking when analyzed by region. In Oklahoma, the Northwest and the Rocky Mountain states the 1980 percentages of educational attainment for Indians were not significantly different from the national average. In the Southwest, on the other hand, more than 36% of the Indian reservation residents had had less than eight years of schooling, a fact to be explained above all by the two large reservations of the Navajos and the Papagos in this region, which had a particularly poor infrastructure for educational

Table 5.10. The educational status of reservation Indians over twenty-five years of age by region, 1980

Region	< 8 years of school (%)	8 years elementary school (%)	4 years high school (%)	> 4 years college (%)
East	17.3	36.7	42.3	3.7
Great Plains (total)	14.1	36.7	44.5	4.7
N. Great Plains	14.8	37.9	42.9	4.4
S. Great Plains	10.5	30.3	52.9	6.3
Rocky Mountains	11.8	35.2	49.7	3.4
Northwest	11.5	33.0	50.7	4.8
Southwest (total)	36.2	24.0	36.8	3.0
Navajo	48.2	17.2	31.3	3.2
Remaining Southwest	22.1	31.9	43.2	2.8
Total reservation Indians	26.4	24.0	40.7	3.6
Total American Indians	16.1	27.9	48.1	7.7
Total U.S. population	10.2	29.2	50.2	16.2

Sources: U.S. Department of Commerce 1980, American Indians, Eskimos and Aleuts on Identified Reservations (unpublished microfilm data); <15>.

access for a long time. The average number of school years completed for Navajos over 25 was 8.4, while the Papagos averaged 9.1 years (see table 5.11), far below the American average.[28] One should not overlook the fact that great progress in education has been made particularly on the Navajo Indian Reservation, however. According to Selective Service records made at the time of the Second World War, 88% out of a total of 4,000 male Navajos ages 18 to 35 were classified as illiterate (quoted in Kluckhohn and Leighton 1962, 146). By the time of the 1950 census the educational situation of the Navajos had improved somewhat, a fact partly to be explained by the high number of its war veterans, yet at that time 57% of the tribal members over 25 years of age still had less than one year of formal schooling (see table 5.11), which placed the Navajos in the lower 3% of the populace educationally. Eighteen years later the authors of the Kennedy Report declared that approximately one-third of the Navajos were still illiterate (Senate Report 1969), a message that was to rouse the conscience of the nation.

The percentage of high-school graduates had increased by 1980, when

28. These 1980 figures for the Navajos and Papagos are by no means the lowest among all reservation Indians, however. On the Cocopah Indian Reservation in Arizona the average number of school years completed was 6.6 and on the Goshute Indian Reservation in Nevada and Utah it was 5.2. The poorest record of all was that of the Miccosukee Indians in Florida, where the average was 0.9. It was not possible to find out why these Indians in particular had had so little schooling. The national average was around 12 years (see table 5.11).

the figure for reservation Indians was only 10% less than that of the American population as whole (see table 5.10). Ten years before, only one-quarter of the adults aged 25 years and older had graduated from high school.

In view of the high expenditures on Indian education—in 1985 almost $1,000 was spent on each federally recognized Indian—the question naturally arises why reservation Indians are still so far behind regarding educational attainment.

The educational deficiencies and poor school attendance were partly caused by the lack of schools, but the inadequate road network and lack of transportation facilities on the reservations also played a role (see figs. 5.15, 5.16). Parents also had economic reasons for not sending their children to school, for in some tribes children were needed to look after the sheep and to help bring in the harvest. One of the principal causes of today's poor education and low school attendance for reservation Indians is to be found in the fact that schools often ignore the unique sociocultural background of the local population and teach few skills that would meet the requirements of daily life on the reservations. This Anglo-oriented educational policy provides little incentive for American Indians and even today leads to their boycotting of the white educational institutions which were forced upon them and which they consider alien to their way of life.

The following statistics on Navajo school attendance will show just how unsatisfactory this situation has been for reservation Indians in this century. In 1911 only 12% of the school-aged Navajo children actually attended school, a percentage which had increased to 46% by 1930 (Johnston 1966, 49). By 1950 only half of the Navajo children attended school. The other 50% either were not able to go to school or their parents did not wish to send them. Ten years later attendance had risen to 86% (Young 1976, 291). This rapid increase between 1950 and 1960 is partly explained by the expansion of the BIA network of dormitories close to the reservation, which enabled about 2,300 students to gain access to public schools in the nearby border towns. Furthermore, special programs were devised in the 1950s to make it easier for Navajo children to attend BIA boarding schools, where about 6,100 Navajos were taken care of by the Indian agency during the school year 1959–1960 (see fig. 5.17). Despite this unquestionable upswing, at the time of the 1980 census 4.1% (about 1,000) Navajo children of school age were still not attending school.[29]

29. With this figure the Navajo Indian Reservation is 0.5% above the average for all reservation Indians, but far below that of some other reservations. The poorest records for 1980 were those of the Shakopee (Minnesota), at 11.8%; Kaibab (Arizona), 14.3%; Soboba (California), 26.7; Sokaogon-Chippewa (Wisconsin), 36.3%; and Miccosukee Indian Reservations (Florida), 40%, where many of the school-age children did not actually attend school. I was not able to find out the reasons for this.

Table 5.11. Educational attainment of American Indians over twenty-five years of age for selected reservations, 1950 and 1980

Reservation	< 1 year of school (%)		< 8 years of school (%)		8 years of school (%)		4 years high school (%)		Ave. no. school years (%)		≥ 4 years of college (%)	
	1950[a]	1980[b]	1950	1980	1950	1980	1950	1980	1950	1980	1950	1980
Blackfeet (MT)	10.8	1.0	37.0	—	19.4	20.4	9.6	49.8	8.1	11.9	0.3	4.7
Colorado River (AZ, CA)	17.4	1.2	39.2	—	15.7	18.0	7.6	52.4	6.7	12.0	0.3	1.5
Crow (MT)	5.0	3.2	42.3	—	24.5	24.8	4.5	52.1	8.2	12.1	1.4	5.8
Fort Apache (AZ)	24.4	1.3	35.6	—	16.0	29.0	4.0	34.0	6.4	10.3	0.4	1.5
Gila River (AZ)	15.0	3.1	40.4	—	12.8	28.6	13.1	38.3	7.2	10.7	0.5	1.3
Hopi (AZ)	19.8	2.8	33.0	—	9.4	17.7	15.1	41.7	—	10.6	0.0	3.2
Menominee (WI)	3.4	1.0	35.2	—	23.4	23.1	6.5	48.7	8.3	11.9	0.8	2.2
Navajo (AZ, NM, UT)	57.1	2.8	18.2	48.2	3.8	17.2	4.3	31.3	0.8	8.4	0.1	3.2
Papago (AZ)	29.2	1.4	47.5	—	7.9	24.3	5.0	35.0	4.1	9.1	0.5	1.1
Pine Ridge (SD)	4.3	1.9	51.0	—	15.8	28.3	7.8	44.8	7.4	11.2	0.4	4.0
Red Lake (MN)	7.3	1.2	42.2	—	28.6	24.6	3.9	43.4	7.7	11.2	0.0	0.6
Rosebud (SD)	4.6	1.3	44.1	—	23.0	23.1	8.6	50.2	7.8	12.0	0.4	4.3
San Carlos (AZ)	8.6	1.6	59.2	—	15.9	25.7	2.6	37.2	6.5	10.5	0.4	1.2
Yakima (WA)	15.2	2.9	32.7	—	17.1	19.0	11.5	51.3	7.8	12.0	0.4	3.9
Zuni (NM)	26.7	1.4	34.4	—	12.7	24.5	9.0	38.3	6.9	10.7	0.5	1.4
Reservation Indians total	20.6[c]	2.3	38.8[c]	26.4	14.7[c]	24.0	7.0[c]	40.7	6.4	10.9	0.6[c]	3.6

[a]The census includes the category of "no school years" completed, which in some cases means that the American Indians in question did not go to school at all.

[b]In the census year 1980 there are no comparative data for the part of the population over twenty-five years of age. The percentages given here relate to children from seven to thirteen years of age, and show how many of them did not go to school at all.

[c]Data regarding the educational status of the total number of reservation Indians were not collected for the year 1950. Those given here relate to the "rural farm" and "rural nonfarm" Indians who, generally speaking, are reservation Indians.

Sources: U.S. Department of Commerce 1950, Population (Special Reports), tables 10 and 21; U.S. Department of Commerce 1980, American Indians, Eskimos and Aleuts on Identified Reservations (unpublished microfilm data); <15>.

Fig. 5.15: Bookmobile of the Navajo Education Center, Window Rock, Arizona, 1992

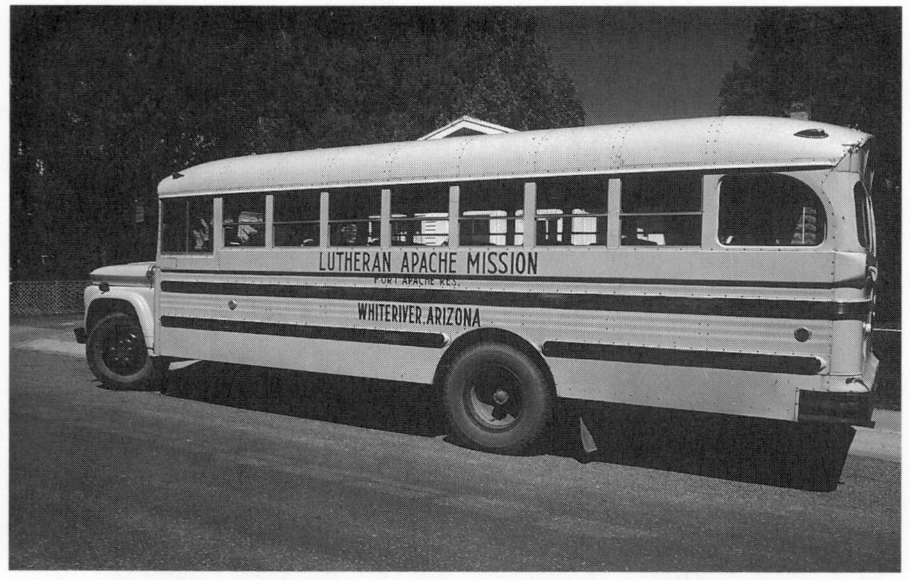

Fig. 5.16: School bus of the Lutheran mission, Fort Apache Indian Reservation, Arizona, 1985

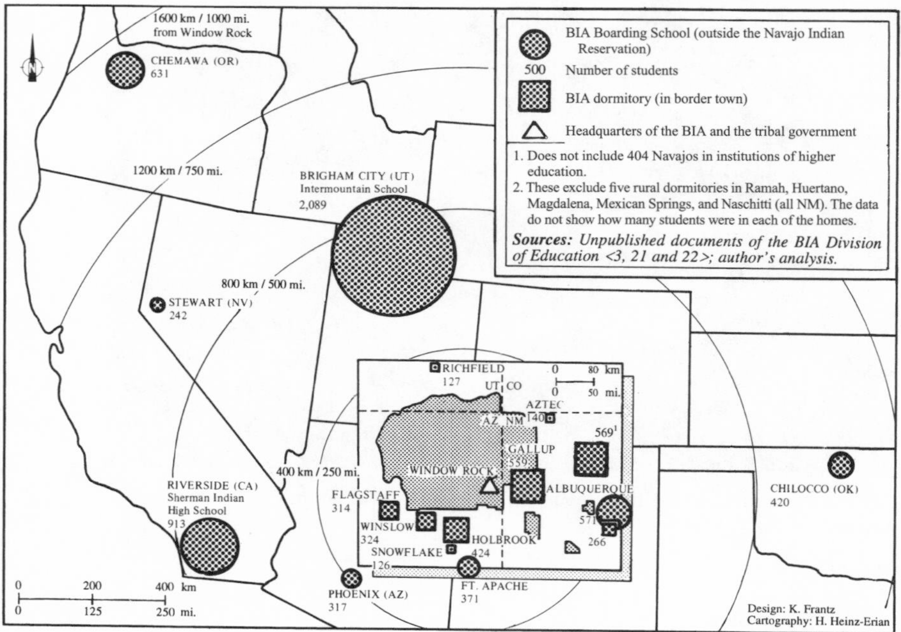

Fig. 5.17: Off-reservation Navajo education, 1959–1960

Another reason for the generally low education levels of reservation Indians is undoubtedly the language barrier which confronts not only many adults, but also children starting school. Unfortunately, it is not possible to make generalizations without research on the English-language competence of individual tribes. It is a fact, however, that for many reservation Indians, English is a second language, which they speak and understand with varying degrees of proficiency.

According to Feest (1976, 198–199), for example, about 30% of Navajo children did not know a word of English when they started school in the early 1970s. Another 39% had such a poor command of the language that they could not follow their teachers' instructions; 21% were bilingual; and 10% spoke English most of the time. These statements correspond with my observations on the Fort Apache <30, 2> and Hualapai <24, 26> Indian Reservations in Arizona. On the Fort Apache Indian Reservation in 1986 three-quarters of the five-year-old children had either no knowledge of English or only a very poor command of the language, and in the same year on the Hualapai Indian Reservation one-quarter of the Indian first-graders had very little knowledge of English.[30]

This situation shows the importance of preschool education, the Head Start Program, and bilingual, bicultural curricula. The Head Start Program was begun at the end of the 1960s, and has produced good results on many reservations—not least thanks to its practice of hiring local, elderly people who served as storytellers and clanship or arts and crafts instructors. Bilingual and bicultural programs, whose legal foundations date back to the 1930s, are still in their initial stages, however, in spite of some promising projects, because school curricula on reservations usually pay little attention to indigenous cultures.[31]

As a reaction against Anglo-oriented curricula dictated by the white man's world, a periodical was published by the National Indian Youth Council in the 1960s, bearing the title *ABC (American Before Columbus).* Yet even today, twenty years after this demonstrative gesture, bicultural and bilingual teaching aids are still the exception rather than the rule. The most serious problem is undoubtedly that there are still far too few American Indian teachers in spite of all the efforts that have been made.

30. The difference between the Apaches and the Hualapais knowledge of languages can also be seen among the adults. Around half the Apaches speak their own language in their daily lives, and more than 90% of the tribe were able to understand this language. At the same time, almost all Apache adults have a command of English, although one-fifth do not speak English well. With the Hualapais, on the other hand, English is the colloquial language of most adults and only an estimated 40% to 75% of them speak or understand Pai.

31. For example, even today the Indian children at some BIA schools are told that America was discovered by Columbus and that the history of the American continent began in the sixteenth century.

Unfortunately, no nationwide data are available concerning the racial or ethnic origins of teachers on reservations, so one must rely on one's own inquiries on specific reservations.

For the seven schools at Whiteriver,[32] for example, the main settlement on the Fort Apache Indian Reservation in Arizona, I found that there were only six Apache and five other American Indian teachers out of almost two hundred teachers in 1986. More than half the white teachers lived outside the reservation, and therefore had little interaction with the Apache community they served on a daily basis. Most of the remaining non-Indian teachers lived in a sort of "trailer ghetto" belonging to the fenced-in BIA compound at Whiteriver (see fig. 3.12g), and they, too, were isolated from Apache society. In order to try to overcome the language problems of the local school children, more than fifty additional Apache teaching aides had been hired during the preceding twenty years.

In Peach Springs, where the only school on the Hualapai Indian Reservation is located, three out of the twelve teachers were tribal members in 1986. Here, too, there were teaching aides, some of whom were elderly Hualapai Indians who kept up tribal traditions and helped to pass them on to the next generation. With their cooperation the three Hualapai teachers made up bilingual teaching materials, which required development of a special kind of phonetic transcription, since their language had been passed on only by word of mouth. While the Fort Apache students have only the standard textbooks used nationwide, the Hualapai Indians have developed their own books, which are based on their cultural values and tribal heritage.

Furthermore, teachers who are members of the tribe, such as those found on the Hualapai Indian Reservation, can serve as role models to encourage Indian children in their own education and future career development. When such models are lacking, young reservation Indians are often inclined to take the attitude that a good Indian should be an academically poor student, because schools are often identified as institutions of the white man's world. Apart from being role models, local Indian teaching staff can also take over important functions outside the school, notably in tribal politics or in cultural fields, in order to help the tribe keep up its identity and support self-determination.

Finally, the level of Indian education is adversely affected by the very high turnover of the non-Indian teaching staff, similar to the high turnover rate of non-Indian physicians discussed earlier. In addition, new teachers have often received little orientation to reservation life and conditions, and

32. Of these seven schools the following four are in the immediate vicinity of Whiteriver: Lutheran Mission School in East Fork (private), Elementary School in Seven Mile (public), Theodore Roosevelt Boarding School at Fort Apache (BIA), and John F. Kennedy Day School in Cedar Creek (BIA).

are thus ill prepared for their jobs. Furthermore, teachers, particularly those in BIA schools, are expected to work almost year-round in contrast to their colleagues outside the reservations, who teach for only a nine-month period. Under these conditions, it is not surprising that the best-qualified teachers are often not to be found on reservations.

Higher education among American Indians is represented by even lower levels of attainment. Nationwide every sixth American of European ancestry and every third American of Asian ancestry aged twenty-five and older has completed at least four years of college (see table 5.1 and fig. 5.4). Among African Americans the proportion is only one in every twelve, and for the American Indians one in every thirteen. The situation is still worse for reservation Indians, where only one in twenty-eight Indians is a college graduate (see table 5.10). Although this proportion has increased considerably since 1950 (see table 5.11), there has been little change since 1970. Furthermore, only 1% of these college graduates have attained postgraduate degrees.

The percentage of people with a college education varies considerably by reservation. While the Hopis and Navajos made great progress between 1950 and 1980 and the percentage of their college graduates among the over twenty-five-year-olds increased from almost zero to 3.2%, higher education on other reservations of the Southwest such as the Colorado River, Gila River, Papago, Fort Apache, San Carlos, and Zuni, played a minor role (see table 5.11). These differences in education levels are certainly not only due to economic reasons, for the Fort Apache, Colorado, and Gila River tribes have resources at their disposal that are at least equal to if not better than those of the Hopis and Navajos. It is more likely that the tribes' attitude toward education, particularly that of their political representatives, plays a critical role. Whereas the Hopis and Navajos have almost always had a college-educated chairman during the past twenty years, on the Fort Apache Indian Reservation, where education was traditionally seen as destructive to family and culture, there can sometimes be an attitude of unconcealed aversion toward it. This attitude seems to be reflected in the tribal funding for scholarships, which in the mid-1980s was more limited than for many other tribes.

In 1984 around 40,000 American Indians studied at the approximately 3,100 U.S. institutions of higher education (Chronicle of Higher Education 1984),[33] and 17,000 of those students received financial assistance from the BIA. This means that in comparison to all other racial and ethnic groups in the United States the American Indians were greatly un-

33. This figure includes community and junior colleges, some of which would not be considered institutions of higher education by European standards but rather as adult education centers or schools for technical or vocational training.

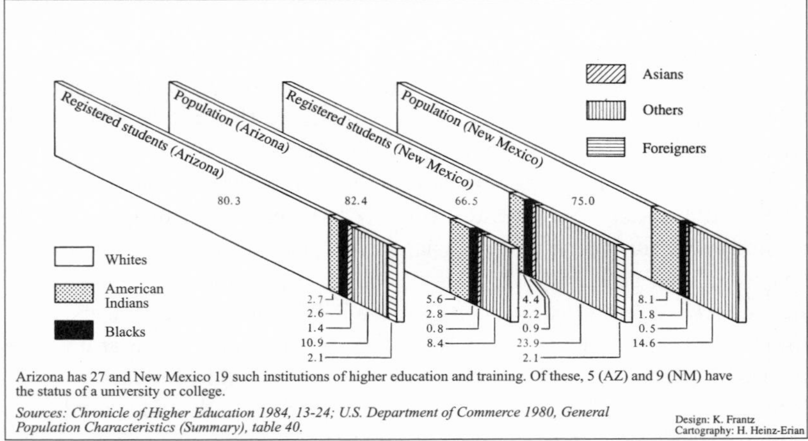

Arizona has 27 and New Mexico 19 such institutions of higher education and training. Of these, 5 (AZ) and 9 (NM) have the status of a university or college.

Sources: Chronicle of Higher Education 1984, 13-24; U.S. Department of Commerce 1980, General Population Characteristics (Summary), table 40.

Design: K. Frantz
Cartography: H. Heinz-Erian

Fig. 5.18: Percentage of ethnic and racial groups in the total population and in higher education in Arizona and New Mexico, 1981

derrepresented. American Indians in Arizona make up 5.6%, and in New Mexico 8.1%, of the total population. Out of 245,000 students enrolled at institutions of higher learning in these states—twenty-seven in Arizona and nineteen in New Mexico—barely 8,300 students (3.8%) were of American Indian origin in 1981 (see fig. 5.18). On the other hand, the proportion of students of Asian or Hispanic origin relative to their populations was most conspicuous.[34]

In view of the very different standing of the various U.S. universities and other institutions of higher learning, it would be appropriate at this point to consider briefly the racial and ethnic composition of the student body at leading American colleges and universities. American Indians are underrepresented at the prestigious private and state universities, a prestige that can partly be seen in the number of foreign students (see table 5.12). With a few exceptions, only a fraction of 1% of students are American Indians in these leading schools. Moreover, these relatively few Indians usually do not come from reservations. In these institutions African Americans are equally underrepresented, whereas there is a high percentage of American students of Asian origin.

It should be pointed out that a number of large American universities offer minority programs that are specially designed for disadvantaged population groups. According to Locke (1978; Chronicle of Higher Education 1984; see fig. 5.19), eighty-seven universities offered American Indian Studies programs in 1978, and fifteen of these schools were each able to at-

34. Students designated with the word "Others" in figure 5.18 are largely of Mexican origin.

Table 5.12. University students by race and foreign origin at leading U.S. universities, 1984

	White	Black	Amer. Indian	Asian	Hispanic	Foreign
PRIVATE UNIVERSITIES						
Brown	87.9	4.8	0.2	2.3	0.7	4.1
Columbia Univ.	83.8	4.7	> 0.1	3.9	3.2	4.4
Cornell	76.0	4.4	0.2	6.4	3.5	9.5
Dartmouth	87.8	5.6	1.0	1.4	0.4	3.8
Georgetown	80.3	5.8	0.2	2.1	3.5	8.1
Harvard-Radcliffe	75.5	5.5	0.5	4.3	3.6	10.6
Princeton	75.7	5.5	0.2	3.4	3.7	11.5
Stanford	69.2	5.0	0.4	5.3	5.0	15.1
Univ. of Chicago	81.8	3.5	0.1	5.0	1.8	7.8
Univ. of Pennsylvania	81.9	4.9	0.3	2.7	1.5	8.7
Yale	80.7	4.9	0.2	4.3	2.3	7.6
STATE UNIVERSITIES						
Univ. of Arizona	86.8	1.3	0.8	1.3	5.2	4.6
Univ. of California–Berkeley	66.7	3.8	0.5	18.4	4.7	5.9
Univ. of California–L.A.	65.7	5.0	0.4	16.5	6.5	5.9
Univ. of Michigan–Ann Arbor	83.4	4.8	0.4	3.1	1.3	7.0
Univ. of Minn.–Minneapolis-St. Paul	91.8	1.5	0.4	2.1	0.7	3.5
Univ. of Washington	81.0	3.0	0.9	10.1	1.4	3.6
Univ. of Wisconsin–Madison	89.1	2.0	0.3	1.5	1.1	6.0
PRIVATE INSTITUTES OF TECHNOLOGY						
Cal. Tech.	66.2	0.8	0.2	10.5	2.1	20.2
MIT	68.0	3.3	0.2	4.7	2.0	21.8
% of total U.S. population	83.1	11.7	0.6	1.3	3.3	—

Sources: Chronicle of Higher Education 1984, 13-24; U.S. Department of Commerce 1980, General Population Characteristics (Summary), table 40.

tract more than 500 Indian students, although only a few of these programs were offered by universities of international standing. The interdisciplinary structure of these programs, which were often set up as a separate discipline intended to meet the particular needs of the American Indians, often creates an aura of isolation around these departments (Churchill and Hill 1979, 45). This causes many Indians to view the programs with disfavor even though the courses are often taught and the programs coordinated and directed by American Indians.

American Indian Studies programs cannot obscure the fact that institutionalized higher education that takes into account the sociocultural background of the American Indians is relatively new and even today is

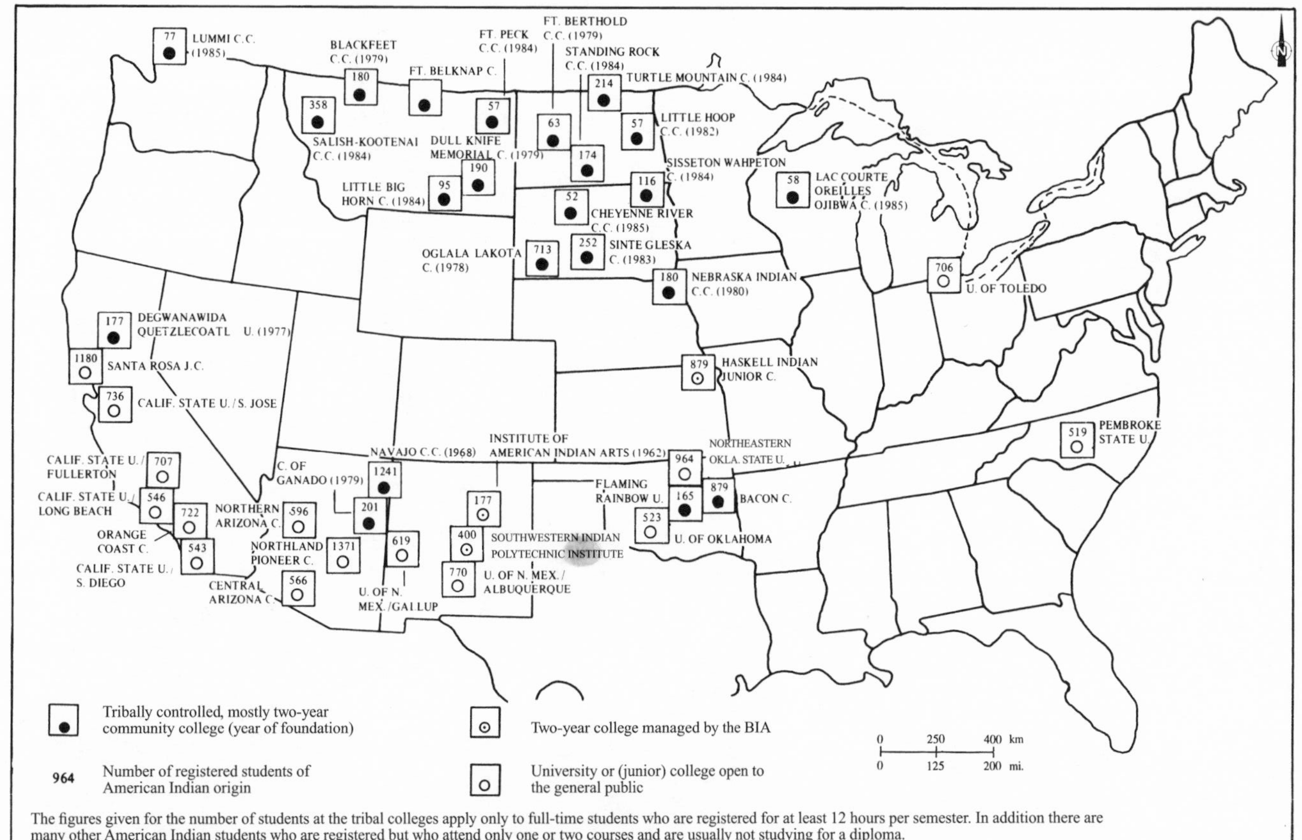

Fig. 5.19: Indian attendance at community colleges managed by Indian tribes or the BIA, and universities and colleges open to the general public with more than 500 registered students of Indian descent, 1985

inadequate. It is true that in the British colonial period there were a few American Indian colleges whose purpose was to civilize neighboring Indian tribes and convert them to Christianity, but these institutions did not have any widespread impact.

The two best-known institutions of this type were Moor's Charity Indian School in Lebanon, Connecticut, and Dartmouth College in Hanover, New Hampshire. Moor's Charity School, an academy for college preparation, was founded in 1754 by Reverend Eleazar Wheelock, who also founded Dartmouth College fifteen years later. Dartmouth, one of America's leading liberal art colleges, and which grew out of Moor's Charity School, was the first institute of higher learning primarily dedicated to the education of American Indians. Soon more white students than Indians were attending Dartmouth, as it met with little response from the Indians, and in this initial stage it then became known as a place where prospective Indian missionaries were trained. In 1984, when Dartmouth offered a small Indian Studies program with a staff of five professors, forty-five American Indian students (1% of the student body) were studying there (table 5.12).

The federal government has not paid very much attention to the higher education of American Indians. The first initiative on the part of the government to found an institute of higher learning for the Indians was to enlarge Carlisle Indian School in Pennsylvania to include postsecondary education, but this plan was soon abandoned. The BIA did create post secondary institutions for American Indians that still exist today: Haskell Indian Junior College in Lawrence, Kansas; Southwestern Indian Polytechnic Institute in Albuquerque; and the Institute of American Indian Arts in Santa Fe, which together had a total enrollment of 1,500 full-time students in 1985 (see fig. 5.19). These "BIA colleges," however, barely manage to survive in the country's world of higher education.

In accordance with the policy of Indian self-determination, based on the Indian Self-Determination and Education Assistance Act, the federal government has helped create and support a number of tribally controlled community colleges on the reservations since 1975. Some of these colleges go back to the late 1960s, but many of them were not accredited as two-year junior colleges until the beginning of the 1980s.[35] Nationwide there were more than twenty such institutions, with around 4,500 full-time students in the mid-1980s, and perhaps the same number of students who registered for only one or two courses (see fig. 5.19). Fifteen of these community colleges were recognized and financially supported by the BIA, yet they are all under the control of the tribes and people whose job it is to keep an eye on

35. Two of them, the Oglala Lakota College and the Sinte Gleska College, also offer four-year courses leading to Bachelor of Arts or Bachelor of Science degrees.

Fig. 5.20: Main building of the Navajo Community College with ceremonial hogan and teepee of the Native American Church, Tsaile, Arizona, 1992

Fig. 5.21: Headquarters of the Oglala Lakota College, Three Niles Creek, Pine Ridge Indian Reservation, South Dakota, 1980s. Photo courtesy Peter Schwarzbauer.

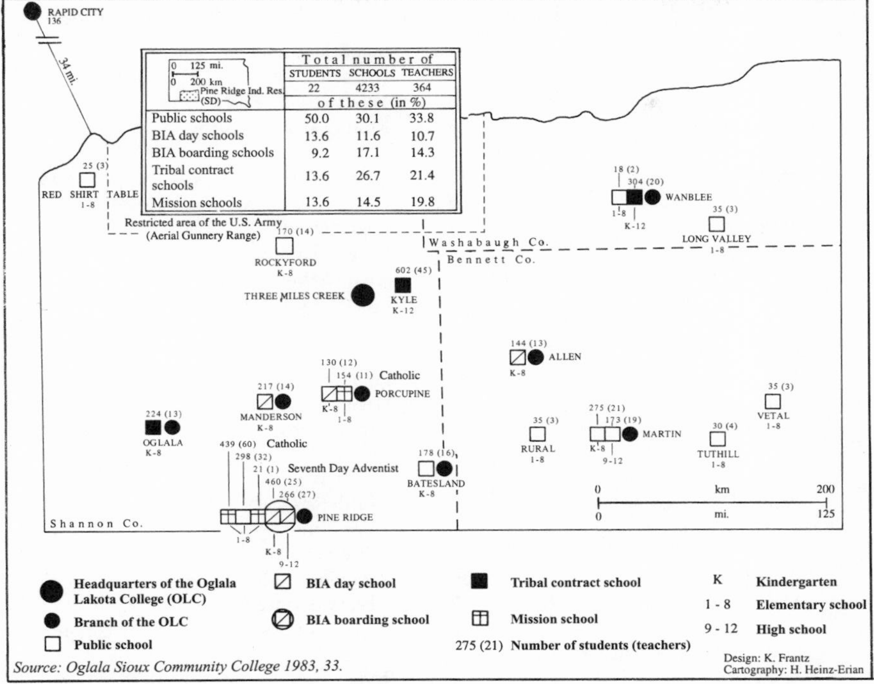

Fig. 5.22: Educational institutions on the Pine Ridge Indian Reservation in South Dakota, 1982

the interests of the reservations. This has meant, among other things, that many of the courses offered are vocational and of a practical nature.

The Oglala Lakota College on the Pine Ridge Indian Reservation in South Dakota is a good example of a tribally controlled community college. Except for the Navajo Community College it is the oldest institution of its kind (see figs. 5.20, 5.21). As on other reservations, the college has several branches (see fig. 5.22) and it is financed by the tribe and various government agencies, as well as by private endowments and gifts. Of the 421 full-time students in 1985 more than 90% were Oglala-Lakota Indians,[36] and there were also around 300 part-time students. About 64% of the teaching staff of twenty-five were white, 32% Indian, and 4% African American. With two exceptions all these teachers were university graduates. Seven of them had doctorates and eleven master's

36. The Lakotas, also known as the Tetons or western Sioux, consist of seven tribes, one of which is the Oglala tribe (see Ballas 1970; Haines 1976; Lindig and Münzel 1985, 141–145). The designation "Sioux" is a French corruption of the Ojibwa word *Na-do-we-isi-w-ug,* which means "smaller adder" (quoted from Schwarzbauer 1986, 10). The Sioux call themselves "the allies" or "the friends."

Table 5.13. American Indian graduate students with BIA scholarships, 1980

Education[a]	414	Soil Science	7
Psychology	66	Forestry	4
History	35	Veterinary Science	2
Anthropology	23	Mining and Metalurgy	1
Humanities[b]	16	SOIL, MINING, AND VET. MEDICINE	14
English	15		
American Studies	8		
Linguistics	5	Engineering	20
Philosophy	1	Urban Planning	4
HUMANITES	583	Architecture	3
Sociology	212	ENGINEERING AND ARCHITECTURE	27
Ethnic Studies	21		
Economics	198	Fine Arts	
Tourism	6	Design	15
SOCIAL SCI. AND ECONOMICS	473	Photography	1
		ART	51
Law	66		
Political Science	3	Health Services[a]	125
JURIS. AND POLITICAL SCIENCE	97	Journalism	17
		Police Science	17
Biology	29	Religious Studies	8
Mathematics	12	Library Science	2
Chemistry	6	OTHER AREAS	169
Environmental Sciences	4		
Zoology	2	Total	1,436
Geography	2		
Computer Science	2		
Geology	1		
NATURAL SCIENCES	58		

[a]These data give no information regarding the number of future teachers or doctors.

Sources: Data found in the U.S. Department of the Interior, BIA, Office of Education; <1>.

degrees. The courses offered at Oglala Lakota College included such subjects as the culture and history of the tribe, handicraft skills, home economics, and health care. Theoretical and practical courses in agriculture as well as in economics, education, and social studies were also included in the curriculum, and from 1974 to 1982, 191 students were graduated from the college. Of these 30 received bachelor of science degrees, (4 master's degrees, and 157 associate of arts degrees. Seventy percent of these graduates found employment on the reservation (Oglala Sioux Community College 1983, 70; <1>).

In addition to the tribally controlled community colleges on the reservations there are comparable institutions outside Indian territory, which, like the BIA colleges, are not intended for only one tribe. Among them is the Degwanawida-Quetzalcoatl (D-Q) University, named after an old cultural hero of the Iroquois as well as a hero from early Mexican culture. This is a university in the vicinity of Davis, California, that is open to all Indians and Chicanos of North and Central America, a product of the pan-Indian movement.

According to a survey of the American Association for the Advancement of Sciences (quoted in Churchill and Hill 1979, 52) in 1976 an estimated 28% of all American Indian students majored in social sciences, 20% in education, and 13% within the field of health services. Only 7% had natural sciences, economics, or engineering as their majors. Although this survey gives no figures for reservation Indians, one may reasonably conclude that at the time not enough experts were produced for the tribes to be in a position to develop their natural and economic potential in their own interest.

An examination of scholarship grants for higher education by the BIA leads to the same conclusion. In 1980 some 17,000 American Indian students received BIA financial support, of whom more than 1,400 were graduate students (see table 5.13). These students were often interested in studying subjects that were not very relevant to the reservations, however, and among students supported by the BIA there was only one student each in geology and mining, two in veterinary medicine, four in forestry, and seven in agricultural science. In view of the abundant mineral resources, as well as livestock, timberland, and agricultural operations on many reservations, this is undoubtedly a situation that ought to induce tribal leaders to introduce new policies in higher education that would benefit tribes.

6 The "Economic Spirit" and Economic Structure of Indian Reservations

American Indian Value Systems and other Regulating Factors

SOCIOCULTURAL COMPONENTS OF THE ECONOMY

In order to understand economic conditions on American Indian reservations it makes sense to consider certain fundamental sociocultural features which are significant elements in the economic life of reservation Indians. A description of these features will also provide a framework for the following chapters.

This approach is by no means new to social and economic geographical research. Alfred Rühl, for example, in his works on Spain (1922), the Orient (1925), and America (1927), has shown that the so-called economic spirit of the people of a particular country or region can have a decisive influence on its economic development. No doubt Rühl's ideas emerged not only from his own research, but from Max Weber's epoch-making studies regarding the economic ethos of world religions published a few years earlier (1920). After the Second World War it was Eugen Wirth, more than anyone else, who followed in the path of Weber and Rühl. In a landmark article on Iraq, Wirth (1956, 44) demonstrated that "basic mental attitudes, predilections, and antipathies, . . . value judgments, and innate dispositions," which altogether, like Rühl, he called the "economic spirit" of a people, could have a decisive influence on their economic activity. Such noneconomic elements have also been considered in more recent research in economic geography. Herbert Popp, for example, a student of Wirth's, completed a detailed study of the changes in an oasis in southeastern Morocco which attached particular importance to the economic spirit of the local population in his analysis of their irrigation economy (see Bencherifa and Popp 1990).

With respect to American Indian reservations, however, this type of research has been widely neglected, with only a few exceptions, primarily provided by nongeographers. This gap in the literature is rather surprising because the value system of American Indians and its consequent effects on the economic life of reservations are so different from those of mainstream American society that they could provide geographers with a very interesting and stimulating field of research, as the following discussion will indicate.

Anglo-American Attitudes as a Standard for Analyzing the Reservation Economy

The cultural independence of Indian tribes and their strong adherence to traditional values have always been regarded by white Americans as the real obstacle to the improvement of socioeconomic conditions on the reservations. This attitude is clearly expressed in the following letter written by Charles Henry Burke, the highest-ranking official for Indian affairs of his time:[1]

> Department of the Interior
> Office of Indian Affairs
> Washington [D.C.]
> February 24, 1923
>
> *To all Indians:*
>
> Not long ago I held a meeting of Superintendents, Missionaries and Indians, at which the feeling of those present was strong against Indian dances, as they are usually given, and against so much time as is often spent by the Indians in a display of their old customs at public gatherings held by the whites. From the views of this meeting and from other information I feel that something must be done to stop the neglect of stock, crops, gardens and home interests caused by these dances or by celebrations, pow-wows, and gatherings of any kind that take the time of the Indians for many days.
>
> Now, what I want you to think about very seriously is that you must first of all try to make your own living, which you cannot do unless you work faithfully and take care of what comes from your labor, and go to dances or other meetings only when your homework will not suffer by it. I do not want to deprive you of decent amusements or occasional feast days, but you should not do evil or foolish things or take so much time for these occasions. No good comes from your "give-away" custom at dances and it

1. The commissioner of Indian affairs, who now holds the title of assistant secretary, has, since 1849, been under the direction of the secretary of the interior, who makes this appointment in agreement with the president. Before 1849 Indian affairs were the responsibility of the Department of War.

> should be stopped. It is not right to torture your bodies or to handle poisonous snakes in your ceremonies. All such extreme things are wrong and should be put aside and forgotten. You do yourselves and your families great injustice when at dances you give away money or other property, perhaps clothing, a cow, a horse or a team and wagon, and then after an absence of several days go home to find everything going to waste and yourselves with less to work with than you had before.
>
> I could issue an order against these useless and harmful performances, but I would much rather have you give them up of your own free will and, therefore, I ask you now in this letter to do so. I urge you to come to an understanding and an agreement with your Superintendent to hold no gatherings in the months when the seed-time, cultivation of crops and the harvest need your attention, and at other times to meet only for a short period and to have no drugs, intoxicants, or gambling, and no dancing that the Superintendent does not approve.
>
> If at the end of one year the reports which I receive show that you are doing as requested, I shall be very glad for I will know that you are making progress in other and more important ways, but if the reports show that you reject this plea, then some other course will have to be taken.
>
> With best wishes for your happiness and success, I am
>
> sincerely yours,
> Charles Burke
> (Commissioner of Indian Affairs)

In this letter Burke specifically attributes the miserable state the reservation economy was in and the poverty of the people living there to the specific Indian way of life and to the amount of time, in his opinion excessive, spent on ceremonies and family gatherings. He also felt called upon to demand that reservation Indians put an end to these customs, which he denounced. He even threatened them with certain measures, not clearly specified, in case his demands were not met.

Although this letter dates from 1923, shortly before the IRA and half a century before the Indian Self-Determination Act, the paternalistic attitudes and opinions expressed are, on the whole, still prevalent today. The pressure which the majority of Americans and the federal government put on the reservation Indians, at that time, to acculturate and to assimilate to the American way of life still exists today. The form of this pressure has changed, but the ideological background remains.

Since the Second World War many American Indians have bowed to this pressure, particularly during the termination and relocation era. Many of these assimilated Indians have either exchanged their way of life on a reservation for a new life in a large city or they have joined the army of civil servants in the Bureau of Indian Affairs (BIA) or the Indian Health Service (IHS), whose members often move from one reservation to another in the course of their careers. Separated geographically, and often also culturally, from their own tribes, removed from the social network of their extended families, and dependent on their employers for their livelihood, a large number of these civil servants now identify themselves with Anglo-American values. Moreover, they often feel superior to reservation Indians they have to deal with, not only because they are often better educated, but also because they belong to another tribe.

Many of these civil servants are so-called mixed-bloods, who have a partly white ancestry[2] and who often, from a sociocultural viewpoint, have little in common with the majority of reservation Indians (Wax 1971, 54–55). There are also local mixed-bloods on the reservations who, in some places, occupy the leading positions in the tribal government and administration and in whose hands the private economic sector, usually very small, is often to be found. These local mixed-bloods, as well as those from other tribes, constitute a class of acculturated Indians who may carry on negotiations for the tribe on economic issues with people of the outside world. It is significant that these acculturated mixed-bloods often live apart from the tribal members not only socially but also geographically.

The fact remains, however, that the overwhelming majority of the local population on the reservations are American Indians whose cultural traditions are still very much alive, even if somewhat restricted and varied depending on the individual tribe. For the majority of non-Indian visitors who only get to know the reservations casually, this might not be obvious, especially from the visible material culture reflected in the cultural landscape. In their eyes most of the reservations appear only slightly different from all the other peripherally located, depressed rural areas of the United States.

Although the cultural landscape on reservations has been modified to reflect the Anglo-American cultural world through the presence of highways, Anglo-style housing and settlement patterns, and so forth, the value systems on reservations remain markedly different. This can be demonstrated especially well with those Indian values which influence the reservation economy. Of course, with the spread of Western civilization, not only various characteristics of Indian material culture but also many aspects of its nonmaterial culture have been eliminated or modified. Moreover, through a selective process of acculturation new values have been acquired. Nevertheless, in spite of the massive and unremitting pressure for assimilation, many Indian tribes have been able to preserve their distinc-

2. The term "Mestizo," commonly used in Central and South America for people of mixed blood, and the term "Métis," used in Canada, are not terms used in the United States with the exception of Métis in three special cases in Montana.

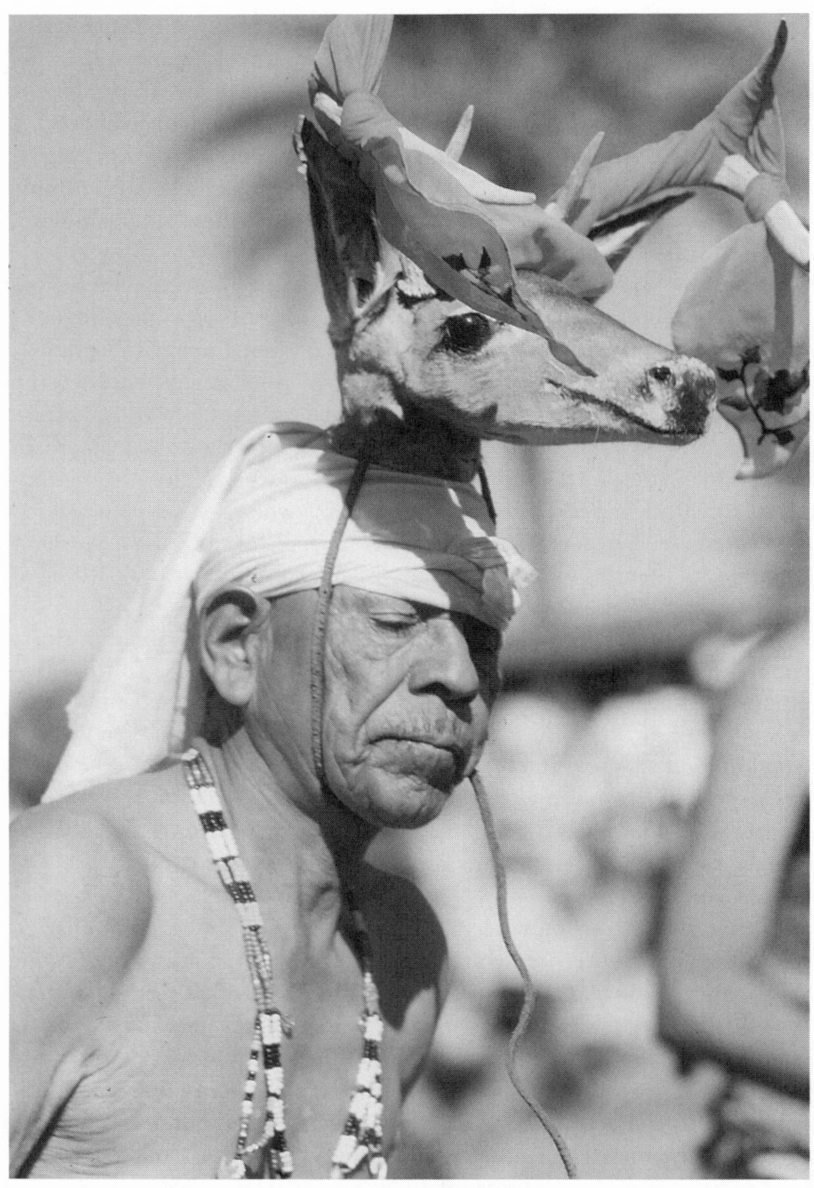

Fig. 6.1: Yaqui deer dancer on Indian Day, Heard Museum, Phoenix, Arizona, 1986

tive sociocultural fabric to an astonishing extent. Parallel to the "New Ethnicity" movement which has taken hold of the United States during the past three decades and has shed new light on the melting-pot theory previously so commonly accepted (see Glazer and Moynihan 1963; Novak 1973), a kind of renaissance of the various ethnic identities appears to be taking place, a renaissance which also encompasses American Indian populations. The ever increasing political self-confidence, the renewed emphasis on sovereignty, the intensified expansion of bicultural methods of instruction, the deliberate promotion of Indian languages, and the revival of numerous religious ceremonies (see figs. 6.1–6.6) are clear signs that American Indians are rediscovering their identity. It can therefore be said that today the Indian reservations in the United States, as well as their varied tribal populations, are also distinctive cultural enclaves within the dominant Anglo-American society, enclaves whose survival for future generations seems to be assured.

The sociocultural spectrum of a tribe should be considered when outside attempts are made in conjunction with local people to promote economic development on the reservation and to adjust the economy to present-day non-Indian conditions. Past experience, however, shows that federal agencies have generally overlooked or ignored this aspect of economic assistance in their programs. The analysis of federal Indian policy in the 1980s indicates that there has been no real change in this attitude (see Presidential Commission . . . 1984). This disregard of the specific sociocultural framework of tribes, as well as the apparently unshakable belief that Anglo-American economic recipes, when combined with capital investment and technology, will provide Indian reservations with the key to successful economic development, has already caused many projects to fail. The numerous empty or run-down industrial parks on reservations across the country are conspicuous examples of this lack of understanding.

All this goes to show, stated bluntly, that the slogan "What's good for General Motors is good for America" cannot be applied to Indian reservations, with the exception of a few tribes that have largely assimilated, and even in these cases the assimilation has not been complete. It would be closer to reality and would give a more precise and discriminating idea of the situation on different reservations to add that "what might be good for the Cherokees may not be good for the Hopis." There seems little chance, however, that this slogan will ever become official Indian policy.

Indian Values as Determinants in Today's Reservation Economy

In view of observations made in the previous chapter it may seem somewhat daring to attempt to define the values of reservation Indians which have a direct influence on today's reservation economy (see fig. 6.7). Because of the

Fig. 6.2: Yaqui Pascola dancer in Hickiwan, Papago Indian Reservation, Arizona, 1985

Fig. 6.3: Crown dancers of the White Mountain Apache, Whiteriver, Fort Apache Indian Reservation, Arizona, 1985

cultural diversity of the various tribes it is undoubtedly true that any generalization must be treated with caution. On the other hand, there appear to be certain values which the majority of reservation Indians do have in common.[3] Whether these values are considered to promote or hinder reservation economic development depends on the ideological viewpoint and the ethnic background of the observer. There is no doubt, however, that according to the official view of federal development agencies, most of the values discussed here are considered to be obstacles to the development of the economy, instead of assets.

An essential element of the traditional value system of Indian tribes is a highly developed collective consciousness, which today still dominates daily life on the reservations. For the white population of America the social status of the individual is determined to a very considerable extent by his or her income and property, which are shown quite openly, whereas on the reservations such individual manifestations, on the whole, count for

3. The following ideas are based on my interviews and observations in 1983, 1985, and 1986 on many reservations in the Southwest, compared with and supplemented by others' research, partly carried out in other regions of the country. See in particular the following: Aberle 1975; Ballas 1973b; Bigart 1972; Bolz 1986; DeMallie 1978; Feest 1976 and 1980; Fonaroff 1964; Gilbreath 1973; Hofmeister 1976 and 1978; Hughes 1983; Jorgensen 1978; Parker 1976; Sorkin 1971; Stanley 1978; Wax 1971; and Weightman 1976.

Fig. 6.4: Giveaway ceremony among the White Mountain Apache, Cibecue, Fort Apache Indian Reservation, Arizona, 1985

very little. Even today, despite certain contrary tendencies, reservation Indians feel primarily bound to their relatives, both to their close relatives and to their more distant ones, as well as to their clans, to the people of their village, indeed, to their community as a whole. Individualistic behavior is regarded as a threat to the survival of the tribe's independence, and people who do not show a sense of obligation to the community are often still ostracized and isolated. In the long run, it is virtually impossible for anyone in this category to remain on the reservation.

This collective consciousness, an attitude of mind that is characteristic not only within tribes of American Indians but also within many other tribal communities worldwide (see Kopp 1978, 90 and 92; Lévi-Strauss 1981; Service 1979; Wesel 1985, 71–107, 232–234, and 252–253), shows itself in a number of different ways. It is to be seen, for example, in the traditional attitude regarding landownership, an attitude very different from that of the overwhelming majority of Americans, yet often found among tribal members even today. It is common knowledge that for many traditional American Indians, land is not regarded as a commodity that can be divided up, bought, and sold. This very different attitude is reflected in the words of the Shawnee warrior chief Tecumseh in his fruitless negotiations in 1810 with William Henry Harrison, who later became president, but then was governor of the Northwest Territory. "Sell a country!" Tecumseh exclaimed,

Fig. 6.5: Sunrise dance of an Apache girl and her friend. In the background are the medicine man and his assistants, Whiteriver, Fort Apache Indian Reservation, Arizona, 1985.

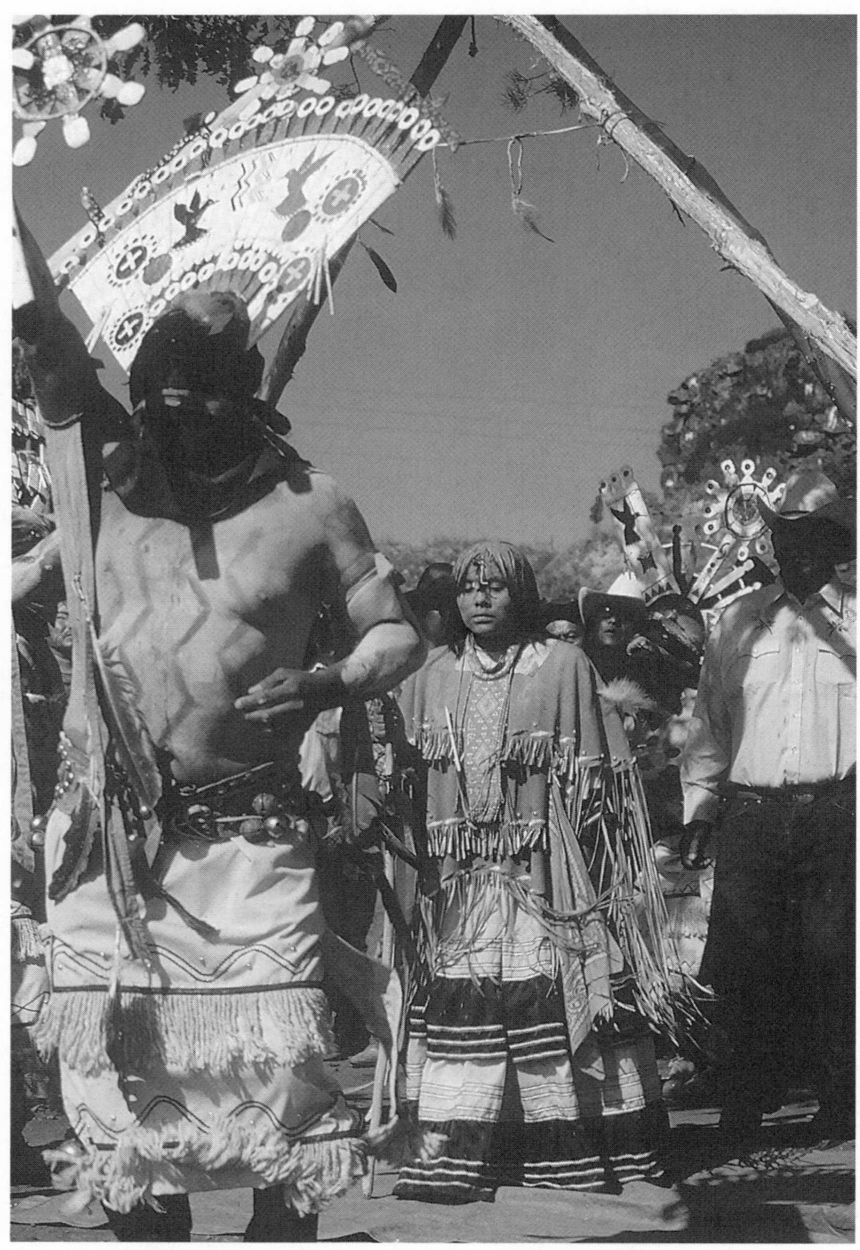

Fig. 6.6: Initiation rite of an Apache girl with crown dancer, Whiteriver, Fort Apache Indian Reservation, Arizona, 1985

greatly vexed. "Why not sell the air, the clouds and the great sea as well as the earth? Did not the Great Spirit make them all for the use of his children?" (quoted in Tucker 1956, 163).

An essential element of the pronounced community spirit, of the "K'é" or "feeling for the other one," as the Navajos call it <34>, is the principle of mutual support and sharing. The origins of this principle are found not so much in religious motives (see Hofmeister 1976, 515; and 1978, 340) as in the Indians' way of life before they lived on reservations. For the nonsedentary tribes whose economy was, on the whole, not adequately developed to keep reserves of food, this cooperative attitude was a kind of insurance for survival in times of need. For them it was therefore simply obvious to share the kill, the food collected in the wilderness, and other natural products as well as occasional crops with their own kin or clan, or with another band or clan with whom they were on friendly terms. Anyone who did not observe this custom had to reckon with sanctions, and even among the sedentary Pueblo Indians, this attitude prevailed. Among the Pueblos it was customary, in case of need, to allow families whose crops were poor to help themselves to what remained of the crops from a neighboring field (Hughes 1983, 61).

The principle of sharing is clearly expressed in the "giveaway" or "potlatch" ceremonies, which are common today at births, funerals, initiation rites, and weddings in many tribes. Even in ceremonies in veneration of a Christian saint this element is present. A family or a particular group of members of the tribe distribute gifts to their relatives and friends, or sometimes to all the people present. These ceremonies often include an invitation to a big, elaborate meal, and it is by no means unusual that the expenses of these giveaways are often so high that a family has to save for several years and slaughter precious livestock to be able to meet them.[4]

It is not surprising, therefore, that in the past the so-called Friends of the Indians, other white reformers, BIA agents, and even Bible-trained Christian missionaries who all had their own ethnocentric viewpoints should

4. I had the opportunity to participate in five Sunrise Dances on the Fort Apache Indian Reservation in Arizona (see figs. 6.3–6.6). These are initiation rites lasting several days in which an Apache girl from 11 to 13 years of age is blessed and psychologically strengthened by the medicine man and his assistants in the presence of her relatives and a considerable number of village residents and friends. According to the Apache's belief, any girl who can endure this impressive but very trying ceremony—which cannot be described in more detail here—will be well-prepared for her future duties as a woman.

Conversations with the participants revealed that the cost of these ceremonies, including the payment of the Crown Dancers, the medicine man and his assistants (see figs. 6.3, 6.6) as well as the costumes, the gifts and food for all people present, can range from $7,000 to $10,000. In most cases the financial support of relatives of the girl's family is necessary in order to shoulder these expenses.

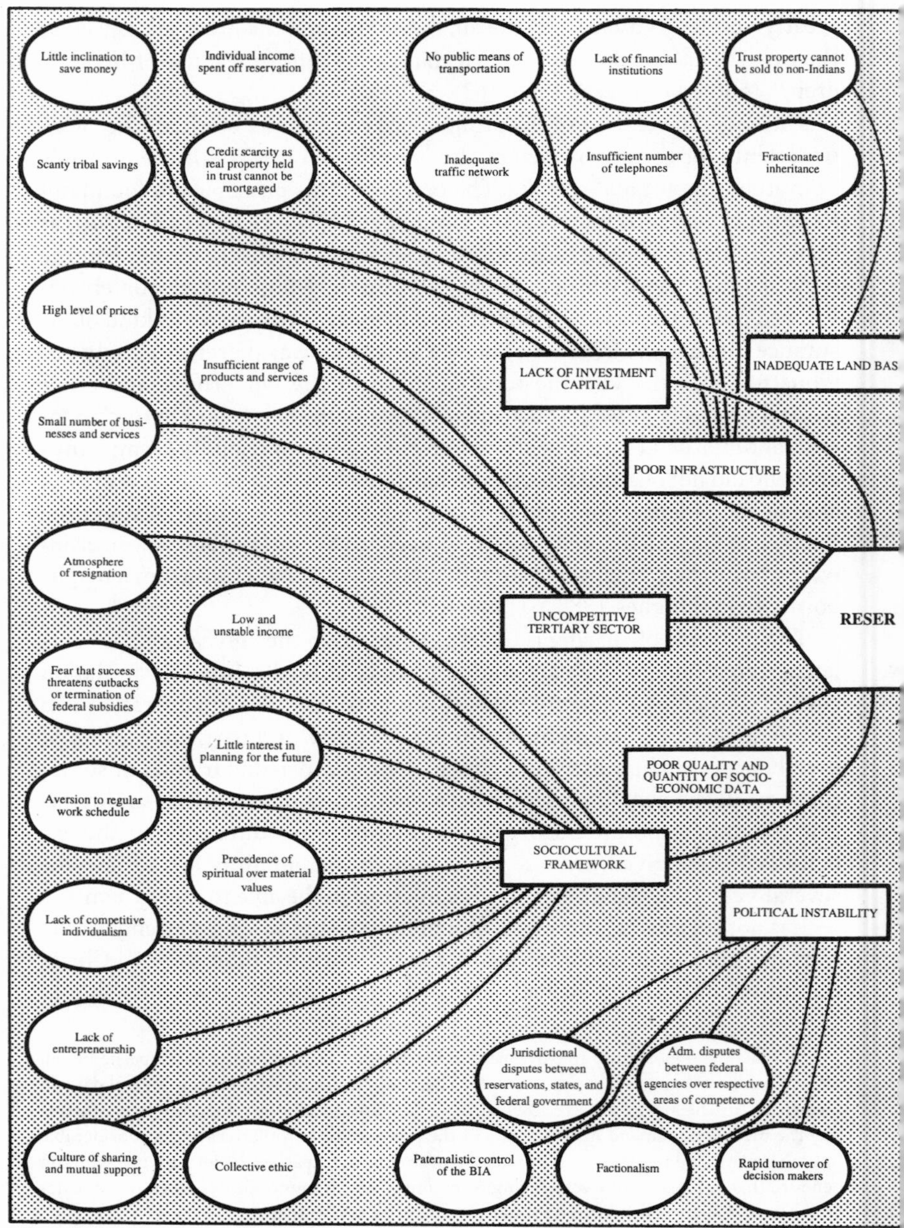

Fig. 6.7: Favorable and unfavorable factors in the economy of Indian reservations, according to the official view

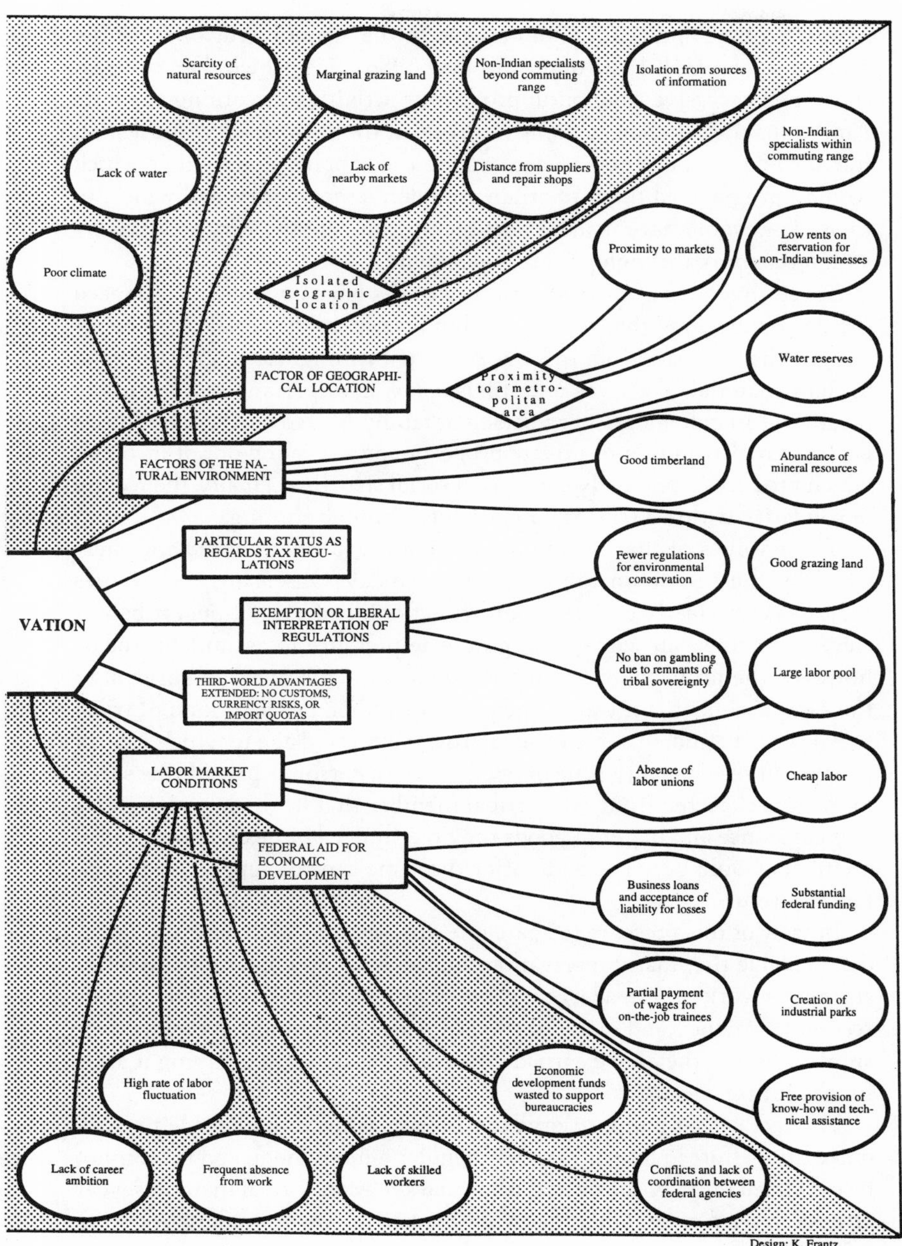

Design: K. Frantz
Cartography: H. Heinz-Erian

have tried time and again to put an end to the "ruinous customs" of the giveaways (see Goll 1940, quoted in Bolz 1986, 171), which they believed consumed excessive time on unproductive activities. According to white Americans these Indian customs were one of the reasons why reservation Indians had not been able to accumulate the supplies and savings which would have enabled them to attain a moderate prosperity, and which, in their view, would have made it possible to stimulate the very weak reservation economy through private investment.

These advocates of the Western way of life have obviously overlooked the deeper meaning that sharing, and in particular, the ritual of exchanging gifts had for American Indians. A mutual willingness to give help and to share material goods was, and still is today, an important measure of the prestige of an individual Indian person or family. Moreover, these traditions lead to a redistribution of private property and a leveling out of incomes, which produces a relatively egalitarian social structure. Finally, the feeling of solidarity within a tribe is strengthened through sharing.

These values of sharing and common land are still honored and cherished by the people on the reservations today. Observing such values indicates very clearly both the sound functioning of a traditional tribal society and success in withstanding outside pressure to assimilate. Today, however, the principle of sharing is no longer a guarantee of survival to the same extent that it was in the past, but rather a specific kind of tribal social security and, in some cases, a provision for old age to which elderly people can resort in any time of need.[5] Once a person is part of this social fabric it is expected from every tribal member that he or she would offer any assistance and material goods they might need or desire. Those tribal members would be expected to offer the same type of support and assistance in return.

In view of this pressure to share one's possessions with others, it is understandable that many reservation Indians have no great motivation for strenuous work to earn a large income or provide themselves with surplus reserves of money and material goods. Instead many Indians rapidly consume whatever they have earned or acquired as opposed to laying it aside for the future.

The fact is that many reservation Indians who would be able to work make little effort to find or maintain regular employment, and this cannot be explained only by the dismal labor market. Many tribal members, year after year, seem to be content with seasonal or an even sporadic work, from

5. Thus, for example, on the Fort Apache Indian Reservation in Arizona, where cattle ranching has played an important role since the turn of the century, a tribal herd was established in 1917 whose proceeds until recently were reserved for the benefit of the elderly people on the reservation. Today this herd is still called the "Old Folks' Herd" <35>.

which they can live for a while. This gives them time to devote to their personal interests without being obliged to submit to the rigorous routine of daily work. Furthermore, they know that in the event of a long period of unemployment they can always count on financial support either from a relative who is working or from a government agency.

It goes without saying that this attitude cannot be considered "proper" by white Americans, many of whom are imbued with the Protestant work ethic (see Weber 1963), and who usually make provision for the future, saving their money with a definite purpose in mind. In their view it is unreasonable that many American Indians make no financial arrangements for their future, and moreover, they regard the custom of sharing as parasitism, something that should be done away with.

These customs and characteristics of Indian culture help to explain why it is that on reservations, savings accounts are seldom to be found. Moreover, those more progressive and better educated Indians who do have such accounts are careful to conceal them from their friends and relatives, a fact which I was able to ascertain from various sources. When Indians aim to become rich, they usually leave their reservations and go their own way, far from the social controls found on the reservation. If they are successful, the idea will hardly occur to invest their own savings on the reservation in view of the prevailing sociocultural conditions.

In the area around Phoenix, for example, there are six very successful Hopi businessmen. In my conversations with two of these businessmen, who lived separated from their tribe geographically and socially, they made it clear that they preferred the anonymity of city life, although they both admitted that they would have preferred to live on the reservation. They realized, of course, that if they lived on the reservation, they would be involved in a network of obligations to their families and relatives which would make it impossible for them to maintain, or even to justify, their prosperity.

This specific sociocultural context with its attendant attitudes goes far to explain why Indians engaged in private enterprise can very seldom be found on reservations, whose residents, generally speaking, simply do not have an entrepreneurial spirit. The consequence is that federal programs designed to stimulate the private sector of the Indian economy have thus far had very little effect.

In view of this pronounced collective consciousness and feeling of solidarity one would naturally consider tribal enterprises and cooperatives to be the appropriate forms of economic organization for reservation Indians, because the work for such organizations can be done by a whole tribe, a kinship group, a clan, or a village community. Tribal enterprises are indeed widespread, and can be seen in the numerous farms and industrial plants owned by tribes in which non-Indians from outside sometimes participate as partners. Cooperatives, on the other hand, which would be in much closer harmony with the social structure of reservations, are comparatively rare, and this lack of cooperatives is one of the limitations of economic development.

Both tribal enterprises and cooperatives are institutions which are impregnated with an ideology radically different from that which predominates in the industry and service-based economies of the Western world. This ideology reveals itself in the fact that as long as the management is influenced directly by the tribe, its primary concern is not maximum financial profit, but rather sociocultural considerations. There are many indicators of this attitude, some of which will be discussed in the following section.

A prominent feature of most tribal enterprises and cooperatives is the number of their employees, which is often much greater than in comparable enterprises outside the reservations. Nevertheless, when measures of rationalization are proposed to reduce the number of personnel, such measures are usually rejected by the Indian board members and tribal leaders who make such decisions because of their feeling of solidarity with relatives and friends. A manager of a tribal enterprise or cooperative who refuses to accept recommended tribal members or tries to discharge them often brings on difficulties which may result in his or her own discharge.

The manager must also have a high degree of sensibility and flexibility when confronted with the often negative attitude of staff regarding regular working hours. This perceived negative attitude can be partly explained if we look back to the prereservation era when Indians simply did not have a daily or weekly routine when engaged in hunting or in agriculture. Traces of this previous way of life, which have survived down to the present time, are often wrongly considered by outsiders to be signs of laziness on the part of the Indians. However, as I shall show, this prejudice is not shared by the people, mostly non-Indians, who have had many years of experience on the reservations as managers of large tribal enterprises.

The Fort Apache Timber Company, for example <36>, has a staff which consists almost entirely (96%) of Apache Indians. The manager of this sawmill, who except for a few short breaks has occupied this position since 1954, speaks highly of the skill and efficiency of his employees, and he is certainly not the only one of this opinion. On the other hand, he criticizes them for not adhering to prearranged schedules, and for frequent absences from their place of work, as well as for their unusually high rate of turnover in the workforce. According to him, tribal ceremonies, family celebrations, a good harvest, the successful sale of young cattle, or the annual traditions to gather acorns in northern Mexico can lead to absences lasting for days, and often no advance notice is given. This manager, who has the prime responsibility for the prosperity of the company, is relatively powerless to discipline his employees in any way for such actions, or to discharge them.

A visitor from outside will find it illuminating to be a customer of one of the tribally operated supermarkets, which usually operate with a deficit, that have been set up during the past few years on many reservations. These supermarkets are intended to take the place of the generally disliked privately run "trading posts" on the reservations, which are generally operated by non-Indians or mixed-bloods. In these new markets it is by no means unusual, as I was able to ascertain on a number of occasions, for poor or elderly people in particular, to take items from the shop without paying. The employees of the supermarket see them take these supplies, but do not interfere, as they are relatives or friends of the customers. Even the management usually tolerates these proceedings, which are in accordance with the strongly developed feeling of solidarity for tribal people and their tradition of sharing with others.

Another characteristic element in the management style of tribal enterprises is that the profits, instead of being reinvested into the business where they are urgently needed, are often used for other, often social, purposes. As already explained, this is also an indication of the priority often given to those values conducive to the general welfare of the community on the reservations, despite the fact that this policy causes the reservation economy to fall still further behind.

Together with the priority given to socially motivated policies, another phenomenon that has a detrimental effect on many tribal economies might best be described as a subliminal fear of success. The causes of this fear can be found in part in the more recent termination policy. During the Eisenhower administration the very reservations that were particularly successful in their economic development had to resign themselves to the curtailment of their subsidies or even "termination." The consequence was that following this experience some tribes were careful not to be too self-sufficient or successful economically because they feared that they might once again lose their status as reservations or be deprived of economic assistance from the federal government.

Another distinctive characteristic of many reservation Indians, a characteristic whose influence is to be seen not only in tribal economies, but also in other fields, including education, is the absence of a competitive spirit, quite noticeable among many tribal members. In America, generally speaking, it is considered desirable to be the best or to become "number one" with all the prestige and status this brings, but this competitive spirit is foreign to the traditional values of most American Indian tribes.

Someone on the reservation who aims to rise to the top at their place of work, for example, or to stand out from the others by conspicuous effort and hard work as a student or a teacher is usually not highly regarded by other tribal members over time. While it is common practice for different groups within the tribes to distinguish themselves collectively from other groups in order to achieve a more advantageous position in case of need, such behavior on the individual level is considered antisocial. Any individuals who continue to behave like a lone wolf must expect that fellow workers and schoolmates will put them under pressure to change their behavior. If this collective attempt to pressure the individual does not produce results, this person and sometimes also their family may be confronted with scorn and distrust. According to a Hopi Indian who works for the BIA outside his reservation, "a community of Indians is like a pot full of crabs. If one of them tries to rise to the top, the others immediately pull them down" <37>.

These forms of peer pressure explain why the highest positions in the management of many tribal businesses are often occupied by whites, or in some cases by Indians of another tribe (see fig. 6.8). Even if local tribal members have the necessary qualifications, they can seldom be persuaded to take leading positions because of this prevalent attitude of egalitarianism. Enjoying good relations with other members of the tribe is held to be more desirable than striving to attain a leading position in the tribal economy, which would mean exposing oneself and running the risk of being branded as a show-off or an outsider.

Bolz rightly points out in his study of the Pine Ridge Indian Reservation (1986, 94 and 142) that every tribal member is enmeshed in a network of relations with kith and kin, and bound by an attitude of mutual respect, including a respect for the elderly, so that only a very limited freedom of decision making remains (see also DeMallie 1978, 295–296). The consequence of these values is that either the authority of a superior is not recognized, which means that in many cases their instructions will not be carried out, or else those in authority are so well-adjusted to the situation and show so much consideration for other tribal members that they never take any decisive actions.

I can testify that the observations of Bolz are valid, not only for the Oglala but for other tribes as well, where the same relations of dependence and mutual consideration prevail. Tribal political leaders are aware of these prevailing values, and therefore they usually appoint people from outside the tribe to top positions in tribal organizations because such people, having no connection with the tribe, can act more independently when decisions must be made. The appointment of a nontribal member has the further advantage that when difficulties arise, the person in question can be dismissed at any time, without provoking hostility within the tribe or cre-

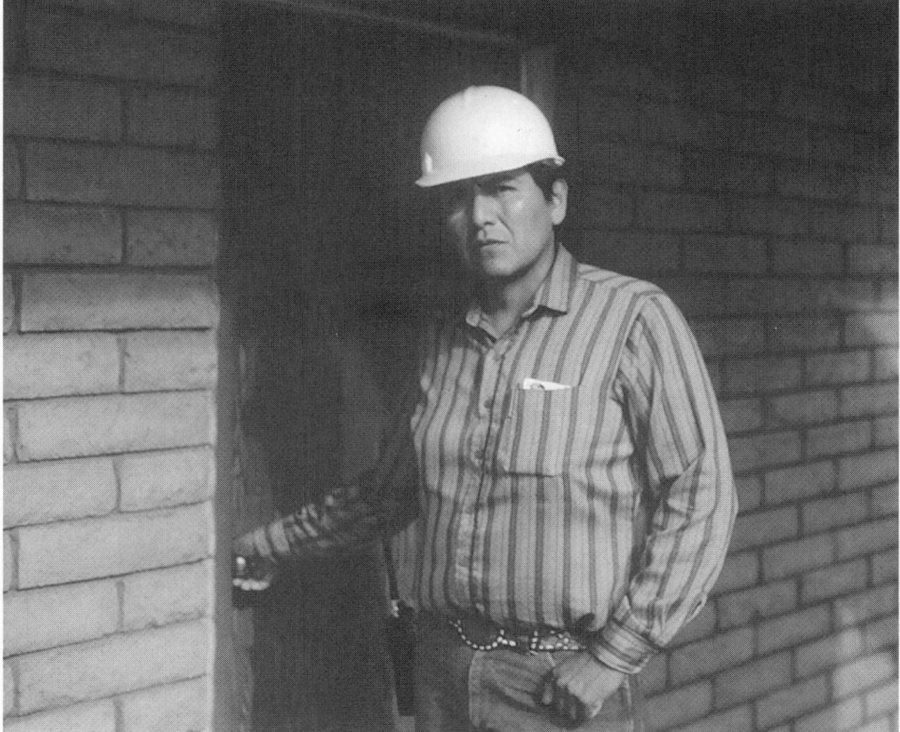

Fig. 6.8: A San Carlos Apache as the assistant manager of the Fort Apache Timber Company, Whiteriver, Arizona, 1985

ating a potential political rival in future elections. Tribal members who have ambitions in the world of business, therefore, are usually compelled to make their way outside the reservation.

The preceding observations shed light on some of the common Indian values and characteristics which have a strong influence on the reservation economy. As already pointed out, these sociocultural values have been and still are generally ignored in the conception and implementation of federal policies for economic development. This was demonstrated once again by measures of the Reagan administration which promoted the intensive development of the private sector on reservations and which aimed to make entrepreneurs out of Indians under the direction of the BIA. Here, too, the responsible officials assumed that reservation Indians were willing and able to bring their own value systems into line with the guidelines of the white programs for development that were proposed for them. These programs would have a much greater chance of success, however, if they were based on the respective sociocultural situation of each reservation in question, rather than the other way around. Such an approach, however, would rapidly put an end to the policy of assimilation that prevails in the BIA and other federal and state agencies. The preconditions for taking this step, despite the IRA and the Indian Self-Determination Act, are still not to be found in Anglo-American society today.

FACTORS THAT HINDER THE DEVELOPMENT OF THE RESERVATION ECONOMY

The urbanization process has been largely completed in the United States and the early agrarian society has given way, first to a society based on industry, and then to one based upon service economy. Indian reservations, mostly peripherally located, were not part of this urbanization process, however, and therefore are now confronted with virtually insurmountable economic problems. These problems and their consequences have given reservations the not very flattering label of "slums in the wilderness" (Hagan 1971, 22). Although economic development in many other distressed rural areas of the United States is hindered by some of the same factors, other problems are endemic to Indian reservations alone. The more important of these factors will be discussed here and summarized in a schematic framework (see fig. 6.7) to help display their complexity.

A closer look at Indian reservations often shows that because of their isolated locations, poor natural resources, and limited economic potential there is a high degree of marginality (see Hofmeister 1976 and 1978; Sorkin 1971 and 1973; Wax 1971). This situation has developed not simply by coincidence, but is predominantly the result of the federal Indian policy previously discussed.

Indian reservations are generally located far from important transportation routes and from urban agglomerations, and often hold only limited prospects for developing an independent and viable economy. Their isolated location (see fig. 6.7) undoubtedly results in a serious handicap for most branches of the economy. A business in the service sector looking for customers with real purchasing power, for example, has only a sparsely populated reservation available, whose many impoverished inhabitants often live in scattered settlements, and therefore do not form an attractive market. The same can be said of businesses which are dependent on transportation, and for which a short distance to the suppliers of raw materials and proximity to markets are of decisive importance. Subsidies or other aid can only partly compensate for the disadvantages of location, and such compensation would be only temporary. It will thus be difficult for the majority of peripherally located tribes to cast off dependency in the future in spite of continuous promotion of self-determination.

When a business enterprise is located on a reservation, it may be isolated from necessary contacts with similar businesses and thus the communication flow is hindered. This isolation can be further increased by generally poor infrastructure: owing to the lack of private enterprises on the reservation, small businesses for repair work and other services are lacking, bank branches remain a rarity, means of communication and sources of energy are commonly both inadequate and expensive, railway connections are rare, the road network is of poor quality and low density, and, generally, public transportation is nonexistent.

On the Navajo Indian Reservation, which includes parts of Arizona, New Mexico, and Utah, and which is by far the largest reservation in the United States, there were, for example, only two good, partly paved roads in 1950, which together were 243 miles in length.[6] The decisive initiative for the construction of roads in this area came during the following thirty years with the exploration and mining of mineral resources on the reservation. This development greatly increased the total number of road miles. The total length of the road network was about 9,400 miles by 1986, according to the BIA roads program, but under 1,900 miles were paved. The remaining roads are unimproved and are partly impassable after it rains.

Despite this great increase in road construction, even today the Navajo Indian Reservation makes a poor showing when compared with similar rural areas of Arizona and New Mexico. If the density of road networks on and off the reservation in Arizona and New Mexico are compared, taken over an area of 1,000 square miles, the road network on the reservation makes up only just over one-third of the equivalent off the reservation. The population density on the Navajo Indian Reservation, however, with fewer than one person per square mile, is much lower than that of rural Arizona and New Mexico.

The development of a road network on the reservations has undoubtedly made daily life for Indians easier in many respects, though this development has also brought with it certain disadvantages for the reservation economy. Parallel to the rise of auto traffic, the expansion and improvement of the road network has led to a higher degree of mobility for reservation Indians. In view of the fact that retail trade and businesses in the service sector are so poorly developed, it is understandable that Indians should make use of this newly acquired mobility to make their purchases in nearby cities outside the reservations where the supply of goods is much greater and the prices lower. Apart from these advantages, reservation Indians can

6. One of these was the road between the two border towns Gallup, New Mexico, and Cortez, Colorado, including a cul-de-sac to Window Rock, the administrative headquarters of the Navajos. The other was the road between Flagstaff, Arizona, and the North and South Rims of Grand Canyon National Park.

enjoy the attractiveness of urban life, in German referred to as "urban glitter," while also escaping, at least temporarily, from the thickly woven network of social control that surrounds them at home.

This tendency to shop in border towns robs the tribe of the benefit of the multiplier effect, because the money earned at home is used to a great extent to "import" merchandise and services. The supermarkets established in recent years by a number of large tribes have not put a stop to the considerable drain of money, as shown by a study commissioned by the Navajos in the mid-1970s. The author of that study came to the conclusion that approximately one-half of a Navajo's household income is spent outside the reservation (Wistisen 1975, cited in Ruffing 1978, 45). On the Zuni Indian Reservation in New Mexico this proportion is even higher, estimated at 84% (Vinje 1982, 94). These conclusions support Hofmeister's view that the competition of the American economy (1978, 342) is generally too powerful for the reservations.

The marginality of Indian reservations is to be seen not only in their isolated location and limited economic potential, but also in their generally poor natural settings, which include the common lack of natural resources, to be treated in more detail in later chapters.

As early as 1976 Hofmeister called attention to the fact that Indian country is generally a very poor, disadvantaged region with regard to its environmental setting and natural resources. Almost one-quarter of these Indian lands are arid with less than 10 inches of rainfall annually (see fig. 6.9). In 1983 about 11% of reservation land was commercially utilized forest (see fig. 6.10), of which one-tenth, down from 40% in 1963, was leased to non-Indian enterprises. On the other hand, croplands represented less than 6% of the total Indian country, and around three-fifths of this cropland, 86% in 1963, was cultivated by non-Indians. The major part of the remaining land was generally used as open range grazing (ED 1986, 114–115).

For many years the federal government tried to make yeoman farmers out of as many American Indians as possible. The extent to which this political program misfired, however, is shown by the fact that today only a small part of reservation land is cultivated, and an exceedingly small number of Indians are farmers. Those tribes who were traditionally successful farmers, the Pueblo Indians of New Mexico and the Pimas of Arizona, for example, have largely lost the skills needed for farming, following controversies over water expropriated by white farmers. The policy that sought to make Indian people individualized, self-sufficient farmers was doomed from the start, since most of the Indian land is not suitable for agriculture without large capital expenditure, owing to unfavorable climatic conditions, poor soil, and inadequate supplies of water.

Fig. 6.9: View over the western part of the Gila River Indian Reservation, Arizona, 1986

Fig. 6.10: Southwestern slope of Mt. Baldy, Fort Apache Indian Reservation, Arizona, 1985

There are still other obstacles to economic development which have little to do with the poor natural environment or the peripheral location of Indian reservations. One of these obstacles has its origins in the legal, administrative, and political structures specific to reservations, structures which underlie the tribe, the BIA, and the respective state, as well as their relationships toward each other.

Outside investors and also the small number of Indian entrepreneurs are therefore often confronted with obscure and intricate administrative practices which act as a deterrent and thus blur the advantages created by favorable tax structures or other benefits. For example, in order to rent a parcel of land in one of the industrial parks on the Gila River Indian Reservation in Arizona to set up a business, it is necessary to enter into wearisome negotiations with the various tribal authorities, and to complete forty-five application forms, many of which are several pages long <31>. Even after these costly and time-consuming formalities have been carried out and various other federal requirements have been met, the veto power of the BIA can still be used to postpone or put a stop to the project in question. The BIA's veto power often has a negative effect on economic initiatives of the tribe, whose members have the feeling that they are constantly being patronized despite the propagated policy of self-determination.

Disputes over the competence of staff and problems of factionalism condition daily life in many tribal governments and among Indian decision makers. These are also factors that create negative effects on the reservation economy and create a climate of political instability, since it is often difficult for the outside investor to know with whom to negotiate or how much importance to attach to any particular negotiations.

First, a business developer is confronted with the tribal chairperson, who is generally up for reelection every two years, but who may be subject to recall if the members of the tribal council give him or her a vote of no confidence before the end of this two-year period. My inquiries have shown, for example, that in the period from 1938 to 1986 there were twenty-one different chairmen on the Hualapai Indian Reservation in Arizona. Seven of them were in office for less than one year, and only three were chairmen for more than one term. The result of this frequent change of leadership is that no one in the tribe acquires very much political power, a principle in accordance with the traditions of many tribes.

The premature expulsion of the tribal chairperson is often accompanied by the simultaneous resignation or replacement of other council members and key tribal staff. The result is that the rules of the game often change in the middle of the process, and that the businessperson, still in the midst of negotiations, even after investments have been made, must abandon the project and leave the reservation.

On some reservations a businessperson from outside who enters into negotiations with the tribal chairperson and staff must also contend with various factions in the tribal council, the "shadow government" in the BIA and, if possible, the traditional tribal leaders, who often do not make themselves known to outsiders.

The political system of the various tribes is very different from that of mainstream society as a whole. Contrary to Western democracies, the principle of representative political leadership and of elected representatives who make decisions for the people on the basis of a parliamentary majority is alien to most Indian tribes. In the past it was the Indian custom to reach a decision only after an extensive discussion among members of a kinship group, a band, a village community, or a tribe, and the problem was discussed until a consensus was reached. Isolated decisions made only by one or more leaders were often not accepted by the people. Even today this Indian style of consensus decision making is widely prevalent on the reservations. The result is that it is often difficult or impossible to reach political decisions that will be accepted by everyone on a reservation whose population consists of many different bands and village communities, and, in some cases, of artificially created fusions of different tribes. Responsible politicians on reservations who make isolated decisions independent of public consensus must then be prepared for the likelihood that their decision will be overthrown or boycotted. Non-Indian businesspersons, once aware of this political environment, can therefore hardly be blamed if they reconsider their plans and decide not to risk an investment on reservation land.

While non-Indian entrepreneurs are free at any time to transfer their business to another place, tribal enterprises are generally bound to the reservation. Their managers are often steered and manipulated by their tribal boards and unduly influenced in their decision making. Unlike the rest of America, where maximizing profits is the primary objective, tribal enterprises are often chiefly guided by social and political considerations.

The complicated and somewhat unclear legal status of the reservation lands and water rights is another deterrent to outside investment. For the arid areas of the West and the Southwest, for example, it is of paramount importance for the manager of a large farm to know the annual amount of water available and how much they must pay for it. Because water is very scarce and in great demand in these regions, and because at the present time more than sixty important cases of litigation are pending between Indian reservations and surrounding communities, counties, or states (ED 1986, 32), the water question remains unclear on many reservations and is a factor that creates great uncertainty for business investors. Complications can also arise from inheritance practices that split the land into extremely small parcels, which makes it difficult for a tenant farmer to cultivate areas of land large enough to earn a profit over time without becoming constantly involved in legal controversies. Another source of legal controversy between the tribe and businesspersons from outside is the question of who should be the owner of improvements and investments made on leasehold land once the lease has expired. Because of such controversies, outside business investors are extremely reluctant to make extensive, nonmobile capital investments and they usually equip their reservation businesses with only the minimum investment necessary.

On some reservations, for a variety of reasons, people are simply not interested in the dynamic development of their economy. It may be that a tribe fears the consequences of a very successful economy, perhaps because the tribal decision makers are still subconsciously influenced by the era of the termination act, when a number of reservations with flourishing economies no longer received subsidies. There are also tribes which seem willing to accept an economic standstill in order to preserve their ethnic identity and to protect themselves from the infiltration of outside change.[7] A similar attitude is to be found among those many American Indians who deliberately remain on the reservation to demonstrate that it is more important for them to live in the security of their tribal society and the homeland they are attached to than to have the prospect of social mobility, which can often only be achieved by leaving their familiar environment.

It has already been pointed out that reservations increasingly serve as places of refuge for those Indians who cannot take part in the migration to the large cities, either because they are too young or too old, or are physically or mentally challenged (see Wax 1971, 71). The consequence of this trend is a brain drain, where many urgently needed and well-educated people move off the reservations, leaving the local economy with a labor force not well-qualified for their tasks.

One of the chief problems for the reservation economy is the lack of investment capital and mechanisms for collateral or loan guarantees. It is true that many tribes have long had trust funds, established by the BIA as compensation at the time when the Indians were forced to cede land to white settlers.[8] Money from the ICC still flows into these trust funds today, if it has not already been distributed to tribal members. Revenues from mining permit fees, royalties, and shared profits along with money from land

7. Thus, for example, the Havasupai Indians have successfully opposed any projects to open their reservation by construction of a cable railway or road, thereby preserving their isolation, which is viewed as idyllic by outsiders.

8. Trust funds have existed since 1819, when for the first time the federal government purchased the territorial claim as well as the agreement to exchange land from an Indian tribe, in this case the Cherokees of Georgia.

leases for agriculture, forestry, and industrial parks is also put into the trust funds. These reserves could be used as capital for new industrial projects, except for two problems. First, the BIA retains the power of veto over the use of these funds and, moreover, has often used this power to limit or stop projects in the past. Secondly, many tribes are not willing to agree to any investments that might put their land at risk.

The tribal capital reserves mentioned above are limited and far from sufficient to meet the need for investment capital. Economic development projects could be carried out using loans from banks and other financiers, but the Indian reservations are not creditworthy, because they cannot offer their trust lands as collateral. The reservation land as well as other tribal property and resource and reparation revenues are held in trust by the United States and therefore cannot be reclaimed in case of bankruptcy.

Only a very few tribal enterprises, such as the Ak-Chin Farms Enterprise at the Ak-Chin Indian Reservation in Arizona, have been successful in avoiding this legal obstacle. At Ak-Chin a skillful tribal chairwoman has set up a flourishing model farm for the tribe using private loans which involved a contract by which part of the harvest would be given to the creditors should the loan not be repaid on time, and in case of bankruptcy the creditors could cultivate the land themselves <32>.

There has been action to create an Indian Development Bank which would follow the example of the World Bank to grant favorable loans to economically underdeveloped Indian reservations, and in fact an American Indian National Bank under the direction of Indians was founded in Washington, DC, in the 1970s. This did not make it any easier for tribes to obtain loans, however.

Tribal enterprises lack investment capital and cannot use land collateral, but this is even more of a problem with individual tribal members and the limitations on entrepreneurship. There are unfortunately no official data or estimates regarding the number of savings accounts held by reservation Indians. On the basis of my conversations with people at the Fort Apache, Gila River, and Hualapai Indian Reservations, as well as my inquiries addressed to managers of local branch banks on or adjacent to these reservations,[9] it would appear that the number of reservation Indians who have savings accounts is exceedingly small. It is not clear, however, whether this is equally true of Indians on other reservations.

9. The Navajo Indian Reservation has had banks for some time, but on all other reservations of Arizona it was only during the past few years that banks were established. These were generally small branches of large private banks found in the state (see figs. 6.11 and 10.4). In 1985 the only bank on the Fort Apache Indian Reservation had around 3,000 accounts, very few of which were savings accounts <33>.

Fig. 6.11: Indian woman with traditional cradle and a baby backpack in front of a bank, Sacaton, Gila River Indian Reservation, Arizona, 1986

As for the creditworthiness of potential Indian entrepreneurs, the situation is essentially the same as for tribal enterprises. Generally speaking, most of their private property cannot be used to secure loans. This applies to land that, although allotted, is still held in trust, as well as income derived from this land. It also applies to housing on the reservations (see La Fontaine 1974, 32). Another reason why lenders are reluctant to grant loans is that they believe that most reservation Indians have had little experience in business and thus represent too great a risk.

The lack of available economic and social data on many reservations may also cause an entrepreneur to decide against setting up a business on a tribal land (Sorkin 1973, 120). This lack puts these Indian reservations at a disadvantage in comparison with many other underdeveloped regions of the United States, where state or local agencies systematically investigate and assemble the data essential for business analysis. The Bureau of Indian Affairs, however, despite its large number of civil servants, has no branch for statistics that aggregates data on a national basis, which means that it often cannot help out with precise data needed by a businessperson.

FACTORS POTENTIALLY FAVORABLE TO THE DEVELOPMENT OF THE RESERVATION ECONOMY

Except for the wealth of natural resources found on some reservations, the utilization of which can be highly profitable for primary economic activities, the factors favorable to the reservation economy are rarely sufficient to compensate for the disadvantages already discussed. For some industries, nevertheless, certain advantages can lead to a decision to locate on Indian lands, at least temporarily.

One attraction for industrial enterprise in Indian country is the large pool of cheap labor (see fig. 6.7) which a businesses can count on under very favorable conditions owing to the high rate of unemployment. This explains why certain industries which may not be competitive in other locations are now moving to the reservations. Here, as in the countries of the Third World, the low wage level (see Sorkin 1971, 98–99) seems assured by the general absence of labor unions and the tremendous unemployment. Although these low wages are also the result of the lack of skilled workers on the reservation, the employer does not bear the full burden of the Indians' lack of training, because the BIA has programs which fund on-the-job training for Indian workers for a certain time, or else reimburse as much as half the wage costs to the employer.[10]

Another advantage that a business may have if it is located on a reservation is that there are fewer environmental and developmental regulations. Furthermore, contrary to the situation in the rest of the country, the implementation of these reduced regulations on reservations is often not strictly enforced by either the tribal governments or the BIA.[11] In this respect conditions in Indian country are similar to those in Third World countries, where the authorities do not want to lose investors because of regulations that would be considered too stringent. There are advantages, however, in locating a business on reservation land instead of in Third World countries, since there are no custom taxes, duties, or import quotas which would hinder business; and since all business is done in dollars, there is no currency risk.

10. The enterprises on reservations that are not owned by the tribe generally pay only the legally required minimum wages, which, on a monthly average, are not much higher than welfare payments, and the workers in these enterprises have little prospect for promotion. This explains why many reservation Indians often display no particular interest in such work, which also means that the available supply of labor is actually not as great as it might appear.

11. For example, the lack of environmental regulation with regard to the mining of uranium on the Navajo and Laguna Indian Reservations (New Mexico) had, in some cases, disastrous consequences, resulting in disease and even death for both miners and others. In some cases the foundations of the houses of diseased people had been constructed with uranium mine tailings (see Peters 1987).

Equally favorable to the investor is the status of tax regulations on reservations and exemptions from some federal and most state taxes. Although some of these advantages may be lost when the tribe itself decides to levy taxes in order to raise additional funds for its budget, the advantage remains. This special tax status reflects the residual sovereignty of the reservations, and is very definitely a locational advantage for business.

Another factor that works to the advantage of the reservation economy is the partial breach of the restrictions on gambling, which has found a place on many reservations since the beginning of the 1980s. While commercial gambling is forbidden in the United States except for Nevada and Atlantic City in New Jersey, Indian reservations maintain that because of their special legal status there is nothing to stand in the way of most kinds of gambling on their territory (with the exception of roulette, which even today is not permitted on reservations). With this opportunity tribes are hoping for an upswing of their economy comparable to that achieved since 1931 in parts of Nevada, where impoverished economic conditions had previously prevailed. The future will show whether these plans can be carried out for the benefit of the tribes and reservation Indians.[12]

Finally there are a number of governmental programs available on reservations which promote and support economic development. These programs and policies involve high costs and include a variety of program activities. In some respects they are one of the few stimuli for the reservation economy and often provide the only reason why businesses are established on reservation land. Of the $2.8 billion of public funds that according to the U.S. Office of Management and Budget were granted to Indians in 1986, an average of $3,500 for every status Indian,[13] somewhat more than 8% was spent to establish and promote economic development projects.

For a long time this promotion of tribal economies was the exclusive responsibility of the BIA, although it paid little attention to this matter until 1934. Until then the Indian agency saw no particular reason to give economic support to the reservations, which, according to the official viewpoint, would simply create obstacles to the process of assimilating and civilizing American Indians. Although this attitude changed with the introduction of the IRA and the reservations then received funds from their trustees to establish tribal enterprises, the fact remains that until the 1960s the BIA did not grant much financial support to the reservation economy.

In conjunction with the so-called War against Poverty, however, as proclaimed by President Johnson in the mid-1960s, other federal agencies,

12. See the epilogue.

13. This amount is only slightly below the 1979 average income for rural American Indians (see table 5.1). Unfortunately, in 1986 more recent statistics on the average income of American Indians were not yet available.

including the Economic Development Administration (EDA), the Office of Economic Opportunity (OEO)[14] and the Small Business Administration (SBA), took the initiative to help reservation Indians. By 1984 there were already more than twenty federal departments or agencies which provided programs to support the development of tribal economies. The total cost of these programs was twice the level of funds granted by the BIA (Presidential Commission . . . 1984, 27). It must be said, however, that these programs, with their various objectives and regulations, were not properly coordinated, and that the major portions of the funds were used for administration.

It was not until passage of the Indian Financing Act (1974) and the establishment of the Indian Loan Guaranty Fund (1984), forty and fifty years after the IRA, that the BIA again seemed to realize its responsibilities for the development of the reservation economy. A basis was now established which made it possible to grant credit not only to tribal enterprises but also to individual Indian businesspersons who, as already explained, would otherwise find it difficult to obtain such credit.[15] At the same time a fund was created with which the BIA would reimburse banks for 90% of the loans to Indians in case an Indian business failed, through a loan guarantee. In view of the large backlog of demand for financing development, the amount of funds available from the BIA, however, was hardly adequate.

Even if federal subsidies are among the major factors which have framed tribal economies, these subsidies are not always helpful to Indian people. Successful overall economic development on reservations is not simply a question of money. Many expensive projects are carried out, not because there is a particular need for them or because they are desired by the tribe, but simply because public money for such projects is available and because non-Indian consultants and lawyers for the tribe acquire a certain prestige by obtaining this money. If some of the tribe's capital had to be spent on such projects, most of the Indian decision makers would undoubtedly be much more cautious. The same can be said of non-Indian businesses on reservation land. Many of them are attracted to the reservations because they can count on operating under large subsidies, and some might be there

14. The OEO, an office within the Department of Health and Human Services, became the Administration for Native Americans in 1977. The Economic Development Administration (EDA), created by the Public Works and Development Act of 1965, is part of the Department of Commerce. The latter agency, besides establishing various facilities to improve the infrastructure and tourism, provided the funding for thirty-seven industrial parks between 1966 and 1973. Only two of these were being used at half their capacity in 1975, eleven had only one leaseholder, and five were empty (General Accounting Office 1975, 10–11).

15. One exception was the Small Business Administration (SBA), which had programs to grant credit to Indian entrepreneurs even before 1974.

because they would probably not have a chance of a business of their own off the reservation. The consequence is that funds intended for reservation Indians often end up in other people's pockets, and once these funds have been used the reservation has lost one more enterprise. Many non-Indian businesses on the reservations close down after only a short time, which demonstrates the result of this policy.

7 The Role of Mining on Reservation Lands

Today American Indians are left with less than 3% of the land that once was theirs (Frantz 1990a, 28). Compared to the rest of the country, however, the value of mineral resources discovered on reservation lands is, in some cases, out of all proportion to the small amount of land that is still in Indian ownership, although estimates of the exact extent of these deposits differ greatly.

The identifiable oil reserves on reservation lands amount to approximately 4.2 billion barrels and 17.5 trillion cubic feet of natural gas, by some estimates, 3% of the known reserves in the United States (see Ortiz 1984, 152; Henkel 1985, 300). Joseph Jorgensen, on the other hand, estimates that these reserves in Indian country may reach from 3% to 10% of the U.S. total (1978, 51).[1]

The situation is similar with Indian coal reserves, which, according to a report of the Department of the Interior, amount to around 70 billion tons, around 5% of the total deposits of the country. The same report reveals that reservation lands contain more than 15% of the nation's low-sulfur, strippable coal (ED 1986, 102; <38>), while other sources suggest that as much as 15% of the total U.S. coal reserves can be found on the reservations,[2] and as much as 33% of the low-sulfur, strippable coal (Ortiz 1984, 152; Bregman 1982, 8).

The uranium resources on Indian lands are also of national importance, although the extraction of these deposits was stopped in 1982 because of the relatively high costs involved compared with other countries, and because of the low world market prices and small demand <38–40>. The uranium supplies discovered thus far on Indian reservations are estimated to be between 15% and 40% of the total U.S. reserves (see Arthur 1978, 1; Bregman 1982, 8; Wiley and Gottlieb 1982, 220), and most surveys assume that at least half of the U.S. reserves are to be found in Indian country (see Cook 1981; Goodman 1982b, 101; Henkel 1985, 300; McDonald 1980, 165),[3] allegedly

1. According to Bregman (1982, 8) and Goodman (1982b, 101) these reserves amount to 4% and 5% respectively.

2. McDonald (1980, 166) speaks of 10%, Goodman (1982b, 10) of 20%; and Henkel 30% (1985, 300).

3. There are even experts who believe that between 65% to 80% of the total U.S. uranium

Table 7.1. Income from mining, agriculture, forestry, and other land leases on Indian lands, 1981 or 1982 (in $1,000)

Tribe	Petroleum and natural gas (1981)	Coal (1982)	Other minerals (1982)	Forestry (1981)	Agriculture (1981)	Rent from businesses, land-use fees (1981)
Agua Caliente (CA)	—	—	—	—	—	6,715
Blackfeet (MT)	7,276	—	—	—	1,017	—
Cheyenne-Arapaho (OK)[a]	8,684	—	—	—	—	—
Colville (WA)	—	—	—	17,703	—	—
Colorado River (AZ, CA)	—	—	—	—	4,341	—
Coeur d'Alene (ID)	—	—	—	—	2,083	—
Crow (MT)	—	2,719	—	—	2,667	—
Five Civilized Tribes (OK)[a]	7,425	—	—	—	—	—
Flathead (MT)	—	—	—	1,452	—	—
Fort Apache (AZ)	—	—	—	3,844	—	—
Fort Berthold (ND)	1,219	—	—	—	—	—
Fort Hall (ID)	—	—	2,299	—	—	—
Fort Peck (MT)	6,601	—	—	—	2,350	—
Fort Yuma (AZ, CA)	—	—	—	—	1,443	—
Gila River (AZ)	—	—	—	—	2,594	—
Hollywood (FL)	—	—	—	—	—	1,195
Hoopa Valley (CA)	—	—	—	2,620	—	—
Hopi (AZ)	—	1,249	—	—	—	—
Jicarilla (NM)	21,169	—	—	—	—	—
Laguna (NM)	—	—	2,191	—	—	—
Kiowa, Comanche, & Apache, Fort Sill Apache (OK)[a]	31,431	—	—	—	4,760	—
Makah (WA)	—	—	—	1,883	—	—
Navajo (AZ, NM, UT)	39,065	7,411	4,408	3,522	—	2,314
Nez Perce (ID)	—	—	—	—	2,332	—

(Continued)

amounting to at least 11% of total world reserves (Ortiz 1984, 152), making total uranium deposits on reservations among the most important in the world.

Compared with the large oil, coal, and uranium reserves, the nonenergy minerals on the reservations are of secondary importance, with the exception of still unexploited tungsten, zeolite, and bentonite deposits,[4] for which

reserves are to be found on Indian reservations (Churchill 1986, cited in Peters 1987, 29; Los Angeles Times, June 10, 1977, part 1, 20).

4. In 1982 these deposits were estimated to amount to 2,000 tons, 137 million tons, and 734 million tons respectively <42>.

(Table 7.1 *continued*)

Tribe	Petroleum and natural gas (1981)	Coal (1982)	Other minerals (1982)	Forestry (1981)	Agriculture (1981)	Rent from businesses, land-use fees (1981)
Northern Cheyenne (MT)	1,329	—	—	—	—	—
Osage (OK)	71,333	—	—	—	—	—
Papago (AZ)	—	—	2,774	—	—	—
Pine Ridge (SD)	—	—	—	—	1,602	—
Quinault (WA)	—	—	—	4,005	—	—
Rosebud (SD)	—	—	—	—	2,105	—
Salt River (AZ)	—	—	—	—	1,652	—
Spokane (WA)	—	—	*	1,127	—	—
Southern Ute (CO)	2,789	—	—	—	—	—
Uintah and Ouray (UT)	13,624	—	—	—	—	—
Umatilla (OR)	—	—	—	—	1,241	—
Ute Mountain (CO, NM, UT)	1,715	—	—	—	—	—
Warm Springs (OR)	—	—	—	11,624	—	—
Wichita, Caddo, Delaware (OK)[a]	23,680	—	—	—	1,189	—
Wind River (WY)	28,128	—	—	—	—	—
Yakima (WA)	—	—	—	19,722	—	—
Total income[b]	194,248	11,379	11,672	65,542	31,376	10,224
% of income of all reservations	91.5	81.3	*	93.0	87.0	26.5

Note: The income from agriculture specified in this table comes entirely from rent. In forestry and mining there are additional proceeds from interest on stock and mining royalties. Only those tribes are listed which in 1981 in the respective lines of business earned more than $1 million.

[a]These tribes, although recognized by the BIA, have no reservation status.

[b]Only tribes with reservation status were considered.

*For the income of this reservation or the total income of all reservations, no data are available.

Sources: ARIL 1982, 56–88; <7>; <16>; Annual reports of U.S. Department of the Interior, BIA Division of Energy and Minerals, and BIA Division of Forestry; U.S. Department of Commerce 1980, (Population).

there will be an increasing demand in the future because of their worldwide scarcity <41> and their many uses in numerous industries. Copper, vanadium, and phosphate sites are also of importance, and there has recently been a significant amount of mining for these minerals on the Papago (Arizona) and the Fort Hall (Idaho) Indian Reservations (see table 7.1, column headed "Other minerals").[5]

5. Besides 1.1 million tons of vanadium, around 2.5 million tons of phosphate were mined on the Fort Hall Indian Reservation in 1982 (ED 1986, 107). The phosphate reserves on this reservation are estimated at around 734 million tons.

A HISTORICAL OUTLINE

That a number of reservations have a wealth of mineral resources today is not without a certain irony, because originally it was the intention of the responsible authorities to leave to American Indians only isolated areas which contained no mineral resources. At the time when the reservations were established some mineral resources such as uranium ore were still unknown, and no one had any idea of the importance that other minerals would have as a source for the nation's energy in the future. In the second half of the nineteenth century, attention was concentrated on gold and silver deposits and later on copper.

These mineral resources were taken into consideration when the boundaries of the new Indian reservations were drawn up. The officers who were entrusted with the creation of the Hualapai Indian Reservation (Arizona) in 1881, for example, doubtless believed that they had carried out their task properly when they informed the authorities in Washington that within the old tribal land of the Pai Indians they had found an area which, in their opinion, was "of little value for grazing and minerals."[6] They were careful not to include the territory to the west of today's reservation in the Cerbat Mountains which is rich in silver deposits.

Despite such precautions, if it was later discovered after a reservation was established that its territory did have valuable mineral resources, the U.S. government frequently resorted to simply changing the boundary. Such land cessions took place on the Great Sioux Reservation, for example, which was established in 1868 (see fig. 2.2). Following the discovery of gold deposits in the Black Hills, this massif was excluded from the Sioux tribal land (Schwarzbauer 1986, 76–77). Step-by-step changes of the boundary also took place on the original Fort Apache Indian Reservation in Arizona, which in 1896 was divided into two reservations of almost the same size, the San Carlos and Fort Apache Indian Reservations (see fig. 7.1). Silver deposits in the vicinity of Globe (fig. 7.2) and the discovery of copper in the immediate neighborhood of Clifton-Morenci, however, led to several land cessions in that area. The reservation of the Chiricahua Apaches in the southeast of Arizona (see fig. 2.5) was abolished altogether when, shortly after its establishment, large gold, silver, and copper deposits were located and potential pastureland was found. These discoveries made it impossible for the Bureau of Indian Affairs (BIA) to preserve the reservation—white ranchers and prospectors would have been up in arms if they had tried—and the land was turned over to non-Indians.

6. The quotation is taken from two letters which Lt. Col. William Redwood Price and Maj. Gen. O. B. Willcox sent to their superiors in Arizona and Washington, DC <43>.

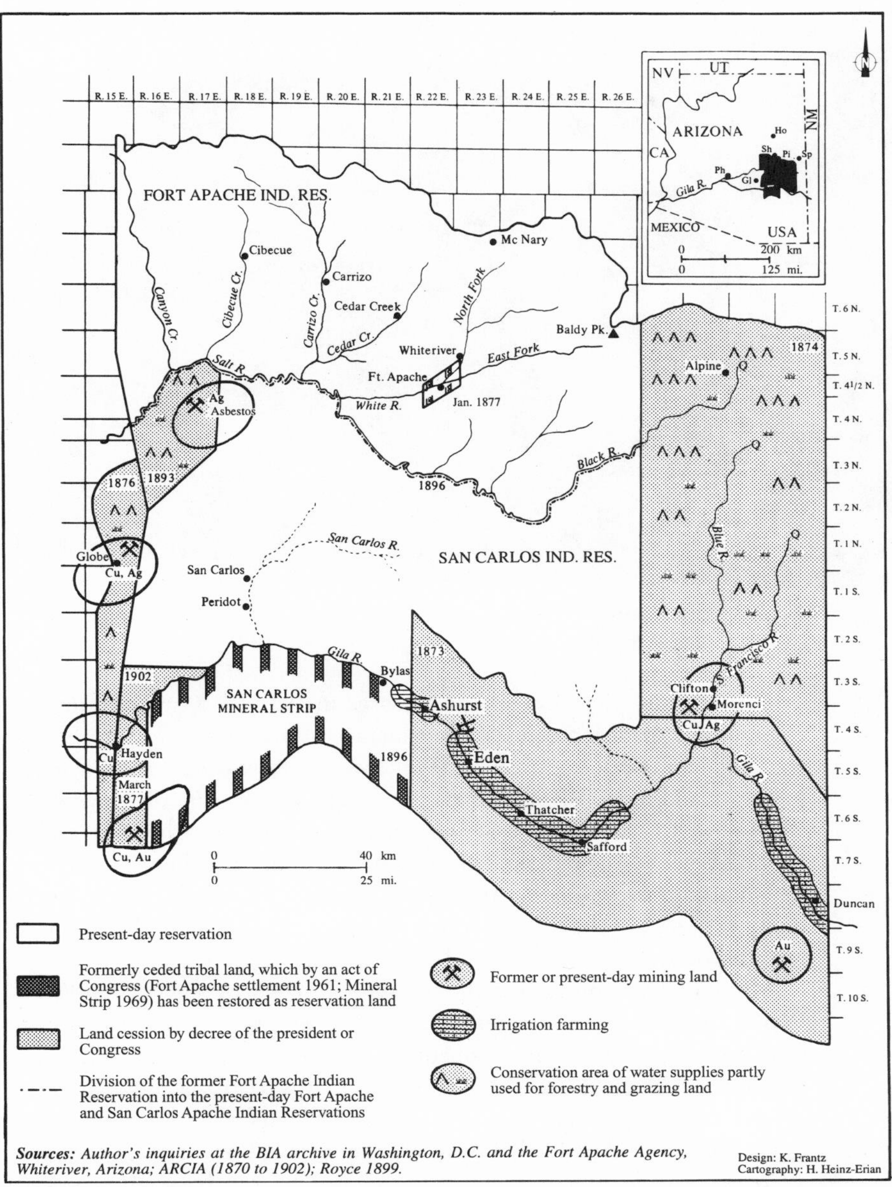

Fig. 7.1: Enforced land cessions on the Fort Apache and San Carlos Indian reservations, Arizona, stimulated by the economic interests of non-Indians

Fig. 7.2: Strip mining of the Magma Copper Company (Pinto Valley Mine) at Miami-Globe, Arizona, 1992

Furthermore, the Mineral Laws, passed in 1972 for a small number of reservations, made it possible for outsiders to have access to the greatly coveted Indian mineral resources quite independently, without bringing in the policy of land cessions.[7] All miners had to do was to communicate with the BIA in order to negotiate leases favorable to themselves which were valid for many years. The outcome of these negotiations was then usually ratified by Congress, while the tribes directly affected by the leases were generally bypassed in the negotiations or simply presented with the results as a matter of form.

More intensive mining on the reservations began after the First World War with the so-called Mineral Leasing Act (1920), which established a legal basis for large mining companies to operate on public land, including Indian reservation lands held in trust. There was particular interest in the oil and natural gas that was found on Navajo lands in the San Juan Basin (New Mexico) and on land of the Osage Indians of Oklahoma in the early 1920s. In order to extract these resources an official tribal council was es-

7. Strictly speaking, the mining of mineral resources in Indian country was already made possible as far back as 1834 by the Act of June 30. This act specified that all leases, including those having to do with mining, required the authorization of the BIA, but at the time few non-Indians made use of this regulation.

tablished on the Navajo Indian Reservation as early as 1921, even before the IRA. The primary function of this council was to sanction highly advantageous contracts with some of the large oil corporations, such as the Midwest Oil Company.

At the end of the 1930s the Mineral Leasing Act, valid throughout the United States, was modified with passage of the Indian Tribal Mineral Leasing Act (1938), which led to standardized mining regulations on all reservations. It was stipulated that the BIA was to advertise for bids for permits, which at first were limited to the prospecting and development of oil. The permit was to be awarded to the highest bidder and was to be valid for at least ten years. Moreover, until its amendment in 1982, this act further stipulated that Indian tribes did not have the right to exploit their own mineral resources. If mining was envisaged, it would be necessary to have a lease with a non-Indian business <42>.

Because of these regulations, as well as the BIA's central role in arranging leases and its close relationship to other agencies within the Department of the Interior and to the board of directors of many multinational mining corporations, American Indians were powerless to free themselves from being taken advantage of and being patronized well into the 1970s. Since the late 1930s long-term contracts had been signed between the BIA and the mining companies, usually with conditions that were most disadvantageous to the interests of reservation Indians. Fixed-price contracts for coal, based on its market value in the 1950s and 1960s, were negotiated on the Navajo Indian Reservation, but by the end of the 1970s the price for a ton of coal had risen 400% without the Navajos gaining any benefit from the increase (Jorgensen 1978, 52). Indian land leases were usually priced much too low and did not take either the quality or the quantity of the reservation's mineral resources into consideration, as is customary in contracts outside the reservations.

It was not until the 1970s, and specifically after the Indian Mineral Development Act (1982), that reservation Indians gained the right to extract their own mineral resources, and it gradually became possible to put an end to this patronage on the part of the BIA in accordance with the much publicized self-determination policy. This new self-confidence of reservation Indians, which also affected activities in the mining sector, was closely linked to the oil embargo of the OPEC countries in 1973, which caused an energy crisis in the industrial countries of the Western world. At that time even the United States, rich in resources, began to realize how dependent it was on other countries for the supply of energy. The nation began a search for new sources of energy that would make it independent of imports, and a number of large American mining corporations directed their attention once again to the Indian coal reserves, which, though already

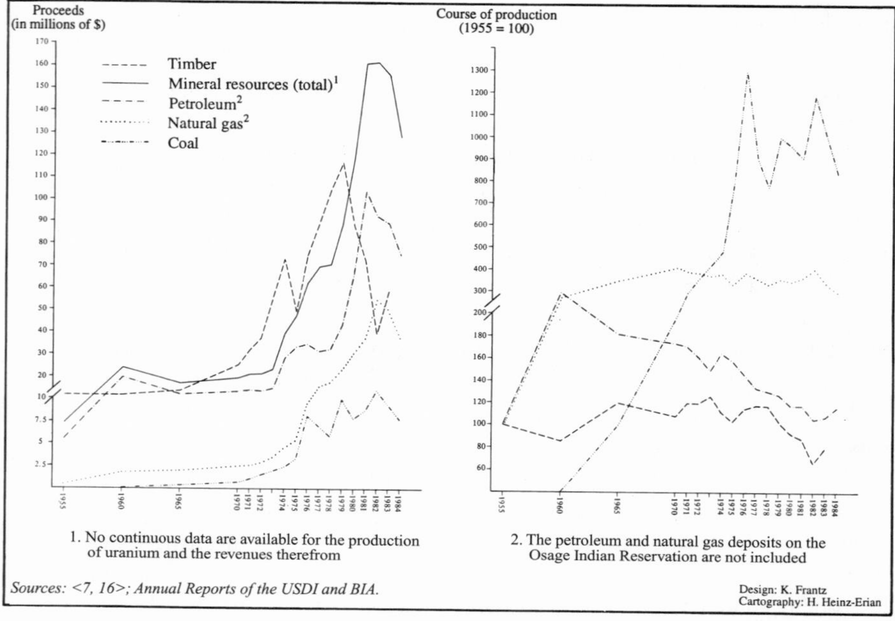

Fig. 7.3: Annual income and production of timber and mining on U.S. Indian reservations, 1955–1984

identified, were still largely untouched. Shortly afterward the U.S. government, on the basis of a study by the National Academy of Sciences (Trink 1989, 93), declared some regions, including land on several reservations, to be areas of national need.

To be better prepared for this renewed onslaught on Indian lands and to cash in on the increased demand for reservation resources, twenty-three of the reservations founded the Council of Energy Resource Tribes (CERT) under the leadership of the Navajos in 1975. This was envisioned as an "American Indian OPEC" which, with the help of its staff of specialists, was to assist the tribes in their negotiations with mining corporations. From 1973 to 1981 the prices of oil, coal, and uranium on the world market rose rapidly, and at the same time contracts were drawn up that were more advantageous to American Indians than they had been in the past, a fact which undoubtedly improved the income of the reservations. During these nine years the output of tribal oil production was cut by about one-fifth, to a 1956 level, while the earnings increased sevenfold, and the natural gas production declined about 4% during this period while tribal revenues from natural gas increased tenfold (see fig. 7.3).

Following a rapid increase in production, mining doubled the amount of coal while tribal revenues increased fourfold between 1973 and 1981. Because of a worldwide surplus of these raw materials, however, extraction declined after 1982 and revenues stagnated.

INCOME, EXPANSE, AND REGIONAL DISTRIBUTION

It is not surprising that large areas in Indian country are reserved for mining and prospecting in view of the significant resources. These mineral development areas, with only a few exceptions, have, however, been leased to multinational mining corporations which have also acquired a majority of the 10,400 prospecting permits, according to my inquiries in the Department of the Interior. The land that has been taken over for such permits amounts to approximately 6,200 square miles, or 7.5% of Indian reservation land in the United States.[8] Oil claims cover almost 90% of this land, a calculation that does not take into consideration the Osage Indian Reservation in Oklahoma, which is extremely rich in oil, but is also one of the few U.S. Indian reservations that will not release relevant statistics. A total of approximately 540 square miles of reservation land has been leased to coal mining corporations. A regional breakdown shows that on some reservations, such as the Fort Berthold (North Dakota), Southern Ute (Colorado), and Uintah-Ouray (Utah), more than 35% of reservation land is used to prospect for and mine mineral resources. On the Fort Peck Indian Reservation in Montana this proportion is 57% and on the Jicarilla Apache Indian Reservation in New Mexico 67% (see table 7.2).

The existence of these mineral resources and the extent of their development should not cause us to overlook the fact that the majority of Indian reservations must manage without this potential source of income. Moreover, even the reservations which have such rich resources very seldom profit from them in the end. The total revenue of reservations from mining, land leases, right-of-way-income and royalties, but not including the wages that Indians earn or the income of certain reservations for their share in production, amounted to about $217 million in 1981 (see table 7.1).[9] If the historic areas of Oklahoma are included, these

8. This leased land is not shown in such a way in the relevant BIA statistics that it can be distinguished from land used as pastureland. If all areas used for both purposes could be listed, the total percentage of leased reservation land would undoubtedly be higher than the 13.6% shown in table 3.1.

9. The figures for income from rent and royalties given in table 7.1 are not entirely accurate, because only a portion of the income from the mining of uranium ore is shown. No figures were available for the Spokane Indian Reservation in Washington, for example. Figure 7.3 does not include the mining of uranium ore in Indian country nor does it include the income from natural gas and oil on the Osage Indian Reservation in Oklahoma. BIA statistics show this income for only a few years.

Table 7.2. Leased land on selected Indian reservations, 1982

Reservation	Area (sq. mi.)	Percentage of tribal land	Percentage of leased land for: Exploitation of mineral resources[a]	Agri-culture	Business and other organizations
Agua Caliente (CA)	37	9.0	—	—	46.1
Blackfeet (MT)	1,470	27.3	28.3	13.5	0.1
Coeur d'Alene (ID)	106	31.0	—	78.0	0.2
Colorado River (AZ, CA)	415	97.8	< 0.1	48.5	3.8
Colville (WA)	1,659	96.2	< 0.01	2.3	< 0.1
Crow (MT)	2,378	26.8	20.7	63.8	0.01
Flathead (MT)	978	92.4	*	19.9	1.5
Fort Apache (AZ)	2,602	100.0	< 0.01	—	< 0.1
Fort Berthold (ND)	651	16.1	38.8	11.1	0.3
Fort Hall (ID)	816	49.8	1.4	31.3	2.4
Fort Peck (MT)	1,432	43.1	56.9	61.2	0.2
Fort Yuma (AZ, CA)	68	80.5	0.1	24.9	1.9
Gila River (AZ)	575	73.8	—	13.1	1.7
Hollywood (FL)	2	100.0	—	—	68.2
Hoopa Valley (CA)	142	96.8	—	< 0.01	0.6
Hopi (AZ)	2,439	99.9	1.3	—	0.1
Jicarilla (NM)	1,201	100.0	66.8	—	0.01
Laguna Pueblo (NM)	712	99.5	3.2	—	0.2
Makah (WA)	42	91.6	—	—	*
Navajo (AZ, NM, UT)	24,368	93.9	4.4	0.2	0.1
Osage (OK)	272	0.4	*	*	*
Papago (AZ)	4,462	98.6	0.3	< 0.01	0.7
Pine Ridge (SD)	2,789	40.0	< 0.01	9.6	< 0.01

(Continued)

earnings exceeded $300 million, 90% of which was derived from oil and natural gas production.

Only a very small number of tribes benefit from mining revenues, however, as already mentioned. Just thirteen reservations were involved in oil and natural gas development in 1983, for example, and test drilling was carried out on nine others <42>.[10] The great producers were the Uintah-Ouray (Utah), Jicarilla (New Mexico), Wind River (Wyoming), Navajo (Arizona, New Mexico, Utah), and Osage (Oklahoma) Indian Reservations, where the tribal revenues were between $13 million and $71 million, as

10. Prospecting and drilling for oil and natural gas was carried out on the lands of 31 federally recognized tribes in 1983. Today they produce 40 million barrels of oil, i.e., 1.2% of the total U.S. production, and 233 billion cubic feet of natural gas, 1.4% of the total U.S. production.

(Table 7.2 *continued*)

Reservation	Area (sq. mi.)	Percentage of tribal land	Percentage of leased land for: Exploitation of mineral resources[a]	Agri-culture	Business and other organizations
Pojoaque Pueblo (NM)	18	100.0	—	—	0.9
Potawatomi (WI)	18	96.6	—	—	0.5
Quinault (WA)	200	5.6	—	—	0.9
Rosebud (SD)	1,492	55.3	—	32.1	0.2
Rumsey Rancheria (CA)	< 1	100.0	—	35.7	6.5
Salt River (AZ)	76	50.4	2.4	35.4	1.4
Skull Valley (UT)	27	99.1	—	—	25.7
Southern Ute (CO)	477	99.1	37.6	9.2	< 0.01
Spokane (WA)	206	78.8	*	4.2	0.9
Swinomish (WA)	5	15.7	—	7.9	23.3
Umatilla (OR)	132	19.5	0.01	37.0	0.6
Uintah and Ouray (UT)	1,596	98.6	36.4	3.2	0.2
Ute Mt. (CO, NM, UT)	933	96.4	28.5	0.01	1.4
Warm Springs (OR)	1,005	91.9	< 0.01	0.3	0.3
Wind River (WY)	2,951	95.0	23.9	2.0	0.2
Winnebago (NE)	42	14.6	—	81.1	0.01
Yakima (WA)	1,800	78.2	0.1	21.1	0.3
Yavapai (AZ)	2	100.0	—	—	21.1

Note: For most reservations information about the area of forestry land is not available.

[a]Areas under lease for the extraction (potential or actual) of mineral resources or for agricultural purposes frequently overlap and therefore cannot be added up to a meaningful total.

*Government agencies have no information about the land under lease.

Sources: ARIL 1983, 4–30 and 56–87; Author's data from the BIA Division of Energy and Minerals, Washington, DC.

well as some tribes in the historic areas of Oklahoma, such as the Kiowa and Comanche Apache in Fort Sill, and the Wichita, Caddo, and Delaware, which, however, have no reservation of their own (see tables 7.1 and 7.3).

Ten mines of the Crow (Montana), Navajo (Arizona, New Mexico, Utah), and Hopi (Arizona) Indian Reservations (see fig. 7.4) achieved a total production of 28.4 million tons of coal in 1983, 3.6% of the total U.S. production that year (ED 1986, 100). On two other reservations, the Southern Ute (Colorado) and the Uintah-Ouray (Utah), exploration concessions for production have been granted. The coal reserves of the Northern Cheyenne (Montana), are estimated at 5 billion tons, but the tribe has withdrawn the exploration rights of the Peabody Coal Company, the country's largest coal producer.

The mining of uranium ore took place on only three reservations at the

Table 7.3: Estimated per capita income from leases for selected Indian reservations, 1981 or 1982

Annual income ($)	Reservation	Number of American Indians	Per capita income ($)	Source of revenues (in $1,000)					
				Petroleum and natural gas[a] (1981)	Coal[a] (1982)	Other natural resources[a] (1982)	Forestry[b] (1981)	Agriculture[c] (1981)	Rent from businesses, land-use fees (1981)
> 10,000	Agua Caliente (CA)	69	97,319	—	—	—	—	—	97,319
	Osage (OK)	4,749	15,214	15,021	—	7	—	153	32
	Jicarilla (NM)[d]	1,715	12,347	12,344	—	—	—	—	3
5,000–9,999	Uintah and Ouray (UT)	2,050	6,875	6,646	—	2[f]	—	168	—
	Wind River (WY)	4,159	6,794	6,763	—	19[f]	—	12	—
	Skull Valley (UT)	13	6,118	—	—	—	—	—	6,118
	Warm Springs (OR)	2,004	5,821	—	—	1[f]	5,800	2	18
	Colville (WA)	3,500	5,102	—	—	6[f]	5,058	28	10
2,500–4,999	Yakima (WA)	4,983	4,911	—	—	5[f]	3,958	798	150
	Coeur d'Alene (ID)	538	4,413	—	—	—	538	3,872	3
	Quinault (WA)	943	4,294	—	—	—	4,247	—	47
	Southern Ute (CO)	855	3,327	3,262	—	—	—	65	—
	Hollywood (FL)	416	2,873	—	—	—	—	—	2,873
	Colorado River (AZ, CA)	1,965	2,563	—	—	27[f]	—	2,211	325
1,000–2,499	Makah (WA)	803	2,469	—	—	—	2,345	—	124
	Fort Peck (MT)	4,273	2,142	—	1,590[f]	2[f]	—	550	—
	Fort Hall (ID)	2,542	2,121	—	—	744[e]	—	1,364	13
	Crow (MT)	3,953	1,764	394[f]	688	7[f]	—	675	—
	Ute Mt. (CO, NM, UT)	1,111	1,586	1,544	—	—	—	—	42
	Yavapai (AZ)	66	1,565	—	—	61[f]	—	—	1,504
	Fort Yuma (AZ, CA)	1,105	1,545	—	—	52[f]	—	1,361	132
	Blackfeet (MT)	5,525	1,507	1,317	—	—	—	184	6

	Hoopa Valley (CA)	411	1,401	—	—	—	1,370	1	30
	Pojoaque Pueblo (NM)	94	1,384	—	—	—	—	—	1,384
	Umatilla (OR)	908	1,380	—	—	—	—	1,366	14
	Spokane (WA)	1,050	1,114	*	—	*	1,073	19	22
500–	Swinomish (WA)	414	856	—	—	—	—	29	827
999	Salt River (AZ)	2,624	830	—	—	68[f]	—	630	132
	Papago (AZ)	6,958	821	—	—	800[e]	—	—	21
	Fort Berthold (ND)	2,640	817	462	—	13[f]	—	342	—
	Potawatomi (WI)	220	743	—	—	—	—	740	3
	Winnebago (NE)	1,140	730	—	—	—	—	730	—
	Rumsey Rancheria (CA)	11	727	—	—	—	—	727	—
	Laguna Pueblo (NM)	3,564	636	—	—	615	—	—	21
	Fort Apache (AZ)	6,880	593	—	—	1[f]	564	—	28
	Navajo (AZ, NM, UT)	105,031	542	372	71	43[e]	34	—	22
	Flathead (MT)	3,771	505	*	—	*	385	93	27
< 500	Gila River (AZ)	7,070	413	—	—	—	—	367	46
	Rosebud (SD)	5,687	371	—	—	—	—	370	1
	Hopi (AZ)	6,593	293	—	268[f]	—	—	—	25
	Pine Ridge (SD)	11,946	138	—	—	3	—	134	1

[a]Income from federal contributions and fees paid by non-Indian enterprises.

[b]Income from interest on the market value of wood that has not yet been cut, in part from non-Indian business.

[c]Income from rent from largely non-Indian businesses.

[d]The tribal oil company is not included in these figures.

[e]Here the mineral resources are phosphate and vanadium (Fort Hall), copper (Papago), and uranium ore (Laguna Pueblo, Navajo). These figures are for the year 1983.

*The reservation gave out no information about its income.

Sources: ARIL 1982, 56–88; ED 1986, 113–114; U.S. Department of Commerce 1980, (Population, American Indian Areas and Alaska Native Villages), 16–20; Author's data from the BIA Division of Energy and Minerals, Washington, DC, and from USGS–Albuquerque Office.

Fig. 7.4: Strip coal mining of the Peabody Company, Kayenta, Navajo Indian Reservation, Arizona, 1992

beginning of the 1980s, the Navajo (New Mexico), Laguna (New Mexico), and Spokane (Washington). As late as 1980, just before mining for uranium on Indian lands ceased, some 24% of the nation's uranium production, approximately 12% of the production of the Western industrial countries (Bregman 1982, 8), came from these reservations. On twenty-two other reservations, including the Acoma (New Mexico), Zuni (New Mexico), and Hualapai (Arizona), field studies and exploratory drilling were carried out by prospectors. Otherwise during the early 1980s only the production of phosphate and vanadium (Fort Hall Indian Reservation, Idaho) and copper (Papago Indian Reservation, Arizona) was of any importance. In each of these cases revenues of more than $2 million were achieved for these tribes.

The Osage, Jicarilla, Uintah-Ouray, and Wind River Indian Reservations are the ones that appear to have profited most if we add up the income from mining permits and land leases to large mining corporations and consider this in relation to the population of these reservations. On these reservations, which account for only 3.7% of all U.S. reservation Indians, a theoretical per capita income of between $6,000 and $15,000 was achieved in 1981 (see table 7.3). These figures are theoretical, of course, because the individual Indians did not, in fact, receive these amounts as a direct payment, as the funds went to each tribal government. An additional 11% of reservation Indians would have received a theoretical income of between $1,000 and $6,000 from mining developments on their respective reservations at that time.

THE ADVANTAGES, DISADVANTAGES, AND PROBLEMS OF TRIBAL MINERAL DEVELOPMENT

There can be no doubt that the forced development of mineral resources on reservations by non-Indian mining companies has brought a number of advantages to the reservations involved, advantages which they would not have had without letting these businesses operate on their land. Although reservations are now permitted to extract their energy resources independently as established by law in 1982, only one, the Jicarilla Apache in New Mexico with its own oil company (Jicarilla Energy Company), has thus far made use of this regulation. Other reservations that intended to start the independent development of their mineral resources have failed so far because of a lack of capital and expertise. They have been able to obtain income only through royalties and rent, or in some cases through the taxation of the licensed mining companies. Joint ventures, such as the one that the Navajos entered in order to share profits from the extraction of uranium, are still the exception.

Income from mining has helped some tribes develop the infrastructure of their reservations, raise the level of education of their members through scholarships and training programs, and reduce the high rate of unemployment by generating new jobs. The non-Indian mining corporations also need a large number of workers, some of whom can be recruited from the reservation population. The total amount of the mining-related wages the Navajos received in the mid-1970s, for example, equaled all other mining revenues (Owens 1978, 55). Similar employment gains were obtained on the Laguna Indian Reservation in New Mexico, where more than 500 Pueblo Indians were active in mining. There can be no doubt that these personal incomes have also benefited the reservation economy.

It still remains questionable, however, whether one can accept the optimistic view that mining would provide the appropriate means to put an end to most of the deplorable socioeconomic conditions on some reservations, for there is a long list of problems and disadvantages. It is a well-known fact, for example, that the prices of oil, coal, uranium, and other mineral products are always subject to great fluctuations which can have disastrous consequences for an economy that is based almost exclusively on these products. This is true not only for developing countries, but also for various Indian reservations in the United States where a substantial portion of the economy is based on mining. Navajo mining revenues, for example, amounted to 94% of the tribal budget in 1954, 50% in 1971, 49% in 1973, and 56% in 1981 (Aberle 1983, 650–651; Hofmeister 1978, 335; <44>).[11] For the Jicarilla Apaches (New Mexico) mining revenues amounted to 81% in 1981, and for the Pueblo Indians of the Laguna Indian Reservations (New Mexico), 92% <45, 46, 39>. These reservations have serious financial difficulties when there is a decline in the price of raw materials on the world market, as shown in the case of the Laguna Indian Reservation with its uranium reserves. The reduction of prices of uranium from $43 a pound to around $20 between 1978 and 1982 <39> brought mining to a complete standstill on this reservation, while tribal budget funds plummeted drastically and more than 500 Indian miners lost their work.

Another problem is that the tribes usually have no real control over the development of their natural resources, even today, because in all important decisions it is still the BIA that has the last word. Apart from the fact that the BIA's very few mining experts are hopelessly overloaded with work, it finds itself constantly involved in a conflict of interests, both with various other federal agencies and with some of the largest corporations of the country. On the one hand the BIA as trustee must represent the interests of the reservations, while on the other hand, as part of the Department of the Interior, it is surrounded by other agencies that have a particular interest in developing the country's natural resources. Such agencies include the U.S. Geological Survey (USGS) and the Bureau of Land Management (BLM), whose views often carry more weight than those of the BIA, which despite its many employees has little influence. The secretary of the interior will usually take the side of the USGS and the BLM in making decisions. Another relevant factor is that outgoing top officials of the BIA sometimes join the board of directors of multinational mining corporations. With their contacts and inside knowledge of tribes they can then smooth the way for their new employers in difficult negotiations with the Department of the Interior. In view of these traditionally close connections and links,[12] it is not surprising that contracts with particularly advantageous conditions are often granted to certain companies or that the extraction of mineral resources is sometimes carried out with minimal federal control.

11. This statement refers to income earned by the tribe itself, to which should be added budget allocations from the BIA and IHS as well as grants and contributions from other federal agencies and private organizations. Income from these sources amount to many times the tribe's revenues.

12. A financial disclosure statement filed by the head of the BIA during Reagan's presidency showed that he owned stock or interest in three oil-related businesses in Indian country (Arizona Republic 1987, 6). After leaving office Reagan's first secretary of the interior and his acting assistant secretary for Indian affairs founded a business consulting firm whose most important clients included mining corporations which either had already obtained the surface leases and permits on reservation land or wanted to obtain them (Washington Post 1985).

It is an open secret that in Oklahoma and other states that are rich in oil, where Indian territory is widely interspersed with non-Indian land, large quantities of oil are illegally drained from under Indian lands by their white neighbors without the BIA taking any steps to prevent it. According to U.S. government estimates, prior to the 1980s, the government and federally recognized Indians lost approximately $5.7 billion in oil reserves in this way (Arizona Republic 1987, 11). The BIA has no precise information about the annual amount of oil and natural gas that the licensed companies take from leased Indian lands, because the meters are read only at very irregular intervals owing to the lack of qualified employees, and quite often the companies even refuse to allow the BIA to do that. The BIA must therefore rely on information provided by the oil producers, which is then passed on in good faith. A congressional report found that a consequence of these conditions was that federal agencies failed to collect as much as $5.8 billion for reservation Indians between 1979 and 1985 (Arizona Republic 1987, 4–6). If this sum is correct, it would amount to roughly four times the total oil revenues to Indian reservations during these seven years.

Moreover, the BIA does not see to it that companies engaged in mining and generating energy observe the precautionary regulations regarding the health of the local population and the conservation of the environment. Such regulations are largely disregarded by these companies because of the expense involved and lack of monitoring. Large quantities of fly ash and sulfur dioxide are emitted onto the Navajo Indian Reservation from the Four Corners Power Plant, which is fueled by Navajo coal and owned by the Arizona Public Service Company in Fruitland, New Mexico. The same happens because of the 2,300-megawatt power plant in Page, Arizona, which is linked by a railway to the Peabody Coal Company mines eighty miles away on the Navajo Indian Reservation (see fig. 7.4). These reservations are left with the pollutants, while a major part of the inexpensive electricity goes to the consumer markets in Phoenix and southern California hundreds of miles away. The Navajos, however, must for the most part buy back the energy that was produced with their coal because they own only 12.5% of the electricity generated at the Four Corners Power Plant. The remaining coal from the strip-mining areas of the Black Mesa (Navajo and Hopi Indian Reservations) is crushed on the spot and, by means of a pipeline that is supplied with groundwater,[13] the coal slurry is shipped to the 1,600-megawatt power plant in Bullhead City, Arizona, more than 270 miles away, where it is first dried and then burned. The energy generated there is also produced for outside markets. Because

13. For this purpose water is pumped up from deep wells at a rate of about 2,700 gallons per minute as far down as 3,300 feet; the Navajos receive little compensation for this groundwater.

no measures are generally taken to recover the strip-mined land, further harm is done to the Hopi and Navajo tribes, harm that is particularly distressing for the local people who have sheep. It will take decades for the prime vegetation on these extremely arid plateaus to return so they can again be used as pastureland, and that may never be possible.

The fact that the BIA neglected its supervisory responsibilities had particularly negative consequences for the reservations on which non-Indian mining companies were prospecting for uranium ore until recently. When mining was begun there in the early 1950s, people were not aware of the attendant health risks. During the next thirty years the health of many American Indian miners suffered as a result and some even died. People living adjacent to these mines were also affected,[14] yet even after the risks became known, the precautionary measures taken by the BIA were inadequate. A visit to the little Indian village of Paguate in the northern part of the Laguna Indian Reservation in New Mexico revealed that at least until the spring of 1983, the piles of radioactive waste from the Jackpile Mine, which was once the world's largest surface uranium mining operation, extended right to the western boundary of this settlement. Other settlements near tailings were found on the Navajo Indian Reservation (New Mexico) at that time. On both these reservations the tailings were used for road constructions and building houses (Peters 1987, 30; <39, 47>), creating an even more permanent hazard to residents.

Without any doubt, the people who really profit from Indian mining live far away from reservations. Except for the Jicarilla Energy Company, the mineral resources in Indian country are mined by white entrepreneurs and then exported. There is no processing of minerals on the reservations. As a result the multiplier effect that could be gained through mineral resources is lost for the local tribal economy. This lost opportunity is reflected, among other things, in the high rate of unemployment to be found even among those tribes for whom mining is of paramount importance and one of the principal sources of income. The unemployment rate of the Jicarilla Apaches, Navajos, and Laguna Pueblo Indians was between 44% and 53% in 1981, when the mining boom was coming to an end. In the case of the Navajo and Laguna tribes this was clearly above the average of 46% for all reservation Indians (see U.S. Department of the Interior, BIA 1982).

In the last analysis, the main profit made from the Indian's natural resources is siphoned off not only by the large U.S. mining corporations and

14. The Shiprock Mine on the Navajo Indian Reservation in New Mexico employed a total of 150 Navajo miners from 1952 to 1970, for example. A survey showed that ten years after the mine closed down, 38 of these miners had died from lung cancer caused by radiation exposure. Another 95 workers suffered from respiratory diseases and cancer (Samet et al. 1984 and Nafziger 1976, cited in Peters 1987, 30).

electric power companies, but above all by the states in which the reservations rich in resources are located. The Navajo case can be used as an example. In 1976 the total income of the Navajos from royalties, land rents, and right-of-way income amounted to $12.8 million, yet in the same year the State of Arizona earned $18.5 million in taxes from non-Indian mining corporations and electric power companies there (Owens 1978, 53), without the Navajos receiving any appreciable sum of money or other service in return from the state.

8 The Critical Issue of Tribal Water Rights

> There has been a lot said about the sacredness of our land which is our body; and the values of our culture which is our soul; but water is the blood of our tribes, and if its life-giving flow is stopped, or it is polluted, all else will die and the many thousands of years of our communal existence will come to an end.
> —Frank Tenorio, Governor of Pueblo San Felipe, 1975

For a long time the relations between white America and American Indians were largely determined by the battle for Indian land. It is generally agreed that this battle turned out to the disadvantage of the indigenous people. Following the shrinking of reservation land, interest concentrated on the reservations' resources, above all on grazing lands and forests as well as the mineral resources, which were exploited by non-Indian business enterprises through long-term leases which were often exceedingly advantageous to non-Indians. However, during the past few decades attention has increasingly concentrated on the Indian water reserves, which are becoming more important all the time for the young, rapidly growing urban agglomerations in the arid areas of the United States.

These present-day conflicts, however, no longer have anything in common with the skirmishes between federal military forces, certain of victory, and hard-pressed tribal warriors, fights that are often still romanticized. Nowadays state and federal courts of law provide the arena for these vehement and costly clashes of interest, and the "modern warriors" are embodied in a host of well-paid lawyers who carry on the struggles for the opposing parties. The extent and the explosive nature of this conflict is shown by the fact that at the beginning of the 1980s the number of public water battles between individual Indian reservations and non-Indian communities or states had already risen to over a hundred (Denver Post 1983, 18), and in 1986 there were nearly sixty major lawsuits, chiefly concerning water, that remained to be settled <48>. These water disputes threaten the very existence of many reservations, for today just under 70% of the reservation Indians and approximately half of the reservations are to be found in arid or semiarid regions (see fig. 8.1).

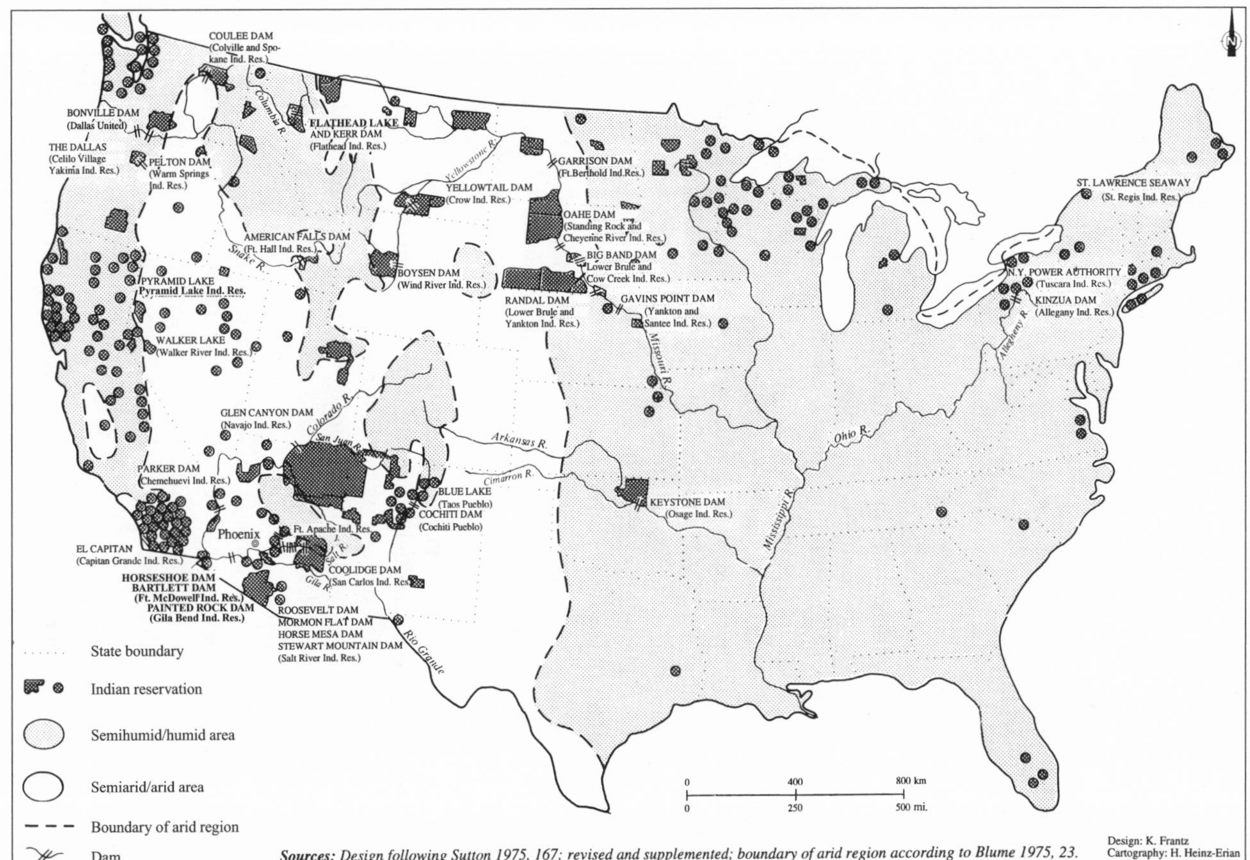

Fig. 8.1: The distribution of Indian reservations between semihumid and semiarid regions of the United States, and dam projects having a direct or indirect effect on the territory of reservations

Much is at stake in these legal actions for the non-Indian parties, as well as for several large cities of the Southwest. Until recently it seemed that there was no limit to the insatiable desire for land and water in the urban developments. Today the situation is different. While it is true that reserves of land for potential development are generally sufficient for these sprawling metropolitan areas, they are nevertheless reaching the limits of their growth because of their excessive consumption of water. Reservation Indians have established secure water rights, and are now beginning to realize that they can keep the water for themselves and successfully defend those rights. There are also increasing difficulties for the intensive irrigation farming that has spread throughout the arid regions of the United States during the past hundred years, owing to the water shortage that will escalate if the Indians' water rights are observed.

In the course of the second half of the nineteenth century aboriginal water rights were largely ignored by the growing number of people who settled in the arid West. This led to a drastic reduction of Indian water resources, which received much official support through the Homestead Act (1862), the Mining Act (1866) and the Desert Land Act (1877). In carrying out these policies, so advantageous to the settlers, the federal government was, however, neglecting its obligations to reservation Indians. Except for a few initiatives and projects, including the construction of small irrigation canals on the Colorado River and Fort Apache Indian Reservations, the government neither took action to uphold specific Indian rights nor made adequate arrangements for tribes to use their water for the development of agriculture on their reservations.

Because the white settlers continued to occupy the land, the local water resources were gradually depleted around the turn of the century in the most valuable portions of the arid regions of the United States. The Federal Reclamation Act (1902) aimed at the reclamation of arid and semiarid land in the West, and sought to counteract this phenomenon with a heavily capitalized program for the development of local water and energy reserves. As a result, major construction projects to build large dams, hydroelectric power plants, and irrigation canals were carried out which at first were intended primarily for local river basins. The reservation Indians were, of course, the people who suffered most from the numerous diversions of the surface water and from the "damming of the American West" (see fig. 8.1), since their legal claim to the water was generally ignored. The vast areas of standing water meant that a great deal of water evaporated. Moreover, through the inundation, the Indians lost another 772 square miles of reservation land. They also lost fishing rights established by treaty: for example, the residents of the Yakima Indian Reservation in Washington lost part of their traditional salmon-fishing area through the construction of the Dallas Dam (Hunt 1970, 98).[1]

Soon even these numerous dam projects, along with the massive tapping of local groundwater, no longer sufficed to satisfy the ever increasing need for water and to sustain the dynamic growth of the non-Indian population and the economy in the growing metropolitan centers of the arid West. The large cities of the Southwest, such as Los Angeles, Phoenix, and Tucson, as well as adjoining areas of intensive irrigated farming, were therefore compelled to construct large-scale canal systems, often hundreds of miles in length, in order to import the water they required. These canals were connected to neighboring river basins. At the end of the 1980s the last of these great projects was inaugurated, the Central Arizona Water Project (CAP), which was to supply the conurbations of southern Arizona with water from the Colorado River. In this case, too, the water rights of reservation Indians were violated. In view of the remarkable boom in these urban regions, which is expected to continue, the water conflict between these metropolitan areas and some Indian reservations will very likely become still more critical. In the future a new distribution of the increasingly scarce water resources will be inevitable.

INDIAN WATER LAW: DIFFERENT INTERPRETATIONS

Indian reservations and their indigenous residents have a unique status under U.S. law. This status, however, appears to be ambiguous in a number of legal areas and is therefore a controversial issue for many Americans. This is particularly true with regard to the question of Indian water law, which is exceedingly complicated, not only for outsiders but even for specialists who have looked into the subject. A definitive clarification of the Indian water law on the part of the U.S. government, something that has long been called for both by the Indian reservations and the states concerned, is still lacking and therefore has become an urgent priority for all parties because of the increasing clash of interests regarding water rights.

In order to understand the disputes which have arisen in connection with Indian water rights and to see them rightly within the general American water use context, it is important to call attention briefly to two legal principles which are of fundamental importance in U.S. water law. They are the Riparian Doctrine (Hofmeister 1976, 513), which is applied in the water-abundant East, and the Doctrine of Prior Appropriation (ibid.), which regulates the allocation of water in the more arid regions of the West.

1. In 1958 the U.S. government made a payment of $4 million as compensation for this loss. This payment, however, did not satisfy some of the Yakima Indians, who would have preferred to uphold their ancestral rights.

The Riparian Doctrine, which evolved from old Anglo-Saxon laws <50> and which at first was applied throughout the United States, states that anyone owning land adjacent to a lake or stream has a right to the use of this water equal to that of all other riparian neighbors (see Canby 1981, 239; Estes 1982, 25; Peyer 1985, 282; Quinn 1968, 110; Sutton 1975, 162–163; Wehmeier 1975, 118). This right, linked to the landownership, does not lapse even if it is not exercised. The only restriction is that each person who has access to the water must consider the "reasonable" use of any other riparian neighbor, and whenever there is a shortage of water must reduce the consumption proportionally.

As the frontier advanced to the arid regions of the West, however, it soon became apparent that the Riparian Doctrine was not suitable for the prevailing conditions. For example, the owners of gold and silver mines located far away from a source of water were obliged to make substantial investments in pipelines and canals through which water could be brought to their sites. The beginnings of intensive irrigation farming were also very costly, as such farming could not always be practiced in the immediate vicinity of a stream. Therefore pioneers who needed water for both mining and irrigated agriculture sought to ensure for themselves continued and adequate supplies of water before they undertook large capital investments (see Canby 1981, 240).[2] In most western states these conditions led to the initiation of the Prior Appropriation Doctrine,[3] a water law which, in a similar way, was also applied in many arid regions of the former Spanish colonies in the New World.

Contrary to the Riparian Doctrine, which expresses a community spirit, the Doctrine of Prior Appropriation developed in an atmosphere of competition. Here the prevailing attitude is "first come, first served," or "use it or lose it" (see Canby 1981, 240). What this actually means is that for landowners, whether they live adjacent to or far from a stretch of water, the important consideration is at what time they first made a claim to the respective state for a particular amount of water, and how far back they can prove that they made use of a fixed quantity of water. According to this legal system, the first user of water from streams and lakes, or groundwater which is found closely subjacent to streams, will always have priority over latecom-

2. It is understandable that the application of the Riparian Doctrine to the arid regions of the West should first have been questioned in California at the time of the gold rush of 1848–1849.

3. The Prior Appropriation Doctrine is to be found in its purest form in Colorado, but it is also found in Arizona, Idaho, Montana, Nevada, New Mexico, Utah, and Wyoming. In nine other states, California, Kansas, Nebraska, North and South Dakota, Oklahoma, Oregon, Texas, and Washington, the Riparian Doctrine and the Prior Appropriation Doctrine are combined (see Canby 1981, 241; Wehmeier 1975, 118).

ers, who can use water only if they do not interfere with the uses of the earlier appropriators (see Hundley 1978, 459). When there is a water shortage due to drought, any later appropriator has to manage with less water or no water at all. A further consequence is that ownership of land is not automatically linked to water rights, a fact which can have a decisive influence on the property value. Furthermore, the appropriators in question must be able to prove that their consumption of water serves a useful purpose (a matter which can be investigated by the responsible authorities in case of doubt) and they must continue to put water to beneficial use throughout the years, or, after a period determined by the state, they risk losing their water supply (see Merrill 1980, 45; Templer 1978, 12).[4]

From all this most of the reservation Indians conclude that it should not be too difficult to defend aboriginal water resources from private encroachments, for today many tribes still inhabit land which has been their home for as long as anyone can remember. Indeed, they were there long before the arrival of the Europeans. Even if the white settlers interpreted the water laws which they themselves established in a way that ignored the prehistoric, that is to say, the pre-Columbian existence of American Indian peoples, it is still an undisputed fact that many reservations in the United States were created at a time when whites had not yet made any legal claims to water.

In accordance with the Doctrine of Prior Appropriation in most cases the water question should have been decided to the advantage of American Indians, provided that they had been able to show that they had made beneficial use of their water. In reality, however, this doctrine was very often ignored when the issue was Indian water. Moreover, this disregard of existing water rights had the full support of all the western states, in spite of the fact that when the states were founded, they were obliged to commit themselves to respect the water rights on federal land, which also included Indian reservations created out of public domain. Therefore, the U.S. government, in accordance with its trust obligations for Indian reservations, found itself compelled, at least officially, to defend by law the Indian interests in the Supreme Court. This was done for the first time in 1908 in the form of the Winters Doctrine, which in its significance can be regarded as a sort of Magna Charta of Indian water rights.

The Winters Doctrine is named after a farmer who settled in northern Montana not far from the Milk River (see fig. 8.1) in the vicinity of the present-day settlement of Chinook at the close of the nineteenth century). In order to practice farming with any prospect of success in this semiarid

4. In Arizona, for example, this claim expires after five consecutive years. The amount of water in question can then be newly allocated.

region, however, he and a number of other farmers in the vicinity were obliged to dig irrigation canals which would divert water from the Milk River to their fields. These canals were built in 1889, one year after the Belknap Indian Reservation was established, a few miles downstream, adjoining the Milk River. In the 1890s the residents of the Fort Belknap Indian Reservation also required an increasing amount of water. These people, the Gros Ventre and Assiniboine Indians, used to be nonsedentary tribes living chiefly from hunting and gathering, but by the close of the century, with the active support of the BIA, their farms were flourishing and they were engaged in successful cattle ranching. The radical transition process from a nonsedentary lifestyle of hunter-gatherers to one of sedentary pastoralism and agriculturalism (see Ross 1981, 199) was suddenly interrupted in 1904 and 1905 when, because of a long period of drought and upstream non-Indian irrigation, they found themselves short of water. The local superintendent of the BIA then applied to the Department of Justice with an urgent appeal to uphold the reserved Indian rights in order to ensure the livelihood of his charges. The Supreme Court in this famous case of *Winters vs. United States* finally reached a decision which in the U.S. legal system, based on judicial precedent, was later to be a guiding principle in Indian water law.

In plain language, the Winters Doctrine says that when an Indian reservation was created, the federal government, as the trustee of reservation land, simultaneously granted this reservation the right to a "sufficient" supply of water and that the waters so reserved are exempt from appropriation under the state laws (Canby 1981, 242–244).[5] It is therefore the position of the U.S. government that in case of litigation, Indian water rights take precedence over those acquired by non-Indians even if the Indian tribe in question does not, in fact, begin to use reservation water until years after the reservation was established and even if the Indian residents had stopped using the water during the interim.

This viewpoint was based on the consideration that one reason for creating reservations was to give the tribes a permanent home and the opportunity to engage in farming and cattle ranching. Since it was a reasonable assumption that most Indian reservations would be unsuitable for large-scale agriculture without adequate water, however, the federal government was to see to it that the tribes would always have access to water once a reservation was established, even if this was not expressly stated in treaty negotiations.

5. This principle is valid not only for treaty reservations but also for reservations established by act of Congress or executive order, although this interpretation of the law has constantly been the subject of violent controversy (see, for example, *Gibson vs. Anderson; Tee-Hit-Ton Indians vs. United States*).

At the time, there were two essential questions in connection with reserved Indian water rights that the Supreme Court left unsettled. The Supreme Court did not specify how to quantify the water rights of a particular reservation, nor did it state precisely the purposes for which the water was to be used, even if agriculture was what it actually had in mind. The Winters Doctrine did not, however, specifically exclude the use of water for purposes other than agriculture.

The BIA, on the other hand, accorded absolute priority to the use of water for agriculture. With this in mind it began a land survey on many reservations in the period between the two world wars with regard to quantifying the irrigable acreage. Furthermore, the Indian agency began to make surveys to ascertain the capacity of Indian lands for keeping more livestock. Both surveys would have provided a basis for determining the amount of water needed, yet they were never used for this purpose.

Despite all the measures the BIA took to improve farming and despite the framework of water rights established for the protection of reservation Indians, the fact remains that the western states generally ignored the Winters Doctrine and continued to allocate large amounts of water to non-Indians, even more water than actually existed. Except for the Department of Justice and some sections of the Department of the Interior which for the most part defended Indian interests, if only in a rather halfhearted way, the policy of these states was actively supported by the federal government despite its role as trustee for reservation Indians. This is shown very clearly by the Federal Reclamation Act, which had unfortunate consequences for many reservations.

The quantification of Indian water rights has been the reason for many controversies between the reservations and the western states and has still not been fully resolved, and little change was made until a 1963 lawsuit, *Arizona vs. California,* more than half a century after the Winters Doctrine had been promulgated. The resulting court order, which is now the second pillar of Indian water law after the Winters Doctrine, concerned five reservations of the lower Colorado River, the Chemehuevi, Cocopah, Colorado River, Fort Mohave, and Fort Yuma.

When making its decision the court considered that it could not have been the intention of the authors of the Winters Doctrine to state that the water required by a reservation in those days would be the same as the water needs of a reservation sixty years later. On the contrary, federal legislation had to consider "future needs," which could arise from expanding populations and economic development. However, to the displeasure of many reservation Indians today, the Supreme Court, just as in the Winters decree, saw the future of a reservation economy only in agriculture. All other future needs—commercial, industrial, or recreational—

including their potential water consumption, were largely ignored. According to the judicial decision in the *Arizona vs. California* dispute there is only one feasible and fair method of determining the maximum water requirements of a reservation, and this is a method based on the measure of "practicably irrigable acreage" (*Arizona vs. California* 1963, 600–601). On the basis of this criterion the above-mentioned reservations, at least on paper, were granted a top-priority diversion right to about 890,000 acre-feet of water annually (an acre-foot is the volume of water covering an acre of land to a depth of one foot, or about 326,000 gallons).[6] This very considerable apportionment does, in fact, amount to almost one-third of the amount that Arizona, the winner of this lawsuit, must be granted (U.S. Senate Report 1975).[7]

In view of the Winters Doctrine and the decree from *Arizona vs. California,* both of which may be considered generally favorable to reservation Indians, it is not surprising that the states which have a high proportion of reservation land and where there is also a water scarcity should have always vigorously opposed these decisions. In these states the question of quantification, when it is settled, will lead to a redistribution of water resources in some areas. At the present time it is often impossible for these states to estimate how much water will be at the disposal of non-Indian users in the near future. Without such quantification it is hardly possible to have comprehensive water management. Therefore, with the help of their courts, the states concerned have been trying to curtail the jurisdiction of the Supreme Court in questions of Indian water rights, and have been seeking congressional actions to further limit tribal rights.

The first landmark of this policy was the McCarran Amendment (McCarran Water Rights Suits Act 1952) passed by Congress in 1952. This amendment, which was aimed against the Winters Doctrine, proposed to limit federal sovereign immunity with regard to its reserved water rights for tribes, which meant that the government, as trustee to Indian reserva-

6. In reality, however, the five reservations mentioned above could at best receive half this amount of water, even if they made full use of their water appropriation. The Supreme Court based its decision on estimates that were made in exceptionally rainy years. According to these estimates the Colorado River ought to have an average annual flow of around 16.4 million acre-feet of water at the gauging station of Lee's Ferry, about six miles to the west of Lake Powell (Comeaux 1981, 252). However, this annual flow is reached only in exceptional years, the last of which was 1922 <51–54>.

7. In this lawsuit the water quotas for the states bordering the Colorado River were partly readjudicated, but if all these quotas were actually used, an additional 10% water (see footnote 6) would have to be poured into the Colorado River and its tributaries, which usually have far less water than estimated. For a theoretical annual allocation of almost 18 million acre-feet of water 26.4% belongs to California, 23.6% to Colorado, 17.4% to Arizona, 10.1% to Utah, 6.3% to Wyoming, 5.4% to New Mexico, 1.8% to Nevada and 9% to the nation of Mexico <55>.

tions, would no longer have exclusive jurisdiction over Indian water rights. The McCarran Amendment gave the individual states the right to pass judgment in their own courts of law regarding future surface water rights, even when "Indian" water was in question. This amendment to congressional legislation was largely granted in 1976 in the lawsuit *Colorado River Water Conservation District vs. U.S.* (1976), in which even the Supreme Court conceded to the state courts the right to share in decisions regarding a new allocation of stream water.

These two verdicts, if they set a precedent, would put an end to the previous sovereign immunity of Indian tribes and their independence from the individual state water laws. They would encroach very seriously on the rights given to the tribes by the federal government, and not only in the domain of water rights. The Indians' previous experiences, both with the state courts and with the state Land Departments, the agencies responsible for water and its allocation, have usually shown that these authorities by no means represent or show any concern for the interests of the Indian reservations.

One of the Indians' main objectives is therefore to insist on their immunity from encroachments of the individual states. Only in this way will they be able to maintain the independence of their reservations.

In the 1970s some states began to set time limits for entering claims regarding water rights with a view to a final settlement for all such claims. These time limits were to be observed by all interested parties within a certain river basin. Many reservation Indians thought, however, that these states had no right to question the rights granted to them by the federal government. They reacted to the planned restrictions, therefore, by filing a large number of lawsuits in the federal courts. Much of this litigation is still pending. The Fort Apache, San Carlos, Payson, Fort McDowell, Gila River, and Salt River Indian Reservations all have taken legal action against the State of Arizona, which upholds the interests of the Salt River Project (SRP) and other major water projects. The SRP provides a large part of the water that is of vital importance for Phoenix, a metropolitan area of more than two million inhabitants, and its surroundings. In view of the hardening front of the parties in this controversy, the outcome of these cases is still uncertain.

Owing to increasing pressures and the possibility that they could lose their court cases, a number of tribes (such as the Fort McDowell, Gila River, Salt River, Ak-Chin, and Papago Indian Reservations), have since decided to settle their water claims outside the courts and to return to the negotiating table in order to find a mutual agreement with Congress. The Ak-Chin and Papago Indians, for instance, which until recently were engaged in legal proceedings against the city of Tucson as well as against 1,700 individual water users in the upper reaches of the Santa Cruz River Valley, have been able to settle their water rights. The Papago Indian Reservation was guaranteed an annual supply of 76,000 acre-feet of water and the Ak-Chin Indian Reservation, 85,000 <60, 32>, which, however, was to come in part from the treated effluent water from the city of Tucson.[8]

In view of the large number of lawsuits that are still pending concerning the allocation of water supplies that are no longer adequate, the U.S. government is not to be envied. In its dual capacity as trustee of the Indian reservations and promoter of the economic interests of non-Indian communities, it has to satisfy demands which are largely irreconcilable. An aggressive policy of congressional action to quantify Indian water rights has been able to avoid extensive lawsuits and provide states clear claims to water while allowing tribes to settle their demands for quantification of current and future water uses.

THE WATER CONFLICT BETWEEN THE STATE OF ARIZONA AND ITS RESERVATIONS: A STUDY OF THE FORT APACHE INDIAN RESERVATION

In no other state of the Union has the conflict between the white authorities and the Indian reservations over the ever increasing water scarcity been carried to the extreme that it has in Arizona. The numerous lawsuits concerning water pending by 1986 are an indication of this conflict (see above, pp. 217–218). Because of the rapid population growth and the booming economy of Arizona, whose future depends on an adequate water supply, there is understandable conflict over this precious resource, which to a considerable extent derives from only a few of the twenty-one Indian reservations in Arizona today. In the period from 1920 to 1980 the population of Arizona increased from 334,000 to over 2.7 million inhabitants, an increase of 814%, and from 1960 to 1980 alone there has been an increase of 362% (Sargent 1988, 138). According to information from the Arizona Water Commission, the Central Arizona Water Project (CAP), and the Salt River Project (SRP), total water consumption has increased more than seven times since 1930 <51, 66>.[9] In 1980 around 4.9 million acre-feet of water were consumed in Arizona. It should be noted, however, that since 1970 the total increase in water consumption has been relatively slight (see table 8.1). This is because irrigated agriculture in Maricopa County, where the metropolitan area of Phoenix is located, previously used huge amounts

8. The rest of the water is either groundwater or is drawn from the Colorado River via the Central Arizona Project (CAP) canal.

9. Exact data for the period before 1930 could not be traced.

Table 8.1. Consumption and water use of Indian reservations in Arizona, 1970

Reservation	Total water consump-tion (acre-feet)	Irrigation agriculture (acre-feet)	Irrigation agriculture (acres)	Live-stock	Mining, energy, industry, public facilities, and households	Rec-reational activities
Ak-Chin	23,041[a]	22,991	11,055	30	20	—
Camp Verde	1,040	1,020	213	*	20	*
Cocopah	1,899	1,889	297	*	10	—
Colorado River	456,322	449,824	34,653	*	1,300	5,198
Fort Apache	6,613	2,255	2,602	1,909	1,219	1,230
Fort McDowell	1,859	1,799	1,285	30	30	*
Fort Mohave	—[b]	—	—	*	—	*
Gila Bend[c]	2,019[a]	1,999	613	10	10	—
Gila River	337,768[a]	336,868	42,748	50	750	100
Havasupai	820	800	173	*	20	*
Hopi	1,609[a]	360	722	840	210	199
Hualapai	1,300	151	82	1,099	50	—
Kaibab	490	60	84	260	20	150
Navajo[d]	31,160[a]	7,908	5,570	14,195	5,158	3,899
Papago	8,787[a]	6,808	4,646	1,879	100	—
Salt River	46,041[a]	45,982	12,479	10	49	*
San Carlos	8,938	3,779	2,298	3,439	1,000	720
San Xavier[c]	5,840[a]	5,820	1,581	*	20	—
Yavapai	—[b]	—	—	—	—	—
AZ Ind. Res. (total)	935,546	890,313	121,101	23,751	9,986	11,496
AZ (total)	4,812,119	4,292,322	1,300,981	—[e]	479,812	39,844[f]

[a]Water is largely obtained from underground reserves. The Gila River Indian Reservation also obtains water from the San Carlos reservoir.

[b]Most of the Mohaves live in Needles, California, and most Yavapais live in Prescott, Arizona, that is to say, outside their reservations. They are therefore supplied with water from the city water system.

[c]The Gila Bend and San Xavier districts are parts of the Papago Indian Reservation but are spatially separated from it.

[d]These data apply only to that part of the reservation that is within the boundaries of Arizona.

[e]No data are available.

[f]Because of the numerous reservoirs and the Colorado River, the correct figure must be considerably larger than that given here.

*Figures are not available, but the amounts are very small.

Source: Arizona Water Commission 1977.

of water,[10] but this type of agriculture has been rapidly declining in the last few years.

This slower growth rate of water consumption cannot obscure the fact that the overall water situation in Arizona is exceedingly precarious. Surface water can supply only 48% of the water needed; the rest must be taken from groundwater. Moreover, the total annual groundwater withdrawal is so great that its replenishment by natural and artificial means amounts to only around 12% of the annual use on the average (Comeaux 1981, 240). In order to limit this overdraft and to make at least a partial recharge of the groundwater possible the CAP was inaugurated at the close of the 1980s. If the rapid urban growth continues, however, this main canal will not be sufficient to put an end to the water crisis in the long run.

The need to resolve the water conflicts in Arizona, already the cause of considerable tension, may become still further aggravated by the legal claims of reservation Indians in the near future unless tribal water rights are quantified. If all the reservations in this state should actually demand and use the amount of water to which they are entitled under the Winters Doctrine and the verdict from *Arizona vs. California,* their annual share of water would amount to over 2.3 million acre-feet (Pfister 1984, 4), nearly 48% of the water consumption in Arizona in 1983. The actual water consumption of all Indian reservations in 1970 amounted to only about 980,000 acre-feet. More recent data are unfortunately not available, but the figure for 1970 represents only about 40% of the amount to which the tribes are entitled. At that time the lion's share, more than 95% of the water consumed by reservation Indians, was used to irrigate about 100,000 acres of cultivated land on the Ak-Chin, Colorado River, Gila River, and Salt River Indian Reservations. These four reservations together had 83% of the irrigated cropland and used 92% of the water that was at their disposal (see table 8.1). The extent of Indian irrigated agriculture at that time is impressive only at first sight, however, because according to the BIA, 61% of it was used on lands leased to non-Indian farmers and around 10% was unused <67>. Moreover, since the 1970s there has been little change regarding land leases.

Indian water consumption has definitely increased in the meantime; yet, according to the estimates of the Indian agency, even in the mid-1980s it still amounted to less than half the amount to which the reservations are legally entitled. Even today a large allocation of water which, at least on paper, belongs to the five reservations which won the *Arizona vs. California* lawsuit is withheld from them, either because of bureaucratic red tape,

10. In the Phoenix urban area the annual consumption of water for agricultural land amounts to five acre-feet per irrigated acre, while urban land requires only around two and a half acre-feet (Wehmeier 1987, 440).

for financial and legal reasons, or because the water has simply not been claimed. (In 1980, when a number of these reservations tried to put an end to some of these obstacles and were partly successful through access to CAP water, the newspaper with the largest circulation in Arizona voiced a widely felt, entrenched attitude toward the Indians' management of their water supply, stating that it was "like giving heroin to an addict" [Arizona Republic 1980].)

With regard to the water question in Arizona, the potential for conflict between non-Indians and Indians is nowhere greater than in Phoenix and the reservations near the city. Although Phoenix has an average annual rainfall of only about 7 inches, the average per capita consumption of domestic water comes to almost 260 gallons, which is 56% more than that of Tucson, the second largest city in Arizona <119, 120>.[11] Despite the aridity of this area, Greater Phoenix is the population and economic center of Arizona because of its dynamic growth. Only 156,000 people lived here in 1950. Thirty years later the population had already increased by 871% to 1.36 million, and by the year 2000 in all probability more than 2.7 million people will be living in this metropolitan area. Parallel to this galloping population growth is the continued urban sprawl. To traverse this conurbation from north to south today one must drive over 30 miles and from east to west more than 45 miles.

Phoenix is a city of swimming-pools,[12] often surrounded by exotic trees and bushes, while people's front and back yards are often covered with lush green grass. There are many golf courses and large areas of green lawns dotted with constructed lakes built to enhance residential land values, but which all require extensive irrigation. The supposedly "highest fountain in the world" adorns a suburb with the appropriate name "Fountain Hills" (see fig. 8.2). All this gives the impression that Phoenix does not suffer from a water scarcity, an impression that seems to be confirmed by the low cost of water, which is only about one-tenth of that in Tucson. The low price

11. One reason for the lower water consumption in Tucson is that there, for some time now, the owners of newly constructed one-family houses have not been permitted to have lawns.

It is interesting to note that, compared to Phoenix, the total per capita consumption of water per day, including water for municipal, industrial, and other purposes, of Riyadh, the capital of Saudi Arabia, a city comparable to Phoenix in terms of population and climate, amounts to around 185 gallons <68>. By contrast the total daily consumption in Innsbruck, Austria, the city where I live, amounts to only 86 gallons per capita a day, and around 50% of this amount is for households only (Municipal Services of Innsbruck 1988, 11).

12. As there are unfortunately no official statistics about swimming pools, the only available information comes from a survey recently published by a local newspaper. According to this article around 20% of households in the metropolitan area of Phoenix had their own swimming pools in 1989 (Arizona Republic 1990, 48).

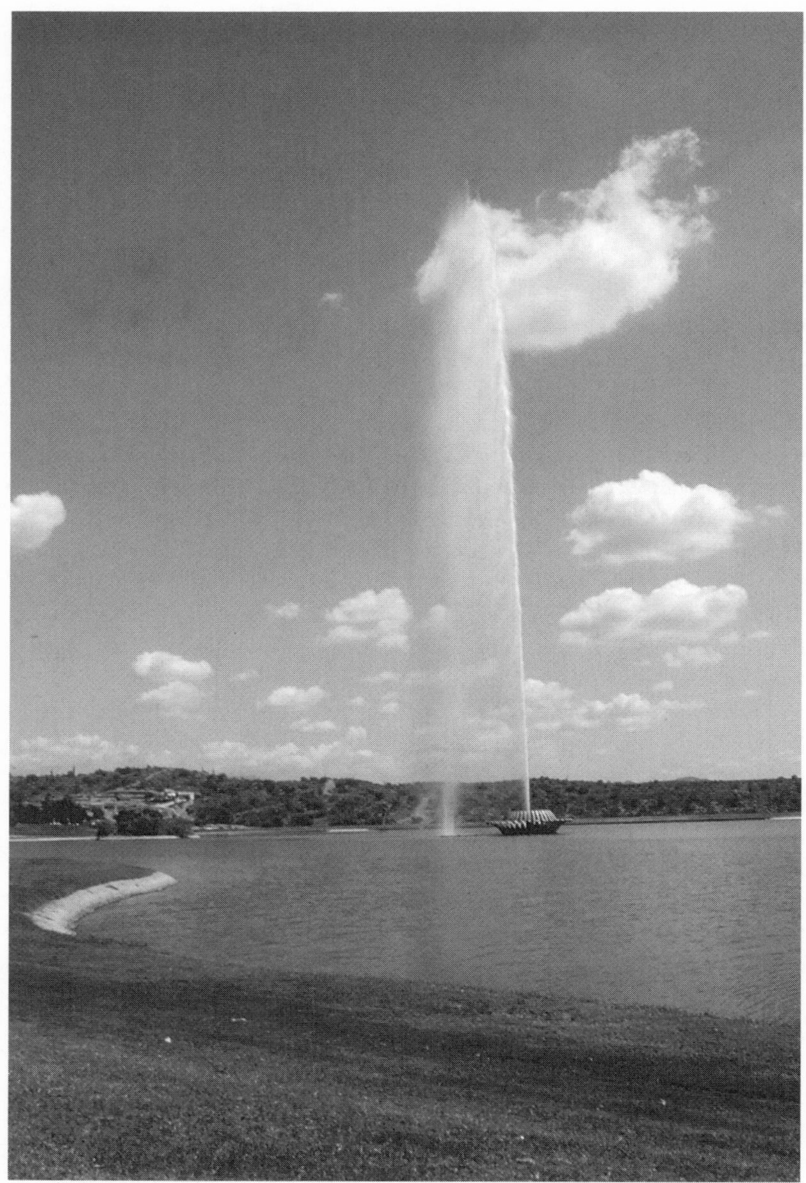

Fig. 8.2: Five-hundred-fifty-foot fountain at Fountain Hill, Arizona, 1992

Fig. 8.3: Earth crack southwest of Chandler, Arizona, 1992. Photo courtesy Elisabeth Bleimschein.

of water, however, is possible only because the SRP, the most important supplier of water and hydroelectric power to the irrigated and urbanized stretches of the Salt River Valley, compensates to some extent by excessive charges for electricity.[13] Moreover, the city is located in Maricopa County, whose crops achieved a yield that at the close of the 1970s put it in fifth place among all the U.S. counties with its agricultural produce from more than 60,000 acres of irrigated land (Comeaux 1981, 259).[14] This intensive irrigation, albeit declining, has not been without consequences. In some places, such as areas northwest and southeast of Phoenix, the groundwater table dropped at an annual rate of up to 10 feet and in some places plummeted to more than 500 feet below the surface by the end of the 1980s. As a result, earth cracks up to 30 feet wide have developed over the past few decades (see Wehmeier 1975, 102–104; fig. 8.3).

The rise of Phoenix as a flourishing city of two million people and of Maricopa County to one of the most important regions of irrigated agriculture in the United States was possible only because the residents of this

13. In 1984 the electric power revenue of the SRP was about $400 million and the water revenue was $7 million. In the same year $24 million from the sale of electric power was used to subsidize the cost of water for consumers <51, 66>.

14. In 1982 the value of all the crops harvested in this county amounted to around $410 million (Arizona Crop and Livestock Reporting Service 1983, 7).

region could always count on a sufficient, inexpensive water supply in the past. The basis of the system through which water was supplied was groundwater withdrawal, which covered about one-third of the yearly requirement, and a chain of reservoirs which, in the wake of the Reclamation Act, were at first constructed along the Salt River (total holding capacity almost 1.8 million acre-feet)[15] and later on the Verde River (317,000 acre-feet)[16] (see fig. 8.1). A network of canals which branches out widely from the Salt River Project within the irrigation oasis of Phoenix and its surrounding area, the Salt River valley, provides for the distribution of water.

In view of their extensive control of the headwaters of the Salt River drainage system the White Mountain Apache occupy a key position in the water supply of metropolitan Phoenix, and over a period of many years an annual average of 486,000 acre-feet of water has come from these mountains. Without taking into account the completion of the CAP in 1995, this amounts to 60% of the city's annual consumption of surface water.[17] So far this strategic position has been of little benefit to the Apaches. For decades the White Mountain Apache tribe on the Fort Apache Indian Reservation did not really benefit from this situation because almost all the water from the Salt River watershed ended up in the reservoirs just outside and downstream from their reservation. Because only a very limited area within the reservation could be irrigated, the State Land Department of Arizona allocated an annual quota of only 36,000 acre-feet of water to the White Mountain Apaches (Lupe 1986, 3; <51, 66>), of which in 1970 they used only 18% (see table 8.1) and in 1985 only 53%. Since the Apaches claim rights to over 200,000 acre-feet of water, they took legal action against Arizona, specifically against Phoenix and the SRP.[18]

The White Mountain Apaches categorically refused to accept the quantification of water for themselves according to the policy which the State of Arizona pursues toward all Indian reservations there. They could not accept the view that their allowance of water should be quantified for all time. Nor could they accept the idea that the amount of "irrigable acreage"

15. Roosevelt Dam (1912): 1.42 million acre-feet (see fig. 8.4), Mormon Flat Dam (1925): 57,600 acre-feet; Horse Mesa Dam (1927): 245,000 acre-feet; Stewart Mountain Dam (1930): 70,000 acre-feet.

16. Bartlett Dam (1939): 178,000 acre-feet; Horseshoe Dam (1946): 139,000 acre-feet.

17. This means that more than 10% of Arizona's total annual water supply and almost 22% of its surface water derives from this one reservation (BIA, Phoenix Area Office 1978, 72).

18. According to a letter which I received in July 1990 from the past tribal attorney of the White Mountain Apaches <48>, the tribe was not successful in this federal court case. The only possibility that still remains to the Apaches is to reopen their case before the Arizona State Court in accordance with the McCarran Amendment. The chances of success in this state court, however, are not very good.

Fig. 8.4: Roosevelt Dam, Arizona, 1985

should be the one and only criterion used in determining their future needs, as cropland plays only a minor role on their reservation. Furthermore, they do not see why they should be expected to remain farmers in an age when for the rest of the country the ideal of America as a nation of farmers has, on the whole, been abandoned (see Windhorst 1987, 480) and when an ever increasing number of workers are turning their backs on agriculture. The Apaches would like to make their own decisions as to how to make use of their own water. They would also like to reserve for themselves the possibility of leasing some of their water rights or selling part of their water for a certain period of time to non-Indian communities, business enterprises, or individuals.

Such a use of water, which has not yet been undertaken by any American Indian reservation,[19] would certainly provide a tremendous additional source of income for the White Mountain Apaches. According to estimates made by CAP experts in the early 1980s, a farmer from central Arizona, if

19. Today many reservation Indians believe that their water, just like their coal and their petroleum, is, indeed, a resource that can be sold and also exported. However, federal courts have always foiled any attempt in this direction, arguing that if a reservation has more water than it needs for itself, it should simply give the surplus to other water users. It is true, however, that up to now it has been customary that Indian water rights are leased to non-Indian farmers along with the lease of Indian lands.

allocated an acre-foot of water, would have to pay up to $100 (Graf 1984, 2). On this basis the Apaches would be losing, at least theoretically, around $49 million per year. This calculation results from the fact that in the period from 1913 to 1975 the average total stream flow of the Salt River measured at a gauging station just outside the Fort Apache Indian Reservation amounted to around 610,000 acre-feet annually (see fig. 8.5). If one deducts from this the amount of water from the uppermost headwaters of the Black River to the east of the reservation, there still remain, as already stated, more than 486,000 acre-feet. So far the Apaches have surrendered this amount of water to the non-Indian share holders of the SRP free of charge.

This conflict over water between the White Mountain Apaches on the one side and the SRP and the city of Phoenix on the other, which has been fought in various federal and state law-courts since 1956, also involves a number of issues which have so far been largely neglected in the specialist literature on the subject. For example, conversations with tribal members as well as a considerable number of old memoranda and letters from the tribal administration and the BIA agency in Whiteriver dating from the period between the 1930s and the 1960s show that since at least 1938, large-scale vegetation manipulations were carried out on Apache land (see fig. 8.5). The Arizona Water Resource Committee, the Watershed Management Division of the Arizona State Land Department, the BIA, the Civilian Conservation Corps, the SRP and the Salt River Valley Water User Association, the municipality of Phoenix, the U.S. Geological Survey, and the University of Arizona carried out Juniper Eradication and Riparian Vegetation Programs, which originally were also supported by the tribal council. The cooperative cattlemen's associations of the Fort Apache Indian Reservation were also originally involved.

All these authorities, as well as other agents of non-Indian interests, assured the Apaches that the juniper eradication and the removal of the riparian vegetation would help to improve the quality and yield of their rangeland. For the Apaches, who at that time were still mainly living off cattle revenues, the improvement of their severely deteriorated natural rangeland, which for decades had been overstocked with Anglo-American cattle, was undoubtedly an economic necessity. Heavy grazing affected the infiltration of water by reducing the shallow-rooted grass and by compacting topsoil by trampling. As a result of this drastic reduction of the original grass cover, which also functioned as an inflammable fuel, the number of wildfires had decreased. This in turn upset the ecological balance, so that the so-called weed trees, in form of juniper-pinyon, were gradually invading large areas of the formerly productive open grasslands. In a letter to his superior in Phoenix (BIA, Fort Apache Agency 1952) the then BIA superintendent of the Fort Apache Indian Reservation reported that despite the programs already under way, almost all previously treeless winter ranges would soon be infested with juniper-pinyon timber growth. As a result of this growth in the preceding two decades it had been necessary to reduce the grazing capacity of Apache rangeland from almost 30,000 to less than 17,000 head of cattle and to virtually put an end to sheep grazing (McGuire 1980, 181).

It is understandable, therefore, that the Apaches immediately and without hesitation accepted the offer of outside aid to stop their grazing land from being invaded by junipers and other species. At first sight, it is not so easy to understand why this aid should be given to them by a municipality such as Phoenix and the SRP or the Salt River Valley Water User Association located there, which have never had the welfare of reservation Indians at heart. A more careful investigation of the matter, however, reveals that the real purpose of this offer was not to improve Indian pastureland but rather to increase the water runoff to the densely populated areas of central Arizona. By clearing the rangeland of pinyon-juniper growth and removing part of the riparian vegetation along the stream banks in the lower and therefore drier areas of the reservation, the water users in Phoenix hoped for larger water yields, even if only temporarily.

Two pilot projects were carried out in the early 1950s to demonstrate how much additional water could be obtained by manipulating the vegetation, water which, from the viewpoint of the interested parties mentioned before, would otherwise remain unused on the reservation. The first project, carried out by the Arizona State Land Department along with the SRP, involved the total eradication of a 5.3-mile-long, 22-acre strip of riparian forest at Carrizo Creek, above the mouth of Corduroy Creek (see fig. 8.5). In this extreme manipulation of the ecosystem all trees, such as willow, ash, cottonwood, and sycamore, were eradicated, in some cases including their deep roots, which had drawn large amounts of water from the subsurface. With this radical approach it was possible to increase the runoff by 42 acre-feet (13.7 million gallons) in the first seventeen weeks, approximately 2 acre-feet of water per acre of riparian forest (Arnold 1962). It is no longer possible to ascertain how long this increased water yield lasted, as the relevant data are missing. The second project concerned runoff experiments carried out to the west of Cibicue, a small Apache settlement. Here on a test area whose size unfortunately can no longer be ascertained, all juniper-pinyon stands were removed. In the first year after clearing, the runoff was 2.7 times larger than in an average year before, and 1.8 and 1.5 times larger in the following two years respectively.

This experimental work was very encouraging for the operating authorities, so that on the Fort Apache Indian Reservation, and on a number of other Indian reservations in Arizona, juniper eradication programs were

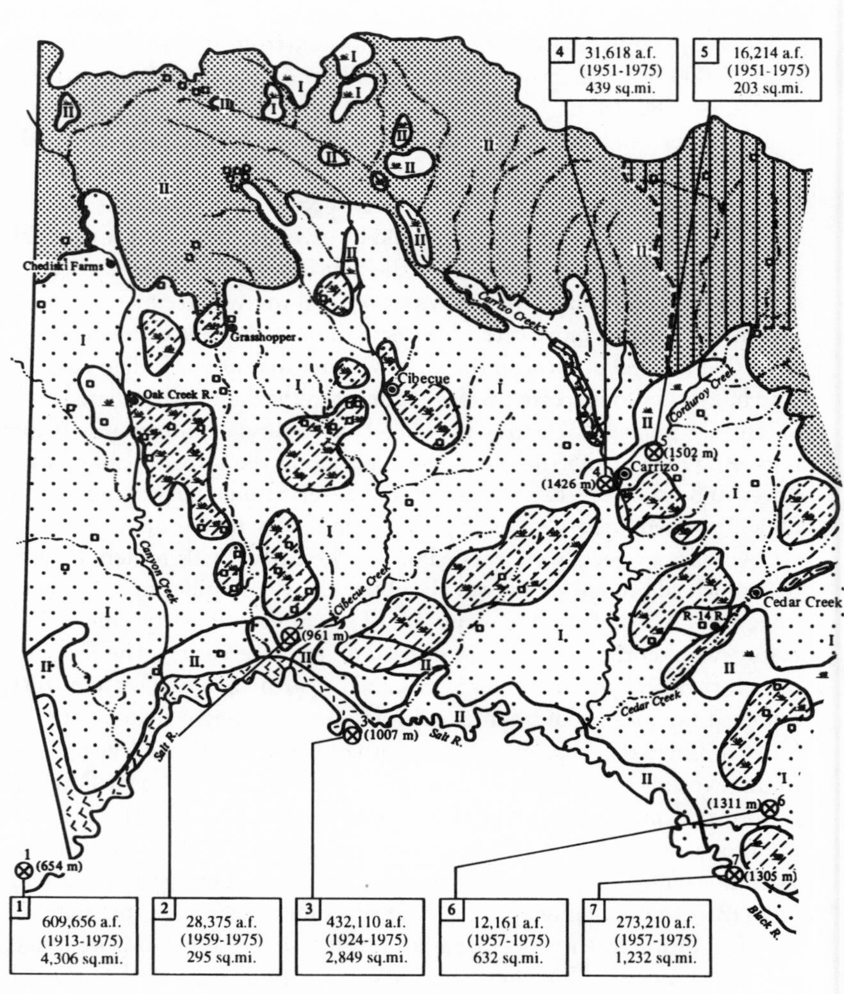

Sources: Map based on the U.S. Department of the Interior Topographic Map 1:100,000, Fort Apache Indian Reservation, Arizona (two sheets), 1977; field research; analysis of aerial photographs taken by the author (summer 1985) and by the BIA (summer 1980); documents of the BIA Agency of the Fort Apache Indian Reservation (Branch of Land Operations); two BIA reports (BIA, Fort Apache Agency 27.8.1963; BIA Phoenix Area Office 1962); and personal interviews <36, 58, 61-65>. The classification of vegetation follows Brown 1982, Klink and Mayer 1983, Rinschede 1984.

Fig. 8.5: Measures to increase the supply of water in accordance with American Indian demands, and direct and indirect efforts of non-Indians to increase the runoff from the reservation: the example of the Fort Apache Indian Reservation in Arizona

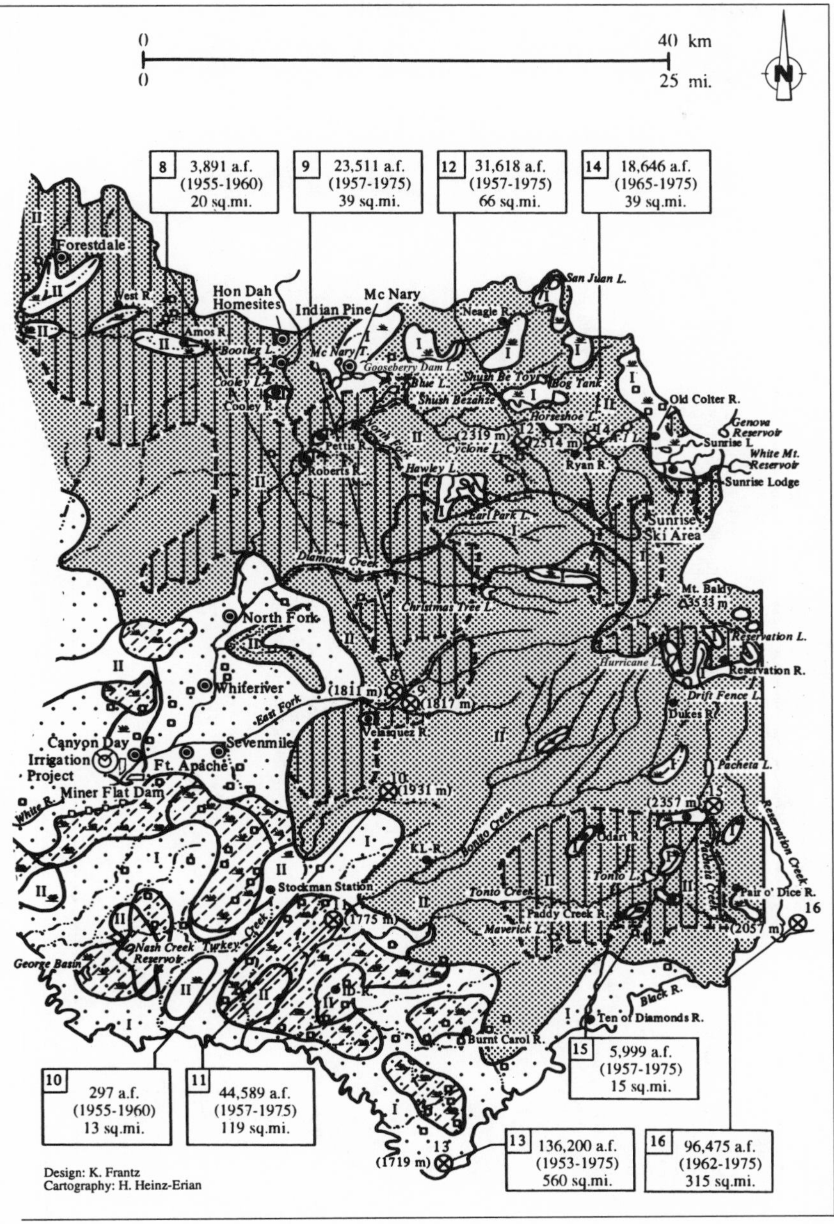

0 40 km
0 25 mi.
N
8 3,891 a.f. (1955-1960) 20 sq.mi.
9 23,511 a.f. (1957-1975) 39 sq.mi.
12 31,618 a.f. (1957-1975) 66 sq.mi.
14 18,646 a.f. (1965-1975) 39 sq.mi.
10 297 a.f. (1955-1960) 13 sq.mi.
11 44,589 a.f. (1957-1975) 119 sq.mi.
13 136,200 a.f. (1953-1975) 560 sq.mi.
15 5,999 a.f. (1957-1975) 15 sq.mi.
16 96,475 a.f. (1962-1975) 315 sq.mi.
Forestdale
Hon Dah Homesites
Indian Pine
Mc Nary
San Juan L.
North Fork
Whiteriver
Canyon Day Irrigation Project
Ft. Apache
Sevenmile
Miner Flat Dam
Stockman Station
Sunrise Ski Area
Sunrise Lodge
Mt. Baldy 3533 m
Old Colter R.
Reservation L.
Reservation R.
Drift Fence L.
Dukes R.
Pacheta L.
Velasquez R.
Burnt Carol R.
Ten of Diamonds R.
Pair o' Dice R.
Design: K. Frantz
Cartography: H. Heinz-Erian

(Continued)

FOREST FORMATION

I Subalpine coniferous forest[1]
II Montane coniferous forest[2]

WOODLAND-SCRUB FORMATION

I Juniper-pinyon woodland[3]
II Interior chaparral[4]

GRASSLAND FORMATION

I Mountain meadow grassland[5]
II Plains and Great Basin grassland[6]

DESERT SCRUBLAND FORMATION

Upper Sonoran Desert scrub[7]

COMMERCIALLY UTILIZED FORESTS

AREA WITH INTENSIVELY TREATED VEGETATION
(Riparian forests and juniper woodlands eradicated by prescribed burning, herbicides, or mechanical uprooting with the intention of increasing forage production and, to a certain extent, runoff).[8]

River

Intermittent stream

Projected dam

Constructed lake

Irrigation project

Large water tank or water catchment basin[9]

Settlement

Ranch

Water gauging station

1	609,656 a.f. (1913-1975) 4,306 sq.mi.

WATER GAUGING STATION
Average total stream flow (in acre-feet per year)
Recorded period
Catchment area

1. *Upper Region:* Engelmann spruce *(Picea engelmannii),* Corkbark fir *(Abies lasiocarpa var. arizonica),* White fir *(Abies concolor)*, etc.

   Lower Region: Douglas fir *(Pseudotsuga menziesii),* lumber pine *(Pinus flexilis),* quaking aspen *(Populus tremuloides),* etc., associated with ponderosa pine *(Pinus ponderosa)*

2. *Upper Region:* Ponderosa pine associated with Douglas fir, etc.

   Lower Region: Ponderosa pine associated with pinyon and juniper

3. *Upper Region:* Pinyon *(Pinus edulis Engelm.),* alligatorbark- and seed juniper *(Juniperus deppeana* and *J. monosperma),* etc., associated with ponderosa pine. Riparian forests along rivers and intermittent streams with black- and narrowleaf cottonwood *(Populus trichocarpa* and *P. angustifolia)* as well as Arizona sycamore *(Platanus wrightii).*

   Lower Region: Utah juniper *(Juniperus utahensis Engelm.),* herbs, and grasses as well as partly riparian forests *(Phreatophytes)*

4. Birchleaf mountain mahogany *(Cercocarpus betuloides),* pointleaf manzanita *(Arctostaphylos pungens),* shrub live oak *(Quercus turbinella),* etc., associated with grasses, succulents, and xerophytes; partly riparian forests

5. Grasses and sedges, e.g., wheatgrass *(Agropyron spp.),* mountain brome *(Bromus marginatus),* and sedges *(Carex spp.),* as well as perennial herbs and forbs, such as asters *(Aster spp.)* and larkspurs *(Delphinium spp.)*

6. Mixed short-grass communities such as Arizona fescue *(Festuca arizonica),* grama grasses *(Bouteloua spp.),* and wheatgrass; partly invaded by juniper stands

7. Creosote bush *(Larrea tridentata),* ironwood *(Olneya tesota),* littleleaf paloverde *(Cercidium microphyllum),* whitehorn acacia *(Acacia constricta),* bush buckwheat *(Eriogonum deserticola),* jojoba *(Simmondsia chinensis),* ocotillo *(Fonquieria splendens),* various species of yucca and opuntia

8. Between 1938 and 1963 all the Juniper woodland and the sporadic stretches of riparian forest were destroyed over an area of 367 square miles on the Fort Apache Indian Reservation. With the help of extant documents I was able to map out approx. 290 square miles. No data, however, exist for the remaining area.

9. There are said to be more than 500 water tanks or water catchment basins on the Fort Apache Indian Reservation. Many of them, however, can no longer be used. Owing to the lack of documents only 145 of these facilities could be included in this map.

(Fig. 8.5 continued)

enforced. For a long time even the tribes contributed to the financing of these eradication efforts; for example, the White Mountain Apaches spent altogether $413,000 up to 1962, a considerable sum for them, on this program. This money, together with funds from the outside, paid for the eradication of all junipers and pinyon pines over an area of 280 square miles, and the riparian vegetation on an additional 16 square miles of land (see fig. 8.5).[20] Under this policy the following procedures were carried out:

- Trees and bushes were chopped or sawed by hand and then burnt. (This method involved a great deal of work and was expensive without being very effective, because some species of tree, such as the alligatorbark juniper, soon grow back again.)
- Prescribed juniper-pinyon stands were burned. (This method was less expensive, but was successful only on denser stands.)
- Tractors uprooted juniper-pinyon trees with a cable attachment, and these trees were then burned or left to rot.
- Powerful bulldozers overturned the species to be removed and massive plows then destroyed the roots in the ground. (These latter two methods were both effective, but in large areas they involved much work and great expense.)
- Herbicides were used in efforts to control junipers. (This was the most expensive procedure and from an ecological viewpoint the most undesirable. On the Fort Apache Indian Reservation it was used only once.)

Certain conditions were imposed on these actions to reduce the negative impact of the manipulated vegetation. For one thing, the ground that had been torn up with tractors, chains, and plows had to be leveled and reseeded with grass in order to increase the forage production for livestock. Very frequently, however, these measures, which appear to be absolutely necessary in a region with extensive snowmelt in the spring and heavy summer thundershowers, were either not carried out or not done efficiently. On the Fort Apache Indian Reservation, for example, only small, often badly affected areas subjected to these treatments were leveled and only 15% to 20% of the affected areas were reseeded. It is therefore not surprising that such a radical change in the vegetation in the upper headwaters of the Salt River and its numerous tributaries caused only a temporary increase in the water yields, but also caused lasting and severe erosion damage which not only the Apaches but others are now fighting. As a result of this extensive denudation process, the operators of the downstream dam projects, the chief

20. Vegetation manipulations were also carried out among the Hopis, Hualapais, Papagos, and San Carlos Apaches, as well as off-reservation land, albeit not to an even remotely comparable extent. Among the Hualapais, for example, the areas concerned amounted to 81 square miles and among the San Carlos Apaches to 125 square miles (Despain 1985).

beneficiaries of the artificially increased water supplies, have had to cope with the problem of increased sediment yield, which reduces the storage capacity of the reservoirs.

Apparently, however, the operating authorities were so well satisfied with the increased water yields induced by these Juniper Eradication and Riparian Vegetation Programs on the Fort Apache Indian Reservation that in the early 1960s, not without a certain piquancy, they conferred what they considered to be a particular honor on one of the projects under way, the Cibicue Watershed Project. Ironically, they declared this region to be the "Watershed of the Year" and they even produced a movie to encourage the continuation of such programs.[21] Today one can speculate as to whether this movie was in reality a reaction to the objections of the White Mountain Apaches to these programs, objections clearly expressed at the time. The fact remains that in 1963 the tribal council was successful in demanding an end to all eradication efforts, having already begun to develop new water claims in the mid-1950s. In both cases the water users of Phoenix immediately fought back, but without achieving any notable success. Furthermore, in 1975, the tribe denied reservation access to technicians from the SRP and the USGS for collecting hydrological data, and shortly afterward they enforced the dismantling of all the hydrological data collection stations, some of which were not set up again until a decade later, and then only with police protection.

It was in 1983 that the escalating water conflict between the Fort Apache Indian Reservation and the metropolitan area of Phoenix came to a head when the tribe filed a water rights suit against the State of Arizona, specifically against the SRP, in Federal District Court. Among the 126 allegations in the suit, which was not limited to the water conflict, were objections to Arizona's policy for measuring the quantity of the water allowance, to the suppression of the tribal irrigation projects, and to the wrongful appropriation of water to benefit non-Indians, which, according to the Apaches, was carried on for decades under the pretence of upgrading grazing and forest land. This last accusation, leveled against the BIA's manipulation of forestry, was first made in the 1980s. It is difficult for an outsider and nonspecialist in this field to determine to what extent it is justified, but the BIA, the agency responsible for the management of Apache forestry, maintains that the measures that it has undertaken in the past all serve to put the forestry of the Apaches on a sound basis (BIA, Fort Apache Agency 1983). These measures include prescribed burning of "dead-and-down" timber and undergrowth, the enforced harvesting of trees in the commercial for-

21. This film was financed by the Arizona Water Reserve Commission, whose members are industrialists, businessmen, and well-to-do farmers of the Phoenix area.

est land, and the regulation of logging methods used for this purpose, as well as the removal of beetle damage in an area to the north of Hurricane Lake (see fig. 8.5) by clear-cutting the Engelmann spruce stands.

The tribal chairman and the legal staff of the White Mountain Apaches totally reject this argument. They maintain, on the contrary, that all these measures were, above all, a ruinous exploitation of the reservation's forest resources, as shown, among other things, by the fact that for decades the annual growth rate of the forest could not keep up with the cutting rate. It is their belief that behind the forest management of the BIA was a strategy, worked out in agreement with the SRP and other institutions concerned, to increase the water runoff from the reservation (Lupe 1986; <35, 48>). The BIA's proposal in the 1950s and 1960s to remove the ponderosa stands in areas of noncommercial value, a proposal that at the time was partly carried out, shows that these suspicions cannot be fully denied.

An account of the water conflict between the Fort Apache Indian Reservation and Phoenix urban area would be incomplete if it considered only the seemingly insatiable thirst of the residents of the Salt River Valley and the methods by which they helped themselves to the additional water they needed at the expense of reservation Indians. Although the water consumption of the reservation up to the present time has been very modest in relation to the water potential, and although the water claims of the Apaches have often amounted to nothing more than declarations of intent, the fact should not be overlooked that since the 1950s the Apaches have sought both to intensify and to diversify their use of water.

At first this more intensive use of water was limited to agricultural requirements. It is true that since the mid-1960s the extent of private irrigated agriculture which many Apache families modestly pursued along stream banks has sharply declined, and many small irrigation canals dating from the time that the reservation was founded were in a dilapidated condition by the late 1980s. Since the Canyon Day Irrigation Project began operations in the early 1980s, however, the water requirements for agriculture on the reservation have more than doubled, increasing from 2,255 (see table 8.1) to a total of 4,621 acre-feet. Although this tribal irrigation project (see figs. 8.5, 8.6), whose enormous investment might not pay for itself, does little more than simply serve the forage production for the tribal cattle herd, it allowed the Apaches to demonstrate successfully that they now need more of their water than before (see fig. 8.7; <48>).

Furthermore, the water use for Apache livestock, in particular cattle, has increased continually since the Second World War. In the IRA era a start was made to construct a network of water catchment basins and water tanks to supplement the cattle troughs provided by nature, which, of course, were not evenly distributed throughout the reservation. This network of stockwater

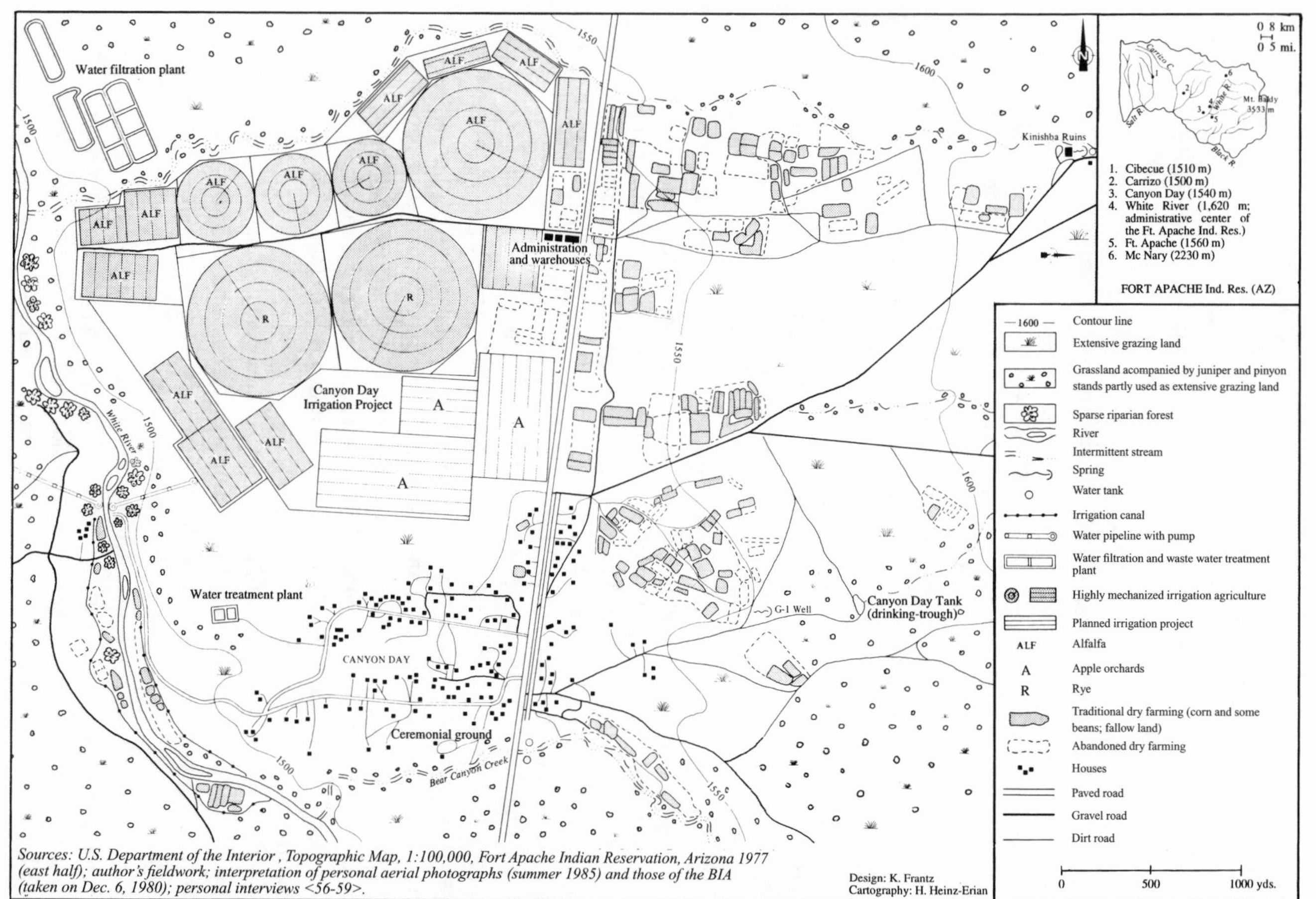

Fig. 8.6: Forms of agriculture on the Fort Apache Indian Reservation at Canyon Day, 1985

ponds, which includes five major ponds stocked with fish in the northeastern and southeastern portion of the reservation[22] was built to support improved and expanded livestock operations throughout the reservation. In the mid-1950s there were more than 200 such ponds. Today there are around 540 stockwater ponds with a total storage capacity of over 3,240 acre-feet of water (see fig. 8.5).

After the Second World War the people of Arizona began to recognize the recreational value of the Fort Apache Indian Reservation. Against the vehement opposition of many Apaches, who believed that their traditional way of life would be threatened by tourism, the majority of the tribal council decided to open certain areas of the reservation land to outside people seeking recreation. Particular attention was given to the northeast portion of the reservation, including the attractive area surrounding Mount Baldy, 11,600 feet high.[23] This area, rich in forests with clean, cool water running year-round, has a particular power to attract people suffering from the summer heat of central Arizona and seeking a cool retreat. In order to capitalize on this recreational potential and to develop it properly, the tribe founded the White Mountain Apache Recreational Enterprise in 1954. Two years later this enterprise, in cooperation with the BIA, constructed an earthfill dam across Trout Creek to create Hawley Lake, 260 acres in size, at an elevation of about 9,000 feet in a very scenic natural area about twelve miles northwest of Mount Baldy. It was ironically named after the BIA superintendent on the reservation at the time (see fig. 8.5).

The creation of this recreational lake, to which eighteen others were added during the next twenty-five years, led to protracted controversies in the courts with the SRP. For the tribal government and its white counselors these constructed lakes were an attraction for summer tourism, and therefore a sensible and welcome addition to the tribal economy, while on the other hand SRP officials saw a threat to their previously uncontested water claims through these developments. Together with the five major stockwater ponds, the mean combined surface area of these recreational lakes is over 1,250 acres and their total average storage capacity is estimated to be about 165,000 acre-feet (BIA, Fort Apache Agency 1985, 173). That is the same volume of water that was lost to the SRP reservoirs' downstream dam projects in the three years it took for the lakes constructed on the Fort Apache Indian Reservation to fill up. Besides this temporary loss of water, which, in fact, amounts to almost 9% of the total capacity of all reservoirs built along the Salt River (Theodore Roosevelt Lake, Apache Lake, Canyon

22. They are the Big Meadows Tank, the Bog Tank, the San Juan Lake, the George Basin, and the Nash Creek Reservoir (see fig. 8.5).

23. The core area of Mount Baldy remained closed to outside visitors because it is a sacred site for the Apaches.

Fig. 8.7: Canyon Day Irrigation Project, Fort Apache Indian Reservation, Arizona, 1985

Lake, and Saguaro Lake), there is, according to the SRP, an additional permanent water loss which results from a doubling of the evaporation rate owing to the artificial increase of the water surface.[24]

The last bone of contention is the conflict between the White Mountain Apaches and the irrigation oasis of urban Phoenix which developed in the mid-1980s over a dam that the tribe wants to build on White River, not far from the Apache settlement of Canyon Day (see fig. 8.5). This proposed Miner Flat Dam, if it is constructed, will be an additional attraction for tourists keen on water sports, provide the tribe with its own source of hydroelectric power, which up until now has been imported and very expensive, and finally, provide an abundant and reliable water reserve for the nearby tribal farm, an irrigation project that already exists. This reservoir would also provide a secure water supply for future industrial projects that the Apaches would like to attract onto the reservation, but which have so far only been discussed.

The financing of this dam project, expected to cost $17 million (White Mountain Apache Tribe 1985–1986, 19), shows the contradictions of American policy toward reservation Indians. The construction of this dam is supported by Congress, and, in particular, by some East Coast senators, who are

24. The evaporation rate is discussed in an unpublished study of the Laboratory of Climatology at Arizona State University in 1975 <69>.

generally supportive of American Indians and who hardly ever have to deal with Indian affairs in their home states. They have already approved $5 million for design of the project. The Economic Development Administration (EDA) of the U.S. Department of Commerce has in the past promoted many projects of reservation Indians with a view to supporting their economic independence in accordance with the Indian Self-Determination Act. It has already promised its financial support for this project as well. On the other hand, the State of Arizona regards the Miner Flat Dam Project as a danger to its water supply and is therefore fighting against it with all possible vehemence. For the time being it is not yet clear whether the dam will actually be built or not.

9 The Importance of Agriculture and Forestry

AGRICULTURE AND GRAZING: THE NATURAL LIVELIHOOD OF RESERVATION INDIANS?

The declared intention of federal Indian policy from the beginning was to make farmers and ranchers out of American Indians. This policy reached its peak in the nineteenth century during the era of the General Allotment Act (GAA) but declined in importance after the Second World War with the first signs of a cash economy on the reservations. The federal government assumed that agriculture and livestock should be the natural sources of income for the majority of American Indians, but they largely ignored the various traditional ways of life of the various tribes and their adaptation to unique climatic and physical environments. In order to make these federal programs successful and to convert Indians to farmers, far more BIA staff and a great deal more funding would have been required.

An assessment of agriculture on the reservations—something that is possible only in very general terms owing to insufficient data—shows that even a hundred years after the enactment of the GAA, the objective of creating sustainable farming in Indian country, except for a few reservations, has generally not been achieved. Today only 5.5% of all reservation land is reported to be arable, and according to statistics, more than two-thirds of this arable land is used for dry land farming (ED 1986, 115–116). Around 57% of the arable land on the reservations is leased to non-Indians. On the other hand, rangeland amounts to more than four-fifths of all Indian country, 10% of which was leased to non-Indian livestock holders in 1983, although twenty years before, allegedly 41% was under non-Indian leases (ED 1986, 115). Some experts consider the surprisingly high proportion of pastureland estimated in the statistics to be exaggerated <67, 77, 97>. Unfortunately there are no exact data encompassing all reservations. However, my own investigations have shown that on some reservations of the Southwest—in particular the Fort Apache, San Carlos, Havasupai, Hualapai, Gila River, and Salt River—a considerable part of the land classified as rangeland is either not used at all for this purpose or has only marginal use. Moreover, many of the pasture areas were classified as forest pastures, and, as such, were also included in forest statistics.

Despite the considerable amount of arable land and pastureland, the fact remains that according to the 1980 census only 6.9% of all gainfully employed reservation Indians were employed in the primary sector (see table 5.8), not including reservation Indians engaged in mining.[1] This percentage was not significantly higher than the 3.5% for the whole of the United States. Furthermore, only 2.8% of reservation households obtained their income from individual farming in 1980, 0.5% less than the U.S. average (ED 1986, 43). These figures show that agriculture and forestry, together with fishing, provide only about the same number of jobs as the building and construction trade. The 6.9% employed in the primary sector cannot, however, compare with the 16% employed in the secondary sector (see table 5.8).

Unfortunately there are very few statistics about the revenues gained from agriculture on the reservations throughout the United States. The data provided by the BIA and other federal agencies are limited to information on rental income on Indian lands and give no information about revenues that the reservations themselves make, whether on a tribal or individual basis. Such figures can therefore be obtained only through inquiries on individual reservations, and even there, no information can be obtained about the revenues of individual Indian farmers and livestock owners, but only of the tribal farms and herds. Furthermore, the BIA statistics, which are incomplete, do not distinguish between rental income from farming and rental income from grazing permits.

Considered as a whole, rental income from agriculture amounted to $31 million in 1981 (see table 7.1) or approximately $106 per capita for the reservation population. This was roughly only half the revenue achieved from forestry and less than one-sixth of the revenue which reservation Indians obtained from mining,[2] yet many more Indian families receive an income from agriculture than from forestry and mining. Agriculture remains important not only for its rental income, but above all for the considerable revenue sometimes obtained from tribal farms and tribal herds. Subsistence farming, on the other hand, plays only a minor role in the cash economy of today's

1. Unfortunately data for primary activities for reservation Indians do not distinguish between those engaged in fisheries, agriculture, and forestry.

2. These comparisons give only an approximate idea of the amount of income earned from the different sources within the primary sector as they are based on incomplete data. While the agricultural data, as already stated, give the rentals only for arable land and, in part, for pastureland, the forestry data are identical with the stumpage fees which all Indian and non-Indian entrepreneurs must pay to the tribe. On the other hand, income from mining consists of rentals from leased land as well as royalties from non-Indian mining companies (see table 7.3).

reservations, and full- or part-time Indian farming and ranching also generally play a rather small role, although they should not be underestimated.

REGIONAL EXAMPLES OF RESERVATION AGRICULTURE IN ARIZONA

That only vague information is available regarding agricultural conditions on the American Indian reservations is due not only to a general lack of relevant official data and records, but also to the BIA's inability to give sufficient attention and care to the agricultural sector. To obtain a deeper insight, it is necessary to rely on one's own inquiries and observations, which is possible only with regard to certain issues and within a restricted regional framework. The state of Arizona with its large American Indian population and its twenty-one[3] reservations will serve as an example.[4]

In 1986, besides three relatively small tribal farms on the Fort Apache (878 acres), Fort McDowell (800 acres), and Kaibab Indian Reservations (250 acres) <57, 78, 79>, there were four large tribal irrigation projects on the Ak-Chin, Colorado River, Gila River, and Papago Indian Reservations which together covered an area of 35,000 acres. Irrigation projects of the Papagos (Mission San Xavier del Bac, San Xavier section of the Papago Indian Reservation) with 1,000 acres and the Maricopas (Maricopa Colony, Gila River Indian Reservation) with 882 acres could be added to these tribal farms. These larger projects were organized on a cooperative basis. Several dozen Indian families participated in each of these cooperatives using their allotted land, while another farmers' cooperative of approximately 250 acres was about to be opened in the Northwest of the Hopi Indian Reservation at the time.[5]

3. Exactly how many Indian reservations there are in Arizona is a matter of interpretation. The territory of some reservations, such as the Colorado River, Fort Mohave, and Navajo Indian Reservations, lies mostly within Arizona. These reservations are, therefore, counted as reservations of this state, in contrast to the Fort Yuma Indian Reservation, of which only a small part is in Arizona. In the case of other reservations made up of several parts, the question arises whether each part should be counted as a separate reservation or not. Thus, for example, the Cocopah and Yavapai-Prescott Indian Reservations each consist of two small plots of land. The Camp Verde Indian Reservation has five such areas, and the Hualapai Indian Reservation is divided into one large and two small parts (Valentine, Big Sandy). Each of these reservations is usually considered to be a single entity. On the other hand, the Papago Indian Reservation, with its three large sections (Papago, San Xavier, Gila Bend) and two small plots, is often counted as three independent reservations.

4. The brief survey of the distribution of tribal and cooperative agricultural operations which follows is based upon my field inquiries, although I was not able to obtain a comprehensive view of the extent to which individual Indian farms and ranches exist, nor can this be found in the relevant literature.

5. The oldest tribal farm in Arizona, the Gila River Farmers Association of the Pima In-

Fig. 9.1: The Navajo's traditional way of keeping sheep and goats, 1956. Photo by Lauren Post, courtesy of Malcolm Comeaux.

Five Arizona reservations—the Fort Apache, Hualapai, Navajo, Papago, and San Carlos—had tribal herds in 1986. These five reservations had cattle associations that operated independently from the tribes, in addition to those located on the Hopi Indian Reservation and the San Xavier section of the Papago Indian Reservation.[6] Interviews with people in charge of Indian livestock operations and data provided by almost all of the tribal governments in Arizona and by the BIA identified approximately 700,000 sheep, 125,000 goats (see fig. 9.1), 98,000 cattle, and 26,000 horses, donkeys, and mules on the Indian reservations of Arizona in the mid-1980s. While raising sheep and goats was almost exclusively limited to the Navajos (700,000/125,000) and to the very much smaller number on the Hopi Indian Reservation (1,250/60), most other reservations had only cattle grazing.[7] The largest cattle owners were the Hualapais (4,200), the Hopis (4,000), the Papagos (18,000), the San Carlos (17,500), the White Mountain Apaches (15,900), and the Navajos (30,000). Among the Pimas on the Gila River Indian Reservation cattle ranching has been in the hands of Mexican leaseholders since 1982 (see table 9.1).

---

dians in Blackwater (Gila River Indian Reservation) was already moribund in 1986. It was founded in the wake of the IRA in the 1930s.

6. It was not possible to ascertain whether such cooperative livestock associations are also to be found among the Paiute Indians.

7. There are no significant livestock operations on the Ak-Chin, Colorado River, Cocopah, Fort Mohave, Pasqua Yaqui, Salt River, and Fort Yuma Indian Reservations.

Table 9.1. Livestock on Indian reservations in Arizona, 1985

Reservation	Cattle	Sheep	Goats	Horses and mules
Ak-Chin	—	—	—	1
Camp Verde	20	—	—	2
Cocopah	—	—	—	—
Colorado River	—	—	—	3
Fort Apache	15,900	—	—	1,000
Fort McDowell	40	—	—	6
Fort Mohave	—	—	—	—
Fort Yuma	—	—	—	—
Gila River	6,100[a]	—	—	15
Havasupai	130	—	—	300
Hopi	4,000	1,250	60	480
Hualapai	4,200	—	—	60
Kaibab	800	—	—	50
Navajo	30,000	700,000	125,000	20,000
Papago	18,000	—	—	3,000
Pasqua Yaqui	—	—	—	—
Payson	—	—	—	3
Salt River	—	400[b]	—	10
San Carlos	17,500	—	—	1,000
Yavapai-Prescott	20	—	—	6
Total	96,710	701,650	125,060	25,936

[a]Cattle ranching on the Gila River Indian Reservation is carried out by Mexican rancheros, who rent the pastureland from the Pimas.

[b]A part of the reservation is leased as winter pasturage for these approximately 400 sheep. They are taken care of by Basque shepherds and belong to non-Indian herdsmen.

Source: These figures are estimates based on the author's own inquiries and information obtained from tribal governments and BIA agencies. Official figures on the economy of Indian livestock activities in Arizona, which would include all reservations, are not available.

Changes and Present-Day Structure of Farming on the Gila River and Fort Apache Indian Reservations

Irrigation agriculture has played a central role since time immemorial among the Pimas, who call themselves "Akimel O'odham" or "People by the River." From their ancestors, the Hohokams, who lived in the Salt and Gila River valleys until the middle of the fifteenth century (see Comeaux 1981, 64–65 and 71–76; Russell 1908, 87), the Pimas took over an extensive, complex system of irrigation canals which caused their agriculture to flourish to such an extent that, besides taking care of their own needs, it produced considerable amounts of surplus crops year after year. This agricultural surplus was sufficient to supply immigrant settlers and was most beneficial to the U.S. Army, which fought against the Apaches in this region shortly after the Civil War. It is reported that the Pimas sold a million pounds of wheat to feed the troops in the Arizona Territory in 1869 (Board of Consultants 1972, 28).

After the Apaches had been pacified, thousands of Anglo Americans settled in southern Arizona. Many built their farms upstream from the Pima villages to the east of the Gila River Reservation, established in 1859, where they diverted water from the Gila River, the lifeblood of the Pima Indians. As a result the agricultural traditions of the Pimas literally dried up, their crops withered, their agricultural base deteriorated rapidly, and they declined to abject poverty (Stout 1872, 167; Ludlam 1880, 4; Russell 1908, 32–34). Already by the 1890s they had to rely on annual deliveries of wheat from the federal government (Hayden 1924, 59). Within just two decades this once self-sufficient tribe became a welfare recipient and the Pimas became dependent on the goodwill of the BIA and the missionaries sent to them.

It was not until the completion of the Coolidge Dam in 1929 (see fig. 8.1) that the reservation was again supplied with a minimum amount of water, though the white farmers in the region were the principal beneficiaries. The Pima generation of that time, however, no longer had the relevant knowledge and skills to revive the flourishing agriculture of their ancestors, partly because they had no tribal role models to copy. Neither a U.S. Department of Agriculture model farm to the north of Sacaton which was installed around the turn of the century, nor a Mormon demonstration farm established in the period between the two World Wars could alter this. An additional problem was that in the wake of the Allotment Act, which was enforced between 1916 and 1921 on the Gila River Indian Reservation, somewhat later than on most of the other reservations in Arizona, the allotments of individual Indian families had gradually been broken up into small parcels of land because of the hereditary system and the transfer of land title to multiple descendants. The consequence was that areas large enough for profitable farming were exceptional.

The development of Pima irrigated agriculture has been ineffective in the decades after their reservation was established (see fig. 9.2), and even down to the present time has essentially shown no improvement, despite the fact that today agriculture is again the most important source of tribal income. Individual Indian farmers for whom agriculture is more than subsistence farming are rarely found today among the Pimas, a situation common to the reservations. Over half the irrigable acreage of the Gila River Indian Reservation has lain idle for a long time, and although today approximately 21% of irrigated farmland is cultivated by individual Pima families, according to tribal records, there are only three families left whose

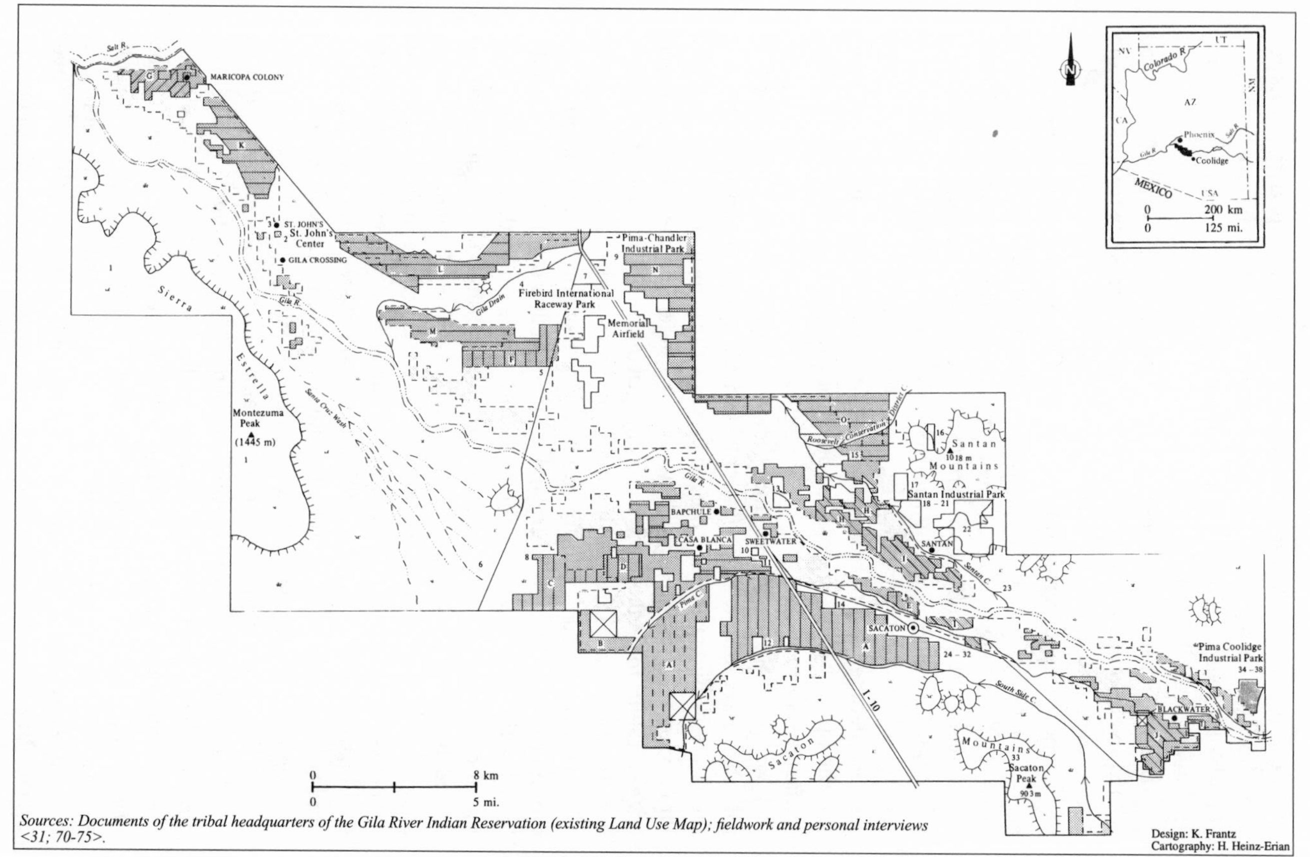

Sources: Documents of the tribal headquarters of the Gila River Indian Reservation (existing Land Use Map); fieldwork and personal interviews <31; 70-75>.

Fig. 9.2: Land use on the Gila River Indian Reservation in Arizona, 1986 *(continued)*

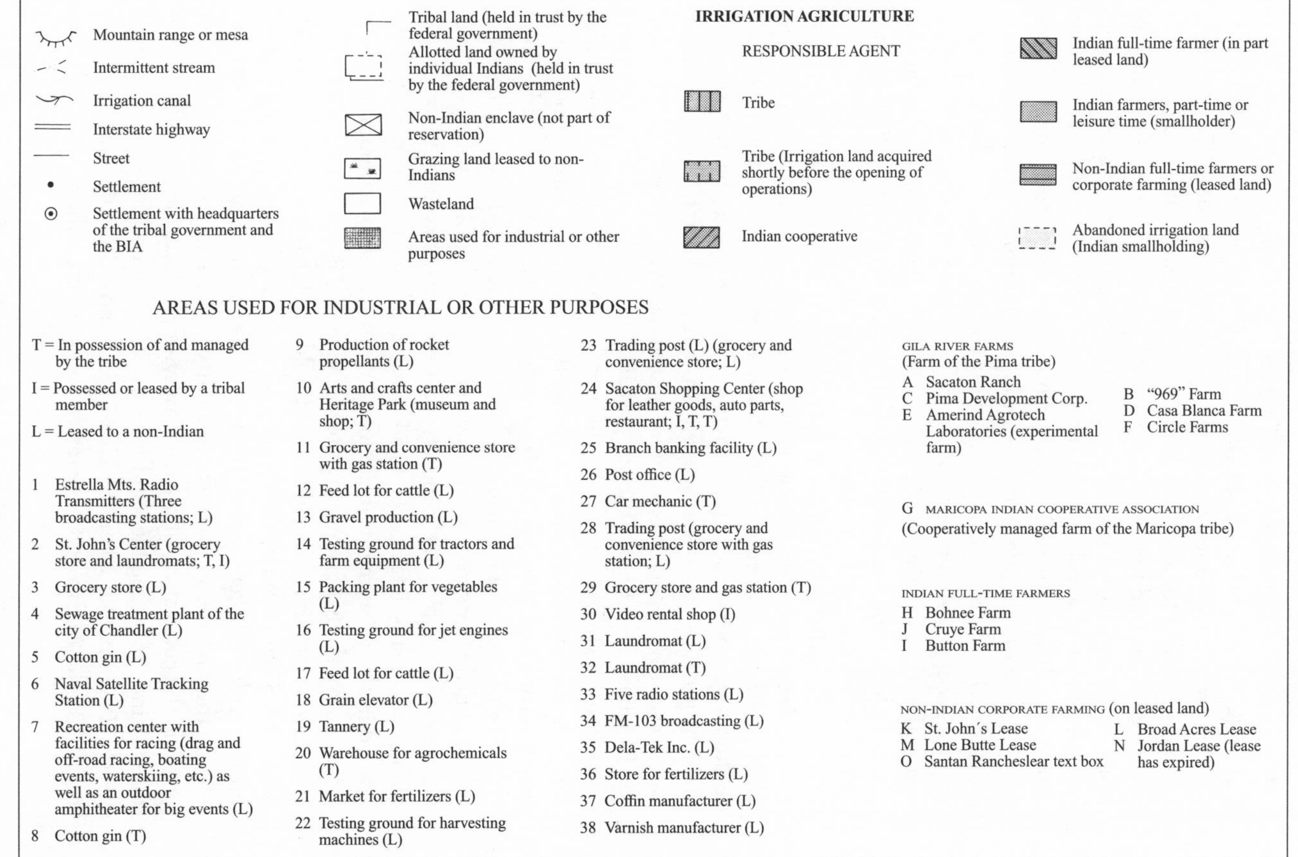
Mountain range or mesa
Intermittent stream
Irrigation canal
Interstate highway
Street
Settlement
Settlement with headquarters of the tribal government and the BIA
Tribal land (held in trust by the federal government)
Allotted land owned by individual Indians (held in trust by the federal government)
Non-Indian enclave (not part of reservation)
Grazing land leased to non-Indians
Wasteland
Areas used for industrial or other purposes
IRRIGATION AGRICULTURE
RESPONSIBLE AGENT
Tribe
Tribe (Irrigation land acquired shortly before the opening of operations)
Indian cooperative
Indian full-time farmer (in part leased land)
Indian farmers, part-time or leisure time (smallholder)
Non-Indian full-time farmers or corporate farming (leased land)
Abandoned irrigation land (Indian smallholding)
AREAS USED FOR INDUSTRIAL OR OTHER PURPOSES
T = In possession of and managed by the tribe
I = Possessed or leased by a tribal member
L = Leased to a non-Indian
1 Estrella Mts. Radio Transmitters (Three broadcasting stations; L)
2 St. John's Center (grocery store and laundromats; T, I)
3 Grocery store (L)
4 Sewage treatment plant of the city of Chandler (L)
5 Cotton gin (L)
6 Naval Satellite Tracking Station (L)
7 Recreation center with facilities for racing (drag and off-road racing, boating events, waterskiing, etc.) as well as an outdoor amphitheater for big events (L)
8 Cotton gin (T)
9 Production of rocket propellants (L)
10 Arts and crafts center and Heritage Park (museum and shop; T)
11 Grocery and convenience store with gas station (T)
12 Feed lot for cattle (L)
13 Gravel production (L)
14 Testing ground for tractors and farm equipment (L)
15 Packing plant for vegetables (L)
16 Testing ground for jet engines (L)
17 Feed lot for cattle (L)
18 Grain elevator (L)
19 Tannery (L)
20 Warehouse for agrochemicals (T)
21 Market for fertilizers (L)
22 Testing ground for harvesting machines (L)
23 Trading post (L) (grocery and convenience store; L)
24 Sacaton Shopping Center (shop for leather goods, auto parts, restaurant; I, T, T)
25 Branch banking facility (L)
26 Post office (L)
27 Car mechanic (T)
28 Trading post (grocery and convenience store with gas station; L)
29 Grocery store and gas station (T)
30 Video rental shop (I)
31 Laundromat (L)
32 Laundromat (T)
33 Five radio stations (L)
34 FM-103 broadcasting (L)
35 Dela-Tek Inc. (L)
36 Store for fertilizers (L)
37 Coffin manufacturer (L)
38 Varnish manufacturer (L)
GILA RIVER FARMS
(Farm of the Pima tribe)
A Sacaton Ranch
C Pima Development Corp.
E Amerind Agrotech Laboratories (experimental farm)
B "969" Farm
D Casa Blanca Farm
F Circle Farms
G MARICOPA INDIAN COOPERATIVE ASSOCIATION
(Cooperatively managed farm of the Maricopa tribe)
INDIAN FULL-TIME FARMERS
H Bohnee Farm
J Cruye Farm
I Button Farm
NON-INDIAN CORPORATE FARMING (on leased land)
K St. John´s Lease
M Lone Butte Lease
O Santan Rancheslear text box
L Broad Acres Lease
N Jordan Lease (lease has expired)

(Fig. 9.2 *continued*)

sole source of income is farming. These are the Bohnee (1,000 acres), the Button (2,500 acres),[8] and the Cruye Farms (900 acres) (see fig. 9.2). These farmers are the only ones who have anything close to the 1,000 acres considered necessary for a full-time farmer in this region of Arizona. The remaining land is said to be divided among the approximately one hundred small-scale Pima farms on the reservation. Except for small garden plots, however, most of which are not intensively cultivated, and indeed sometimes not cultivated at all, I could find only a very limited number of these minifarms. It can therefore be assumed that the number of minifarm owners has declined dramatically in the past few years. On the basis of most inquiries it appears that these small-scale landowners prefer to lease their land rather than cultivate it, despite the fact that leasing it gives them only a very little income, because of the small area and low rent. In 1986 the rent including water rights was between $100 and $150 per acre.

The only other alternative to leasing farmland, leaving it unused, or farming it on a subsistence basis is for the small landowner to join together with others in a cooperative farm. Such a cooperative of "allottees," who could not survive as farmers on their own, makes it possible to farm more efficiently and to have easier access to investment capital. Despite these advantages, such cooperatives are rare on the reservations, not least because the BIA has given them little support. The Maricopa Indian Cooperative Association (MICA), located in the northwest of the Gila River Indian Reservation, has experienced this lack of support. This cooperative, which began farming in 1954 with only two members and 30 acres of land, increased in size in the following years, and in the mid-1960s the cooperative obtained a bank loan for the development and mechanization of its operations, although at the time reservation Indians were seldom able to obtain such loans. Because of a bad harvest, however, the cooperative defaulted in the repayment of the loan, which led to its temporary closing <84>. Today the MICA has a white manager and is operating again. It has twenty-six allottees as members and holds 1,600 acres of land, of which around 975 acres are farmed. Alfalfa, cotton, and wheat are the only crops grown, with a market value in 1983 estimated at more than $900,000 <74, 85>. In the past, the superior water quality of the adjacent Gila River made possible the cultivation of a wide variety of produce, including many different kinds of vegetables and melons, which found a ready market in nearby Phoenix. But when the Gila River almost dried up near the end of the 1920s, the Maricopa Cooperative had to rely on the strongly alkaline

8. The manager of Button Farm is an Anglo American who has married into a Pima family. It is the rule rather than the exception that whites or so-called mixed-bloods are the largest "Indian" farmers or ranchers on a reservation.

Fig. 9.3: Tribally owned farm at Colorado River Indian Reservation, Arizona, 1986

groundwater in the region to irrigate its fields. This is why only field crops can be cultivated today.

The Indian farms that are operated on an individual or cooperative basis, similar to those on the surrounding reservations, appear to be very modest when compared with the tribal farms or with the seven relatively small irrigation units and the five large ones which are leased to individual non-Indian farmers or to large agribusinesses (see fig. 9.3). For twenty-five years most of the cropland on the Gila River Indian Reservation has been under the control of large, heavily capitalized modern farm corporations, following the general trend of U.S. agricultural production toward corporate control and large-scale, mechanized farming practices. This trend toward large corporate farms on the reservation is expected to continue in the future in view of the fact that after years of litigation a water settlement has been reached, so that the Pimas can now count on a substantial appropriation of water from the Central Arizona Water Project (CAP).[9] With this increased water supply the tribal government now intends to use another 25,000 acres of land for agriculture. While land under cultivation in Maricopa County is diminishing because of urban sprawl, the realization of such a project would mean that the Gila River Indian Reservation, as well as the

9. According to the court ruling the Gila River Indian Reservation will now receive one and a half times as much water as metropolitan Phoenix from CAP.

Fig. 9.4: Growing lettuce at a non-Indian tenant farm with Mexican migrant workers, Gila River Indian Reservation, Arizona, 1985

Salt River Indian Reservation, besides being the "vegetable garden" for Phoenix (see fig. 9.4), will also become by far the leading agricultural area within south-central Arizona.

The tribal Gila River Farms, whose origins go back to a model farm of the Department of Agriculture and a BIA farm of the postwar era, was founded in 1968. It encompasses five smaller farms as well as an experimental agricultural station,[10] which altogether comprise 17,375 acres, approximately 38% of the tillable land on the reservation (see fig. 9.2). A major portion of the Gila River Farms is tribal land, while the remainder (1,768 acres) consists of a large number of small and highly fragmented landholdings. The majority of the approximately seven hundred landowners have agreed either to sell their land to the tribe or to lease it for $10 per

10. These smaller farms are the Circle Farms (1,600 acres), the 960 Farm (1,008 acres), the Casa Blanca Farm (360 acres), the Sacaton Ranch (11,925 acres), and the Pima Development Corporation (2,080 acres). The experimental farm has an area of 400 acres.

acre. A considerable number of these landowners, however, many of whom do not live on the reservation, have never actually expressed their veiws regarding selling or leasing their land. Some of the absentee owners, whose tiny plots (less than 226 square feet each) lie idle, have refused to cooperate <67, 72, 75, 86>. A highly controversial decree, at present being contested before the U.S. Supreme Court, has so far prevented these landholders' protests from being heard.

The tribal Gila River Farms, which started with a deficit and has always had a white manager, succeeded in obtaining a $14 million loan from the U.S. Bureau of Reclamation in 1977. This substantial sum made it possible to acquire additional machinery and to improve and extend the network of irrigation canals. Some of these funds were also used to transform semidesert areas into arable land. The result of this investment has been that the Gila River Farms made a profit of $1.2 million with total earnings of $8.5 million in 1984 <72>, creating 250 seasonal and 146 full-year employees, 70% of whom are Pimas. This tribal company is the fourth largest employer on the reservation, following tribal administration, the BIA, and industrial plants. The one drop of bitterness in all this is that almost all of the few managerial positions are taken by white people. Furthermore, Mexican migrant workers hold most of the seasonal jobs here as they do on the reservation farms which are leased to Anglo-Americans, even though 30% of the Pimas able to work are unemployed. The fact that very few of the managerial positions in farming are taken by tribal members will probably not change in the future, not only because of the sociocultural factors discussed in chapter 6, but because not one tribal member has yet studied agricultural science at a university level.

Today on Gila River tribal farms cotton is grown on 38% of the land, alfalfa on 19%, and barley or wheat on another 19%. The remaining arable land includes 1% for citrus fruit and pistachios, 4% for millet, and 5% for various kinds of vegetables and melons, which are produced mainly for the nearby Phoenix market.

Gila River's tribal experimental farm (Amerind Agrotech Laboratories), under contract with the Department of Defense and Firestone Tire and Rubber, operates a plantation to grow guayule and jojoba bushes, but there have been experiments with red squill (a natural rat poison), tragacanth gum (a thickener), and various kinds of beans. Since the Second World War the roots and stems of the guayule plant *(parthenium argentatum)* which are rich in latex, have been used as a commercial source of rubber. This plant, native to the Chihuahua desert, grown over 32,000 acres in the United States, is still being investigated for its potential as a desert crop. Experiments are also being made with another native species, the bushlike jojoba plant *(simmondsia chinensis)*.[11] The Jojoba seeds contain a liquid wax with qualities similar to sperm-whale oil that is widely used in the cosmetics industry (see Frantz 1992; <73>) and as a high-temperature lubricant. The director of Amerind Agrotech Laboratories, which employs forty people, is a Choctaw-Cherokee.

The Gila River Farms, like the tribal farms of the Ak-Chin (5,000 acres) and the Colorado River Indian Reservations (8,500 acres)[12] are trying to vertically integrate their agricultural operations. These farms follow the trend of U.S. American agriculture (see Windhorst 1989), and with 45,850 acres they are Arizona's largest cotton producers. They grow cotton of superior quality which is further treated and refined in cotton gins, some of them partly owned by the tribe.[13] A good deal of this cotton is exported as "Swiss cotton" and is used in Europe to make quality shirts. Aside from their relatively limited capital resources, the tribal farms mentioned here have great advantages over similar, non-Indian companies because of their location. They have no expenses for rent or water, and no taxes to pay.

The remaining 18,108 acres, approximately 39% of the irrigated acreage of the Gila River Indian Reservation, are utilized by five non-Indian large-scale tenant farmers and a number of smaller leaseholders. The largest of these farms is the Santan Ranches (4,000 acres), in which a former secretary of the interior who was responsible for the BIA holds a controlling interest. The other large-scale operations are St. John's (2,800 acres), Broad Acres (3,075 acres), Lone Butte (2,950 acres), and Jordan Lease (3,113 acres) (see fig. 9.4). These large-scale non-Indian tenant farmers, whose leases, running from twenty to twenty-five years, give them an abundant water supply free of charge, are all located north of the Gila River, which facilitates access to metropolitan Phoenix. Together with the smaller leaseholders they earned $13.5 million in 1981 (BIA, Pima River Agency 1981), over half of the value of agricultural products earned on the reservation in that year.[14] For the same year these tenant farmers paid approximately $2.6

11. On the San Carlos Indian Reservation in Arizona, where the jojoba plant grows wild, these seeds are gathered in large quantities.

12. On the Ak-Chin (which means "mouth of the arroyo") Indian Reservation the tribal farm takes up all the agricultural acreage. In 1985 the farm employed 120 tribal members, 70% of all gainfully employed tribal members, and made a profit of approximately $1 million <32, 87>. The Colorado River Tribal Farm (CRTF) takes up only 10.5% of the reservation's agricultural acreage. The rest is divided among non-Indian leaseholders (78%), six Indian full-time farmers (9.5%), and many small-scale landowners (2%). The CRTF's profit of around $700,000 in 1984 was earned by a staff of 126 employees, 84% of whom were tribal members <88–90>.

13. Fifty-one percent of the cotton gins on the Ak-Chin Indian Reservation and 50% of those on the Colorado River Indian Reservation belong to the tribes.

14. According to estimates, 73% of the market value of the harvest on all reservations in Arizona came from non-Indian leaseholders at the beginning of the 1980s.

million (see table 7.1) for their leases. At that time, only the white leaseholders of the Colorado River Indian Reservation in Arizona ($4.3 million)[15] and the Crow Indian Reservation in Montana ($2.7 million) had to pay a higher rent.

There has not been much farming on the Fort Apache Indian Reservation, in comparison with the other Arizona reservations. The Western Apaches, descended from the Athapaskans, were hunters and gatherers. Additional needs, whether for food or other products, were met by raiding. Shortly before the Spanish conquest they migrated from the north to what is now the Southwestern United States, and it was not until then that they became acquainted with the agricultural techniques of the Pueblo Indians inhabiting this region, techniques which they partly adopted (see Buskirk 1949; Basso 1983, 465). Even then, farming played only a minor role as a means of livelihood in their new home, the White Mountains and the surrounding area. The Apaches had only small plots along the valley bottoms and Apache women worked the land by flooding the fields in spring using simple ditches, plowing the land, and seeding it.[16] During the heavy summer thundershowers the crops grew quickly and were harvested at the end of August or in early September. The simple method of irrigation is still practiced today in a similar way and is mainly used for growing various kinds of corn, beans, and pumpkins. Although the Apaches were also acquainted with dry farming, only about one-quarter of their food requirements were covered by the crops they harvested (Goodwin 1935, 61).

In the years after the reservation was established, the subsistence economy of the Apaches, including their limited agriculture, broke down altogether, not least because of the enforced resettlement of the Indians to newly founded, concentrated settlements. It was only toward the end of the nineteenth century that the Apaches took up agriculture again, when the local BIA agency encouraged them to grow crops on small plots of land. With this purpose in mind, the BIA set up a "school farm" at Fort Apache, which at first was only five acres in size. The agency also promoted the construction of numerous irrigation canals, but without envisioning the extension of the allotment policy to the Apaches. As a result of these measures the irrigated agriculture of the White Mountain Apaches spread, even if only modestly compared to the size of the reservation. Between 1899 and 1911 the irrigated acreage of arable land increased from 1,250 to over 4,000 acres (Crouse 1903,

15. The market value of the reservation's entire harvest, $66 million, was almost two and one half times greater than that of the Gila River Indian Reservation and 84% of this amount was controlled by non-Indian tenant farmers.

16. All of the farming was essentially done by women, except for some help from children, the elderly, and the sick. The men were out hunting and raiding.

147–151; Clotts 1917, 27).[17] With the rise of wage labor, agriculture declined to 2,885 acres by 1956, and eight years later it amounted to only 1,855 acres. According to a report of the BIA area director (BIA, Phoenix Area Office 1965), however, only 868 acres were actually cultivated, chiefly for corn (50%), alfalfa or grass (30%), and fruit trees (9%). Today, most of the barely one-hundred-year-old irrigation canals have deteriorated, and the irrigated agriculture of the individual Apache families has shrunk to 575 acres (BIA, Fort Apache Agency, Natural Resource Information System 1983). At the same time some White Mountain Apaches are engaged in dry farming, but on only 300 acres of the entire reservation.

The tribal irrigation project at Canyon Day belongs to another category altogether. In the case of this little Apache village, at an altitude of 5,000 feet, two utterly different worlds face one another. A fully mechanized and heavily capitalized farm is located in the midst of small, privately owned parcels of land, on which mainly dry farming is to be found (see fig. 8.6). The Canyon Day Irrigation Project with its elaborate sprinkler systems was established in 1980 at a cost of $5 million, 80% of which was paid by the EDA and 20% by the tribe. This tribal farm of 878 acres provides the needed winter forage for the whole Apache tribal cattle herd and for the herds of the cattlemen associations, but at the same time it is a project that serves as a demonstration of tribal water rights (see fig. 8.7). If the Apaches had been obliged to purchase the amounts of fodder (alfalfa, corn, rye, wheat) that were grown here, it would have cost $450,000 in 1984. That is approximately the same amount that this irrigation project cost during the same year for seed, operating expenses, loan repayments, and the services of the white manager as well as his seven Indian employees <57>.

In view of these comparatively high current expenses as well as the peripheral location of the Canyon Day Irrigation Project and its rather disadvantageous natural environment, the question arises as to just how competitive this project can be in the long run. A similar farm on the Gila River Indian Reservation, for example, in addition to enjoying a central location, could count on having from eight to nine instead of from three to four alfalfa harvests per year. The Apaches are well aware of their disadvantages and are therefore setting their hopes on a diversification of their agriculture, which, besides forage production, would make growing more competitive products possible. The tribe is considering expanding their experimental apple orchards (see fig. 8.6), because in the Sulphur Spring valley in southeastern Arizona, which lies at a similar altitude, there have been excellent harvests.

17. According to Crouse's estimates, there were approximately 2.5 acres of agricultural land per inhabitant of the Fort Apache Indian Reservation in 1903.

Livestock Grazing on the Fort Apache Indian Reservation

In the prereservation era hunting and gathering were the basis for the Apache's way of life. To supplement their means of subsistence, they also carried out raids on sedentary neighboring tribes during the summer months, and these raids first brought them into contact with agriculture. In this respect the relationship between the Apaches and the Pima, Papago, and Pueblo Indians was similar to that between the nomads and the oasis peasants in the arid regions of the Old World. Raising livestock was at first apparently unknown to the Apaches and their Indian neighbors, and, indeed, to most North American tribes.[18] It was not until the arrival of the Spanish explorers and the gradual infiltration of white missionaries, ranchers, miners, and soldiers into the Southwest that American Indians in this region finally came into contact with horses, cattle, sheep, goats, and other domesticated animals of the Old World. While many sedentary tribes, like the Christianized Pimas,[19] were being instructed in livestock breeding by the missionaries and building up small cattle herds near the missionary outposts, the Apaches were learning the value of these domesticated animals in their own way by specializing in rustling. These raids were a constant threat to the Spanish rancheros, a threat which increased considerably as the Apaches learned how to use horses. The result of this newly gained mobility was that a broad stretch of land reaching far into present-day Mexico was denounced as *ranchos despoblados,* frequently designated as "Apacheria" on maps, because the region was largely depopulated and no further attempts were made to colonize it.[20]

When the White Mountain Apaches were compelled to become sedentary,

18. There were, however, a few exceptions, including a few prairie tribes and tribes in the Southeast and the Southwest of the present-day United States. In the pre-Columbian era these tribes were already familiar with breeding dogs, which, before the introduction and widespread use of horses, sometimes served as draft animals and were also used in hunting (see Lindig and Münzel 1985, 113; Russell 1908, 84). Domesticated turkeys were also found in some places.

19. The various Pima bands lived in a region which extended from the Gila River in the southeast of present-day Arizona to the Rio Sonora, in the northwest of Sonora, Mexico. The boundary to the west was the Gulf of California and on the east the San Pedro River. This region, called Pimeria Alta, was where the Jesuit missionary Eusebio F. Kino worked from 1687 to 1711. Kino, who came from Nonsberg (Trentino, Italy), was educated at the Jesuit theological college of Hall in Tyrol, Austria. Together with his brethren he established 16 missionary stations. One of them, the San Xavier del Bac Mission (see fig. 9.5) on the Papago Indian Reservation in Arizona, is supervised by the Franciscans today (see Bolton 1986; Burrus 1954; Smith et al. 1966; <91>). Kino was not only an enterprising missionary but also a talented geographer and cartographer of this region who discovered that today's Baja California was not an island but a peninsula.

20. In the period from 1820 to 1825, 100 settlements were destroyed by the Apaches in this region, 5,000 Mexicans were killed and another 4,000 driven away (Haase 1968, 13). Information on Apache losses is not available.

Fig. 9.5: San Xavier del Bac, Father Eusebio Kino's northernmost missionary church, San Xavier section of the Papago Indian Reservation, Arizona, 1983

the raiding expeditions came to an abrupt halt and their regular supply of cattle was interrupted. This situation did not change until the last quarter of the nineteenth century when the military stationed on the reservation began to distribute cattle to the Apaches in small numbers. The Apache band leaders, nominated either by tradition or by the commanding officer of the fort, were given cattle from which they and their band members were expected to build up small herds. The cattle were labeled by tag bands stamped with the letter assigned to the leader of the respective band. At first grazing was confined to the valley bottoms along the rivers and to the vicinity of the newly established nucleated settlements (see fig. 9.6). In short, this tag-band system, which is still maintained today by the cooperative cattle associations on the Fort Apache Indian Reservation, goes back to the beginnings of cattle ranching on this reservation. According to a report of the BIA superintendent of the time, the White Mountain Apaches owned 521 head of cattle and 760 head of sheep in 1878 (ARCIA 1879). Fifteen years later 5% of Apache families owned cattle, 1,861 head of cattle altogether (ARCIA 1893, cited in Goodwin 1969, 583). Nothing more was reported about sheep, which later disappeared from the reservation because the Apaches, in contrast to the Navajos and the Hopis, had no interest in keeping sheep.

The rangeland of the Apaches could have carried much more livestock, and was far from being overstocked at the time. This fact was known to

Crouse, the BIA superintendent who shortly after the turn of the century reduced the food rations below the amounts customary and, with the funds thereby saved, provided each of the approximately 400 Apache families of the reservation with one head of cattle (Crouse 1903 and 1905). Crouse soon had to face the fact that not much later only one-fifth of the Apache cattle were still grazing the rangeland, because the tribe either neglected or slaughtered them for food, which was scarce among the tribal members. Crouse's primary objective, however, was to legalize the unlawful livestock operations of the non-Indians on many reservations, which he achieved by issuing them permits with remarkably generous terms. As Hofmeister has pointed out (1967, 52), this led to grossly overstocked pastures and a sharp decline in the grazing capacity on reservations. During Crouse's term of office (1902–1914) a large area of rangeland was leased to white cattle owners (see fig. 9.6). Thus in 1904, besides unlawful occupancy and encroachment, which the few BIA officials were powerless to prevent, there were thirty-two non-Indian ranchers with grazing permits who officially had 8,000 head of cattle on Apache rangeland throughout the year (McGuire 1980, 81). By 1912 the number rose to almost 13,000 cattle. In addition there were 12,000 head of non-Indian sheep on the summer ranges and 28,000 head of sheep year-round (ARCIA 1913).

When the United States entered the First World War the white ranchers on the reservation became even more influential, partly because Crouse's successor was said to have particularly close relations with the major cattle owners of Arizona. This helped him to become manager and joint owner of one of the largest ranching operations of the entire Southwest, the White River Land and Cattle Company, after his retirement from the BIA. On the Fort Apache Indian Reservation this company had more leased rangeland than any other cattle owner at that time. Before the IRA was passed in 1934, there was little change in this leasing policy, and although the number of non-Indian cattlemen leasing land from the Apache tribe had declined to fourteen, the best rangeland, 1,062 square miles (40% of the reservation land), was still in the hands of non-Indians, with around 10,000 non-Indian cattle and 29,000 sheep (see fig. 9.6; McGuire 1980, 148).

Changes eventually took place among the Apaches who, though at first indifferent, gradually became interested in ranching. Supported by the BIA, they began to develop livestock operations shortly after the turn of the century. This can be seen not only from the percentage of cattle owners in the tribe, which amounted to 12% in 1910 and rose from 19% to 27% between 1920 and 1930 (BIA, Fort Apache Agency, Grazing Files 1910–1934), but also from the total number of cattle, which increased to approximately 16,500 by 1930 (McGuire 1980, 148). Then, as today, however, these cattle were not evenly distributed among the Apache cattle owners,

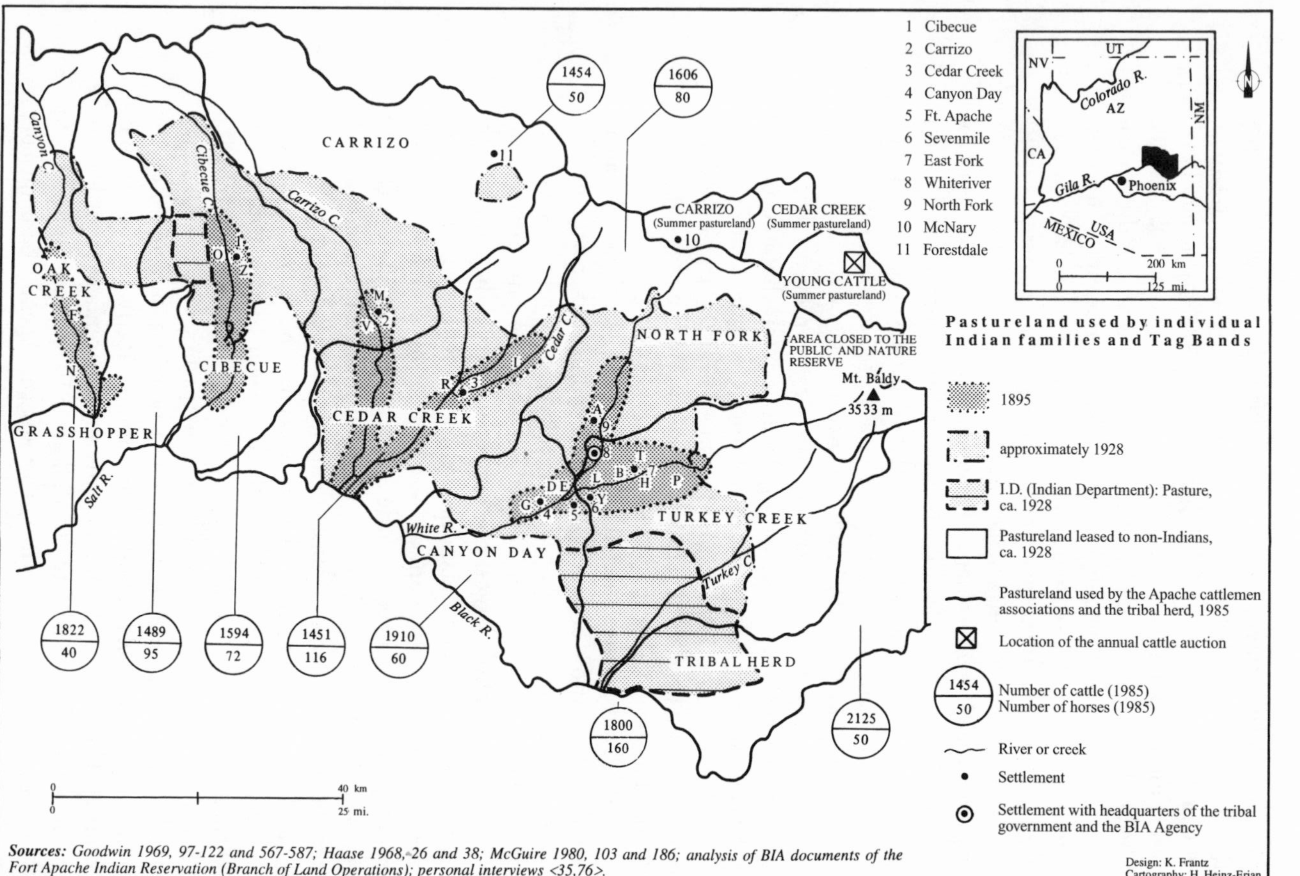

Sources: *Goodwin 1969, 97-122 and 567-587; Haase 1968, 26 and 38; McGuire 1980, 103 and 186; analysis of BIA documents of the Fort Apache Indian Reservation (Branch of Land Operations); personal interviews <35,76>.*

Fig. 9.6: The pastureland economy on the Fort Apache Indian Reservation in Arizona, 1895–1985 *(continued)*

Fort Apache Indian Reservation Census, 1893

TAG BAND[1]	NUMBER OF MEMBERS	CLAN MEMBERSHIP[2]	SETTLEMENT AREA	BAND LEADER
A	295	1, 15, 16	North Fork, Forestdale	*tsájú* "Swollen One" or Alchesay
B	101	1, 2	East Fork, Turkey Creek	*mbàkèna·gic*
C	120	4, 5, 14, 15	Cibecue	*t'à·ìlè·* "He Joins Them" or John Taylor
D	113	1, 8, 10	East Fork Canyon Day Flat Cedar Creek	*nànt' áìłkè'* "Chief Scattered-About"
E	64	1, 7, 12, 13	East Fork Canyon Day	*tcòxùjé* "Wrinkled Testicles"
F	41	5, 9, 11, 17	Canyon Creek	*hàckí·bá' – ò sìndì'* "Angry in Good Condition" or Willy Lupe
G	89	1, 8, 13	Canyon Day Flat	*nánògò·hń* "Coming Home Weak from Wounds and Privation"
H	64	1, 12, 13	Seven Mile Canyon East Fork	*xàtsò·γé* "Weak One"
I	71	1, 8, 12	Cedar Creek	*hàckí·ýkè·d* "Angry He Asks for It"
L	84	1, 13	East Fork Turkey Creek	*ts'òį̀zlì·'* or Chino
M	253	14, 15, 18	Carrizo	*bé·cbìγò'án* "Metal Tooth" or Sanchez
N	100	5, 6, 10	Canyon Creek	*tc'à'ndè·zń* "Long Hat" or John Dayzn
O	45	15, 5, 6	Cibecue	*tł'é'gùlkì·jń* "Spotted on Inside of Thighs"
P	21	1, 13	Turkey Creek	*hàcké·łdà'dìdìldì·* "Angry He Raises His Hand"
R	80	1, 2, 8	Cedar Creek Crossing	*hàckí·yà·bà·dzà·* "Angry He Goes out to Fight
T	72	1, 13	Turkey Creek (originally Chiricahua) Apache	Belinte
V	151	14, 15, 18	Carrizo	*tsí·stc'íln* "Curly Haired"
Y	61	3, 7, 13	East Fork	*'ìckį̀·łgaì* "White Boy"
Z	36	5, 6, 15	Cibecue	*nà·łkéhíjǫ̀·d*

1. When the U.S. Army established this reservation and began to count the population it gave each band leader a letter and the number 1. Accordingly the chief of the North Fork Band was designated as A_1. Male members of his band who were married were designated as A_2, A_3, A_4 etc. At the time all Apache affected by this policy were obliged to wear necklaces showing their tags in order to obtain food rations. The cattle allotted to them were likewise given the same tags consisting of letters and numbers.
2. A band consisted of several clans which, through previous fragmentation, might be found in several different groups. To simplify, the clans in this list have been numbered consecutively. The resulting figures are not identical with those given for the Apache in Goodwin's research (1969).

Source: Goodwin 1969, 583.

(Fig. 9.6 *continued*)

a situation symptomatic of many reservations. In 1920 the Velasquez, Cooley, Amos, Pettis, and Altaha (known as R-14)[21] families, who all had their own rangeland (see fig. 8.5), controlled altogether 165,000 acres of pastureland assigned to them (12% of the ranges on the reservation) and shared approximately 3,500 head of cattle (BIA, Fort Apache Agency, Grazing Files 1910–1934), which amounted to one-third of all Apache cattle.

The background of these families seems to be characteristic of Indian ranching on many reservations. The paternal grandfathers of the Velasquezes and the Cooleys in the 1930s were white. During the early reservation years, they performed valuable services for the military at Fort Apache, one as an interpreter and scout, the other as a meat supplier. They both married daughters of respected Apache families. In the case of C. E. Cooley, he actually married two daughters of the tag-band chief Pedro. With these marriages, which gave them the label "squaw men," they gained the same rights as a tribal member. Their special status, which gave them access to the decision makers within the Anglo as well as the Apache society on the reservation, made it easy for them to build up impressive cattle herds on the best pasturelands in just a short time. Cooley, for example, settled in the North Fork region near Forestdale (see fig. 9.6), where Chief Pedro's clan always spent the summers in the prereservation era. The second and third generations of the Cooleys, two daughters and a granddaughter, married sons of influential white pioneer families who lived as ranchers north of the reservation boundary. In this way the reservation was opened to three more white settlers who created the Amos, Pettis, and West Ranches (see McGuire 1980, 60–69; <93>).

These developments illustrate a widespread phenomenon on Indian reservations that is to be found not only in livestock operations but in all branches of the economy. In the private economic sector the mixed-bloods, who are often situated both spatially and socially at the periphery of the reservation populations, are often distinguished by their exceptional economic success.

If one counts all the cattle, sheep, and horses which were legally grazing on the Fort Apache Indian Reservation between 1910 and 1930, the annual average was around 60,000 head in addition to the thousands of cattle that were herded on the Apache rangeland year after year without a permit. The result was overgrazing, which left scars on the Apache ranges. The reduction of forage resources, the invasion of weeds and

21. R-14, or Wallace Altaha, was the successor to the old tag-band chief R-1 or "Angry He Goes out to Fight" (see Fort Apache Census 1893, fig. 50, cited in Goodwin 1969, 583). He was the only full-blooded Apache on the reservation to build up a large cattle herd, which later was incorporated into Cedar Creek Cattle Association, one of the autonomous cattle associations on the reservation.

woody perennials such as juniper and pinyon pine, the deterioration of the natural cattle troughs, and the massive erosion of the soils all progressed rapidly as a result of overgrazing. This led to a considerable reduction of the carrying capacity of the Apache rangeland during the period between the two World Wars. Influenced by the policies of the IRA, the local BIA agency became aware of this alarming situation and realized it was time to react if it was to remain at all credible as trustee of Indian country and its resources. The consequence was that everywhere on the reservations the livestock industry was radically reorganized and partially improved. The improvements included controlling the growth of juniper and pinyon pines, expanding and maintaining the network of water tanks and cattle troughs, not renewing non-Indian grazing permits, reducing the number of cattle drastically, upgrading the breed, promoting cooperative cattle associations, and above all adopting a rotation system which guaranteed a more balanced use of rangeland.

On the Navajo Indian Reservation at the beginning of the 1930s it was chiefly the herds of sheep and cattle belonging to the Indians that were reduced,[22] but on the Fort Apache Indian Reservation around the same time it was mainly the non-Indian ranchers that were affected by the reduction of stock. The BIA did not renew many of the non-Indian permits, so that in 1953 the last white ranchers had to leave the reservation. From the early 1950s the Apaches and their approximately 16,000 head of cattle had around 2,200 square miles of pastureland at their disposal, including the newly acquired summer ranges at higher altitudes and the sparse ponderosa forests which are quite suitable for pasturage. Since then, even with this relatively small number of cattle, overgrazing might still have continued if not for a reorganization of Indian livestock management. Until this time it had been customary, for the Apaches, like the Navajos with their "hogan grazing,"[23] to let the cattle graze on the ranges located near their wickiup camps. The result was that these pastures soon deteriorated. After the Second World War, however, Apache herding improved because range units were fenced in, a rotation system attuned to the natural environment was adopted, and self-supporting cattle associations established in the 1920s became more prominent as more members joined and the number of cattle increased.

The opposition to this new BIA policy of livestock management was at first so great, however, that transitional regulations had to be introduced.

22. Within just a few years the number of Navajo sheep was reduced from 1.5 million to half that number (Aberle 1983, 642).

23. This expression designates a particular kind of grazing practice whereby the Navajos and other Indian tribes put their livestock out to graze in the close vicinity of their homes. In the evening the animals are brought back to the corral near the hogan.

Thus, for example, some of the important Indian cattlemen, including the above-mentioned Altaha, Amos, Pettis, Velasquez, and West families, as "Individual Indian Rights Cattlemen," were at first not inclined to join the cattle associations because they feared that they would have too little say and that their profits would be too small. At the beginning of the 1950s, however, a tribal council decreed that these cattle owners choose between joining one of the cattle associations or removing their stock from the reservation. While Velasquez and the descendants of R-14 decided in favor of the "Indian" way, choosing cooperative herding, some of the mixed-bloods preferred to sell their cattle and leave the reservation.

Today the rangeland and the cattle industry are still an important part of the Apache economy, as on many other reservations in Arizona (see table 9.1), although overall this industry has declined since the 1950s, as has individual farming, because of the rise of wage labor. Much higher incomes can be gained from working for the tribal timber industry and the tribal sawmill. Almost one-third of all Apache families still receive a modest supplemental income from ranching <35, 59, 76>, however, and ranching is carried out with dedication and enthusiasm because it is recognized as prestigious and brings social success. Outside visitors who are present when the cattle are driven together in the fall, who watch when the calves are branded or the cattle are auctioned, or who attend the annual rodeo will observe that many male Apaches once accustomed to hunting and raiding have not found it difficult to adjust to the more limited freedom of the cowboy (see fig. 9.7).

Like other Indian reservations in Arizona which boast a substantial number of cattle, the White Mountain Apaches have several cattle associations and a tribal cattle herd. Indeed, the reservation has eight such associations,[24] and each has from 40 to 70 members and 1,500 to 2,000 head of cattle (see fig. 9.6). It was not possible for me to find out how many cattle were owned by individual members of the associations, because the eight managers responsible for their respective herds were not prepared to give precise information concerning numbers of cattle. Conversations with tribal cattle owners on the reservation made it clear that the cattle are not distributed evenly among the White Mountain Apaches.

My inquiries among Hualapais in Arizona (see fig. 9.8) yielded a more accurate picture of the conditions of ownership among reservation Indians. In 1985 there were 93 individual cattle owners on the Hualapai Indian

24. The San Xavier section of the Papago Indian Reservation has one such association, San Carlos has five, and the Hualapai and Papago Indian Reservation have four each. I had no opportunity to find out about the numbers of livestock associations among the Hopis, the Navajos, or the Paiutes.

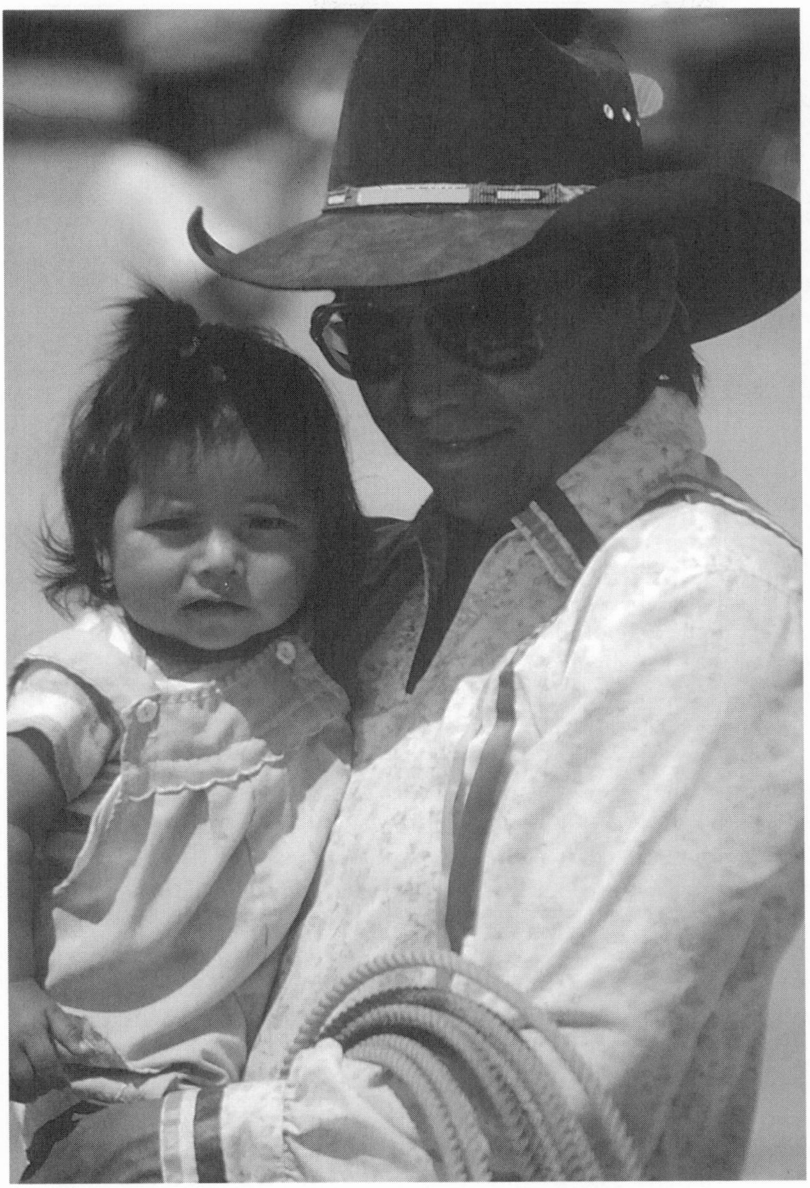

Fig. 9.7: Apache Rodeo champion with his daughter, Rock Point, Navajo Indian Reservation, Arizona, 1985

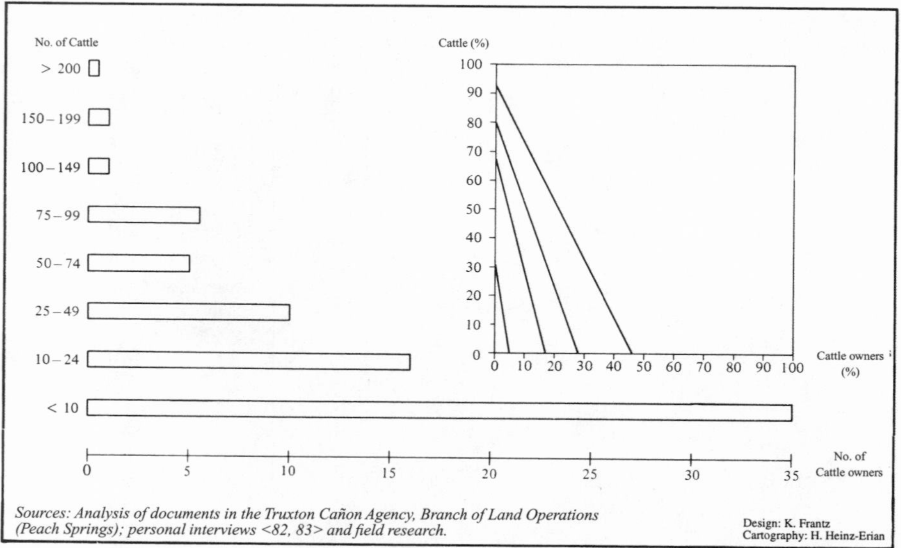

Fig. 9.8: Cattle ownership on the Hualapai Indian Reservation in Arizona, 1985

Reservation, 13% of the tribal members living on the reservation. These Hualapais were divided up among four cattle associations and owned around 4,200 cattle. Five people (5.4% of the cattle owners) owned approximately one-third of the cattle, while 78 people (83% of the cattle owners) owned another third. At the time of my inquiries, however, fifteen cattle owners had just sold their stock. This uneven distribution of the ownership of cattle is found not only among the Hualapais and the White Mountain Apaches, but also on the Hopi, Papago,[25] and San Carlos Indian Reservations. Such conditions would appear to be incompatible with the collective consciousness of reservation Indians and their principle of sharing, which is indeed the view of most of the people on the reservations concerned. The consequence is frequent strife between those who own many cattle and those who have only a few.[26] Thus the cattle associations are

25. One of the few surveys of the distribution of cattle among reservation Indians was conducted in 1959 on the Papago Indian Reservation. At that time around half the families of the tribe had cattle, and two of them had more than a thousand cattle, five families more than 500, and nine families more than 100 (Metzer 1960, cited in Manuel et al. 1978, 527).

26. On the Hualapai Indian Reservation, for example, almost three-quarters of the cattle owners have fewer than 25 head, but in order to earn enough to live from ranching in Arizona, it is estimated that one must have around 200 head. There is only one rancher on the reservation, however, who has this many cattle (see fig. 9.8).

dominated by the large cattle owners, and it is they who determine which tribal members may be admitted to the associations and how many cattle they may own. Furthermore, they often even decide whether the children of a deceased cattle owner will be allowed to take charge of the herd or compelled to sell it.

The cattle associations have their own managers and employees, who see that the cattle have enough water and winter fodder and that the rotation between summer and winter ranges, which makes it necessary to put up fences, works properly. These full-time employees also organize the cattle auctions that take place regularly. They collect membership dues and fees, and distribute the profits. The cattle associations are relatively autonomous in their decisions, but they must observe certain tribal regulations, and they must also pay the tribe use fees corresponding to the size of their herds.

In addition to the cattle associations, the Fort Apache Indian Reservation has a tribal herd that was founded by the BIA in 1917 (see fig. 9.6). This tribal herd or "Old Folks' Herd" has been making profits for a number of years, which have long been used for the benefit of elderly members of the tribe. The herd employs a white stockman and six Apaches who take care of approximately 2,100 cattle. Besides the tribal herd comprised of Hereford cattle, the stockman has also been breeding Beefmaster cattle since 1985. This breed is particularly suitable for the conditions on the Fort Apache Indian Reservation and promise substantially higher profits because of its very tender, low-fat meat. Beefmaster cattle (crossbreeds of Hereford, Shorthorn, and Red Zebu Brahma cattle) were registered in 1954 by the U.S. Department of Agriculture as a distinctive breed, and $1.6 million was allocated for the development of this additional, special herd for the tribe, a sum which made the purchase of 600 heifers and 75 registered Brahma bulls possible. The tribe funded 70% of this amount, and a BIA Economic Development Grant funded the remaining 30%. The plan would generate a growth of the herd to 2,200 animals in seven years <35>.

Despite all these investments and the large number of cattle owners on the Fort Apache Indian Reservation, neither the tribal government nor the BIA seems to have made much of an effort to encourage tribal members to study veterinary medicine. There has not been one tribal veterinarian among the White Mountain Apaches, a shortcoming that also applies to the Hualapais, Papagos, and San Carlos Apaches.

FORESTRY IN INDIAN COUNTRY

For a long time the federal policy to make self-sufficient farmers and ranchers out of reservation Indians overshadowed the commercial utilization of tribal forest lands. The first systematic logging in Indian country during the second half of the nineteenth century must also be seen in this context. Timber harvests on Chippewa lands in northern Minnesota in 1861 and the Tulalip Indians' land in northwest Washington in 1867 were primarily clearing operations, so that more territory could be opened up for agriculture. The BIA, however, did not take any steps toward the development of Indian forest resources at that time. Indeed, with the exception of these clearing operations contracted to non-Indians, which had to be authorized by the president, logging was forbidden in Indian forests until the beginning of the twentieth century under an order of the Supreme Court (*United States vs. Cook,* 1873). This order, however, did not prevent white entrepreneurs far from the federal capital from profiting from the commercial exploitation of Indian timber resources without the local BIA agencies taking any preventative measures to protect tribal timberlands. Indians on the reservations were at first restricted to taking only dead-and-downed timber from their own forests in accordance with an act of Congress in 1889. It was not until two decades later that Congress decided to permit reservation Indians to carry out controlled logging and the sale of Indian timber, and a forestry branch was established within the BIA (Iadarola 1979, 8–9). This branch had the reputation of being particularly responsive to the wishes of the owners of outside sawmill enterprises, however, and, at the same time, neglecting the interests of its own clientele. This situation did not change until the IRA era, a time more accommodating to reservation Indians, when the BIA took the first steps for the management and utilization of the reservation forests and when tribal forestry enterprises, which used Indian timber resources, were permitted.

Even today the responsibility and management of Indian forest resources lies very largely in the hands of the BIA. Despite the Indian Self-Determination Act, whose purpose was to give economic autonomy to tribes, there has been little change in this respect. The BIA's responsibility includes care for the forests, reforestation programs, and fire protection. The BIA also assesses the annual allowable cut and stumpage fees and authorizes or allocates usufructuary rights. According to the law, the tribes could assume authority over their own forests, but either they are not willing to manage their timber resources, or, as is usually the case, they are not yet able to do so because on the reservations that are rich in forests there are few tribal members who have forestry degrees. Most of the reservations are compelled to have their forestry programs operated by the BIA, for which, according to a decision of Congress, it is authorized to deduct 5% to 10% from timber sale contracts as an administrative fee. Because of a shortage of experts and lack of funds (despite the fee),[27] however, not even today's division of forestry within the BIA is in a position to carry out adequately the tasks entrusted to it. For example, forest management inventories are generally either lacking or outdated, yet without such inventories a sustained yield management of the forests is not possible. In order to ensure sustainable management, which prevents the ruinous exploitation of forest resources, there needs to be a better balance between the annual amount of timber cut and the annual timber increment over a period of years.

Neither on the Quinault Indian Reservation in Washington nor on the Fort Apache Indian Reservation in Arizona, both rich in timber, do forest management inventories exist, which means that the assessment of the annual optimum cut is not based on sound silvicultural principles but only on estimates. The White Mountain Apaches today believe that because of the inadequate operation and management of Indian forests the BIA has neglected its trust responsibilities over the years and has awarded timber sales with excessive volumes of timber cuts to non-Indian companies and to the tribal sawmills. It is their view that this has led to a ruinous exploitation of their timber resources. This accusation against the Indian agency is also part of a mismanagement claim by the tribe against the federal government. The Quinault Indians, on the other hand, are of the opinion that in their case the BIA's assessment of the annual allowable cut has been too small and that the agency has failed to enforce minimum annual cut provisions. Furthermore, they believe that the BIA has neglected proper thinning practices, which has resulted in high tree density and declining growth rates of their forests. According to Iadarola (1979, 2) such mismanagement conditions can be found on other reservations as well. The estimated end product of this mismanagement was that these tribes suffered a financial loss of $25.5 million between 1970 and 1974.

Difficulties in the management of the reservation forests cannot be entirely explained by the lack of adequate funds and experts, however; they also arise because of fractured ownership patterns of reservation lands that resulted from the GAA. The configuration of landownership is so intricate and complex in some places that meaningful, efficient forestry operations

27. Because of this lack of funds the BIA spends only $1.60 on average for the care of an acre of forest. Compare with this the U.S. Forest Service, which spends an average of $2.30 for the same area of forest; individual owners, who average $5.50 per acre; and the BLM, which spends as much as $7.40 <16>.

are not possible. On the Colville, Washington; Hoopa Valley, California; Spokane, Washington; Warm Springs, Oregon; Yakima, Washington; and Quinault, Washington, Indian Reservations, for example, a considerable part of the forests are on allotted land (see fig. 2.3). Certain parts of this allotted land—often the areas with the best forests—later came into the hands of interested white parties. The remaining allotments were divided up into ever smaller tracts of inherited land in the course of successive generations, a situation which poses a serious obstacle to forestry development. On the Quinault Indian Reservation, for example, with a land area of only 200 square miles, the GAA is responsible for the fact that today (1985) only 4.6% of the reservation area is tribal land, while 46% belongs to the largest lumber producer of North America,[28] and the remaining 160-acre allotments distributed to tribal members in 1905 had an average of about thirteen owners eighty years later <16>.

From an economic point of view reservation forestry is of far greater importance than reservation agriculture. Today about one-quarter of all Indian country is forested (20,000 square miles), of which 41% is classified as commercially important for timber production. According to Iadarola (1979, 1) this commercial forest area contains 40 billion board feet of marketable timber, representing 1.5% of the nation's total timber-growing stock volume. Although the timber is located on more than a hundred reservations in twenty-two states, it should be noted that almost three-fifths of this estimated timber volume is concentrated on only five reservations: Yakima, Washington; Colville, Washington; Fort Apache, Arizona; Warm Springs, Oregon; and Flathead, Montana (BIA, Fort Apache Agency . . . 1982, 74).

With these renewable natural resources American Indians are able to make handsome profits which, however, are subject to great fluctuations depending on prices and volume of production linked to timber demand (see fig. 7.3). When timber prices peaked in 1979, stumpage revenues (that is, the price purchasers paid for standing timber) amounted to $116 million. By 1981, owing to the rapid decline in timber prices, these revenues had plummeted to $75 million, an amount which was then still twice as high as income from agricultural leases (see table 7.1). Income from forestry is not distributed among as many reservations as income from agriculture, however,[29] and 90% of the reservations producing timber are in Washington, Arizona, Oregon, New Mexico, California, and Montana. Washington tribes have the highest timber income and Montana the lowest. When the

28. This is the Weyerhaeuser Company in Tacoma, Washington, which, with its 32 sawmills, produced more than 3.1 billion board feet of sawn timber (softwood) in North America in 1983 (Forest Industries 1985, 26).

29. Fifty-seven Indian reservations received income from forestry in 1985 <16>.

Indian timberlands are viewed in a cost-benefit context, whereby income is balanced against the running costs which the BIA and the tribal governments must meet to maintain forestry, it becomes clear that in reality only seven Indian reservations make a profit from utilizing their forests <16>. These are the Yakima, Colville, Warm Springs, Quinault, Fort Apache, Hoopa Valley, and Flathead Indian Reservations, which represented almost four-fifths of the total forestry income from all American reservations in 1981 (see table 7.1). Forestry was by far the largest source of income on these reservations and an important renewable tribal resource.

Among the four reservations in Arizona with harvestable timberlands—the Fort Apache, Navajo, San Carlos, and Hualapai—the Fort Apache Indian Reservation is by far the richest in forest resources. Nearly 1,100 square miles, more than three-fifths of this reservation, are forested, while 65% of this is regarded as commercial timberland. With these forest stands, some of which are of very high quality, the Fort Apache Indian Reservation takes a leading position in timber production, not only among reservations, but also on a national level. Only a few reservations in the Northwest produce higher quality timber.

The White Mountain Apaches, who derive their name from a mountain range in the northeast of the Fort Apache Indian Reservation, owe this richness in forests to the privileged location of their tribal lands. Within the predominantly arid state of Arizona, the White Mountains (11,590 feet) and their foothills form a sort of "green roof" (see fig. 6.10) for the state; the only area of comparable altitude and beauty is around the San Francisco Peaks north of Flagstaff. The climatic conditions at these altitudes are conducive to the growth of vast coniferous forests. The highest region of the reservation lies between about 9,000 and 11,000 feet, and is covered by subalpine coniferous forests (see fig. 8.5), which consist predominantly of Douglas fir and Engelmann spruce. Around one-fourth of the reservation's manageable forests are located in this area. Directly below these subalpine forests are extensive, open ponderosa stands, which are mingled with pinyon pines in the lower altitudes (6,000–6,500 feet). Two-thirds of all timber cuts come from these ponderosa pine forests that are unrivaled in the Southwest. Below these pine forests lie the juniper-pinyon woodlands, which take up the largest area of the tribal lands, but which have had little commercial value up to now.

Until well beyond the middle of the nineteenth century White Mountain Apache timber resources remained untouched. Even in the first years after the reservation was established there was little change in this situation except for two rather small, basically noncommercial sawmills, one of them built by the U.S. Army at Fort Apache in 1870, and the other by the

BIA twelve miles to the north of Whiteriver in 1880 (see fig. 9.9) which were intended to supply the army with timber and to provide both employment and building materials for reservation Indians. According to a report of the BIA superintendent in Whiteriver at the time, even a quarter of a century after their construction the timber processed annually at these mills amounted to only 250,000 board feet (McGuire 1980, 86). Before the First World War, however, no real measures had been taken to utilize the tribal forests, although, even at that time, some southwestern foresters were impressed by the considerable timber potential of the Fort Apache Indian Reservation.

It was not until the First World War that these resources were developed, owing to the great demand for timber in the United States during the war. The price of timber rose considerably and in 1917 attracted a large-scale investor, a banker from Flagstaff called Tom Pollock, to the reservation, which was then still quite remote. In the north of the reservation he constructed the Apache Lumber Company, allegedly the largest and most modern sawmill in the entire Southwest (Matheny 1976, 246), around which he built a small company town named Cooley after a pioneer family that was greatly respected at the time. By means of a branchline he also established a railway connection between this newly founded settlement and the Atchinson, Topeka, and Santa Fe Railway at Holbrook, about seventy miles to the north of the reservation. This branchline, intended solely for timber transports, was typical of the "railroad logging era," which in the U.S. came to an end in the 1930s. With investments in the order of $3.5 million (ibid., 245–246), however, Pollock had taken on too much. His business went bankrupt in 1923 and was bought up by J. McNary and W. Cady, two timber contractors from Louisiana. They changed the name of the business and named the company town McNary, which it is still called today (see fig. 9.9). Some 500 African American lumberjacks and sawmill workers whom they brought to the reservation from Louisiana all settled in McNary, segregated from the houses of white skilled workers and from the camps of the Apache and Navajo laborers, which were also separated from each other.

This sawmill, which later changed ownership twice more and finally belonged to the Southwest Forest Industries,[30] acquired ten logging contracts for over $6.5 million on the Fort Apache Indian Reservation between 1917 and 1965. This guaranteed the company roughly 1.3 billion board feet of roundwood (see fig. 9.9), which meant that until 1968 it was by far the largest timber producer on the reservation. During this period it cut and

30. Among the large sawmill enterprises of North America, the Southwest Forest Industries with ten plants and an annual production of 314 million board feet of sawn timber (1984) ranks 33d (Forest Industries 1985, 26).

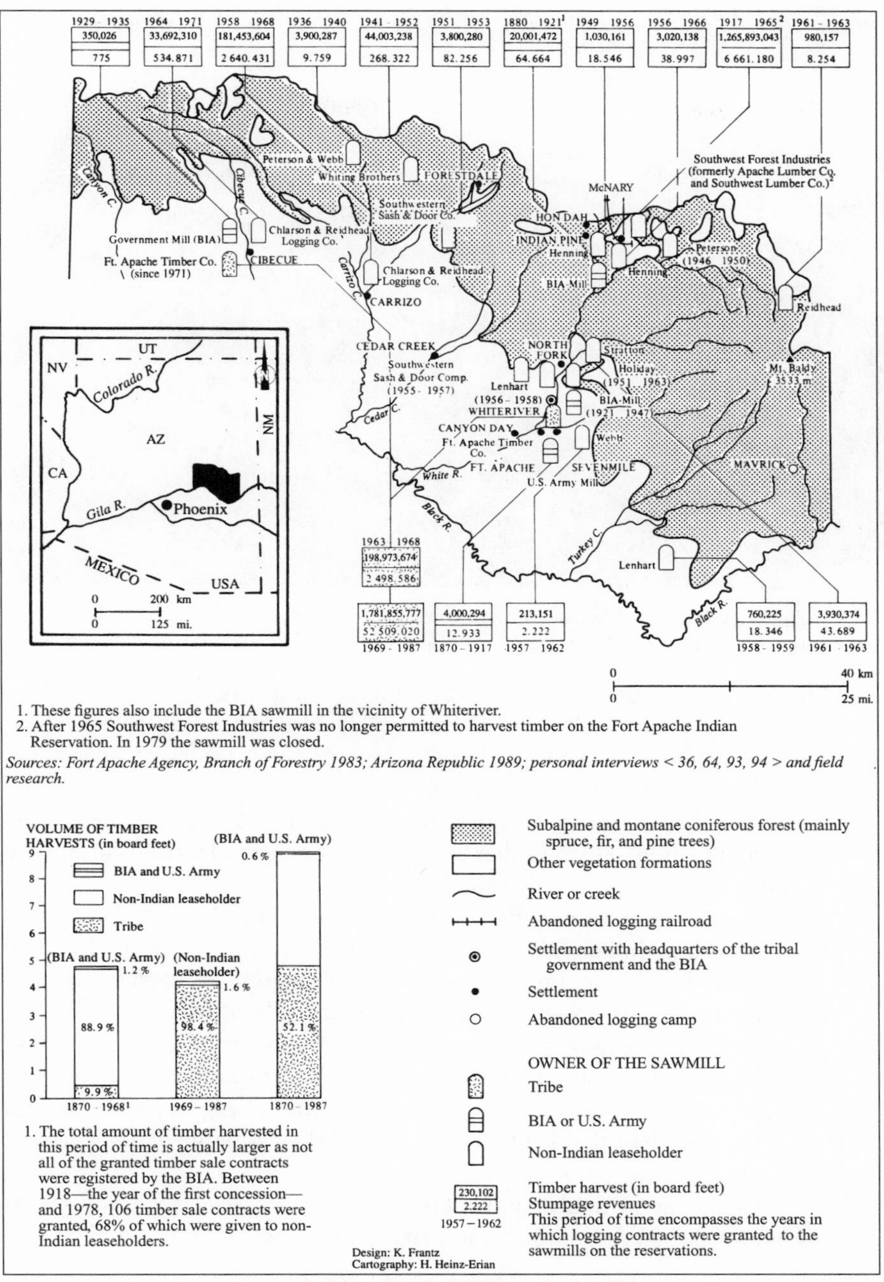

Fig. 9.9: The forest sector on the Fort Apache Indian Reservation in Arizona, 1870–1987

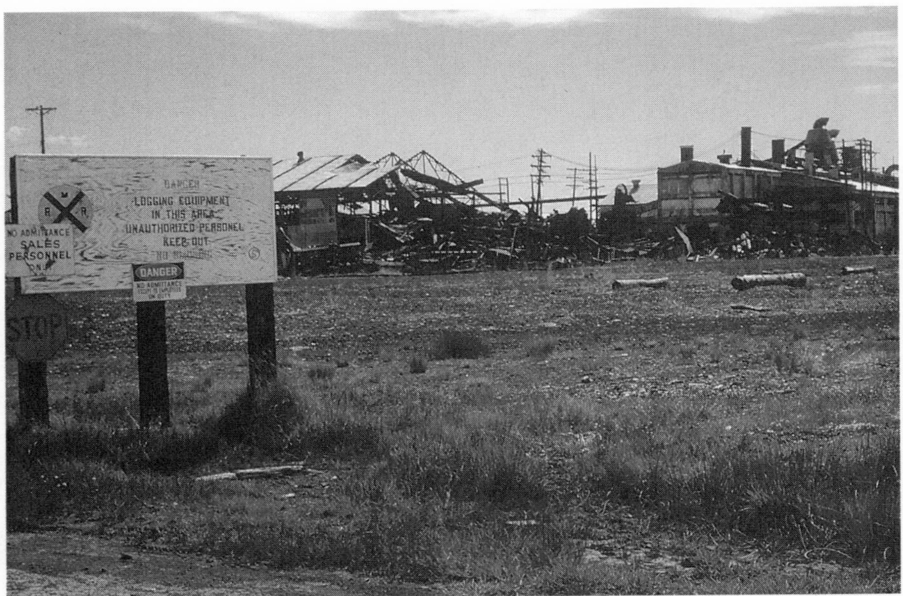

Fig. 9.10: A burnt-down sawmill of the Southwest Forest Industries, McNary, Fort Apache Indian Reservation, Arizona, 1985

processed no less than 63% of the timber harvest on the reservation—almost two and a half times the timber volume of the remaining eleven white logging contractors and sawmill proprietors of the Fort Apache Indian Reservation together, and more than six times as much as that of the tribal timber company founded in 1963 (see fig. 9.9). These facts show clearly that non-Indian timber entrepreneurs benefited most from Apache timberlands until the end of the 1960s.

This situation did not change until 1968 when the Apaches decided not to grant new timber sale contracts to outside businessmen. At the same time they began to put pressure on the sawmill proprietors who were still around to leave the reservation. The plant of the Southwest Forest Industries held out the longest, yet from 1965 on, this sawmill was compelled to purchase all the timber that it processed on the reservation from outside. In 1979 the plant was the object of an arson attack by local people (see fig. 9.10), an incident that led to its closure and the discharge of 250 employees two years later.[31]

A glance at the total amount of timber harvested (see fig. 9.9) shows that after 1968 the White Mountain Apaches were able either to cut almost all

31. McNary, a former site of the Southwest Forest Industries, is a half-deserted settlement with around 240 residents today. Apart from the Apache residents it has a population of around thirty African Americans and thirty whites <95>.

Fig. 9.11: Fort Apache Timber Company, tribally controlled sawmill in Whiteriver, Fort Apache Indian Reservation, Arizona, 1985

of their timber themselves or to have it cut by contracts with non-Indian logging companies. To demonstrate their own need for their forest resources, they constructed a sawmill south of Whiteriver, which with the support of federal EDA funds ($4 million) was gradually expanded (see fig. 9.11). In addition, they acquired another smaller sawmill in Cibecue, a small village in the western part of their reservation. Between 1969 and 1987 these two plants processed about 1.8 billion board feet of timber, which produced $52.5 million in revenues for the tribal government (see fig. 9.9), an annual average of over 93.6 million board feet with a market value of $2.8 million. With this timber production the Fort Apache Timber Company (FATCo) attained the first rank among all tribal forestry enterprises in the U.S. by far,[32] and also placed itself in the ranks of the hundred largest timber companies of North America (Forest Industries 1985, 27). Since 1987 FATCo has gone still further to acquire publicly advertised logging contracts in the Apache Sitgreaves National Forest, east of the Fort Apache Indian Reservation, and formerly a part of the reservation. A tribal resolution made it possible for the FATCo managers to reduce the timber cut on

32. The next ranking tribal sawmill enterprises are those of the Navajos (Arizona, New Mexico, Utah), Warm Spring Indians (Oregon), and Menominee Indians (Wisconsin), which processed between 14.4 and 45.3 million board feet of timber in 1983 (Forest Industries 1985, 27).

the reservation by almost 30% through these contracts <96>, and, in the view of the Apaches, this will help their own tribal forests to regenerate. This measure means that the tribal government accepts an annual loss of revenues of around $770,000 from stumpage payments, for timber that has been cut in the past, and at the same time the leases cause great resentment among neighboring non-Indian sawmill proprietors who believe that they are disadvantaged in their bidding, because of their tax situation and the unfair distribution of federal assistance (Arizona Republic 1989).

The income which this tribe of approximately 8,000 people obtains directly or indirectly from its forestry operations is not limited to stumpage payments. A great deal of it comes from profits that FATCo makes from producing milled lumber. I was not able to ascertain the amount of these profits, but an indication can be gained by looking at the revenues, which amounted to around $20 million in 1985 <36>. The above-mentioned profits and the income from the sale of stumpage are the most important assets in today's White Mountain Apache economy. The wages which FATCo, the BIA, the tribe, and the non-Indian logging contractors on the reservation pay to their over 500 full-time employees, however, are equally important.[33] In 1985 this personal income from timber-related employment totaled almost $7 million, a sum that does not include the wages of the approximately 150 seasonal workers. In view of the great importance that the tribal forests have for the Apache economy, it is surprising that the tribal government has so far been content to leave the management and administration of these forests entirely to the BIA instead of taking a more active role themselves.

33. Of these employees, 59% worked for FATCo in 1985, 28% for the BIA, and 10% for non-Indian businesses. Only 3% received their wages from the tribal government.

10 Manufactures and Services on the Reservations

TRADING POSTS AND THEIR NEW COMPETITORS

For a long time trading posts, as well as the BIA, the U.S. Army, and religious communities, were the only institutions on the generally sparsely populated Indian reservations which provided goods and services.[1] These trading posts were small businesses, and most of them were owned by non-Indian private individuals who sold or exchanged consumer goods for Indian livestock, craft work, and various natural products to satisfy the daily or long-term needs of Indian people.

Until recently these isolated trading posts, authorized by the BIA and distributed to avoid competition with one another, enjoyed an extensive trade monopoly on the reservations, which made it possible for them to manipulate prices to their own advantage. Reservation Indians seldom had cash income and had to rely on whatever terms the merchants set to their wares, and on obtaining credit from the merchants. The consequence was that the local residents grew to rely on trading posts and found that it was difficult to free themselves from this dependency. Lacking any other source of supplies, they could make their purchases only from these creditors, who charged exorbitant prices. In order to be able to afford consumer goods it was often necessary to pledge income which had not yet been earned. Months before a livestock auction, for example, the hoped-for proceeds from the auction may have been guaranteed to some trader, or the Indian purchaser may have promised to make a certain amount of jewelry, rugs, or articles of basketry for them. Often the trading post also functioned as

1. Today quite a number of trading posts still function as post offices, places to cash federal welfare checks, and, as will be explained, banks. In connection with the last two functions some trading posts reserve the right to retain a portion of social welfare remittances when the Indian recipients are badly in arrears with their financial obligations.

a bank which provided the reservation Indians with urgently needed money for their wares.

Even in the first years after the Second World War trading posts still had a virtual monopoly on goods and services, which they later gradually lost with the increased use of the automobile. More reservation Indians could then make their purchases in border towns near the reservations, where they found a wider range of goods and lower prices. In addition, more competition developed on the reservations when a number of tribally owned supermarkets were built toward the end of the 1960s. Granting credit more liberally was the only policy that prevented the loss of customers and kept the attendant loss of business within limits.

It is not surprising, therefore, that as a result of this competitive pressure there has been a marked decline in the number of trading posts during the past two decades,[2] though they have not entirely lost their importance in the small, more peripherally located communities. In larger settlements and in the vicinity of border towns where the local people have numerous opportunities to purchase what they need, it has become increasingly difficult for the owners of the trading posts to make a profit, as they have not been able to adjust to the new price levels in these areas. The customers of these remaining trading posts are, for the most part, people who have no regular source of income or means of transportation, and who therefore still rely heavily on the trading posts for loans when they are in financial difficulties (see figs. 10.1, 10.2).

The Fort Apache and Hualapai Indian Reservations in Arizona demonstrate these conditions, and to compare prices, I put together a basket of goods in various shops on and off these reservations (see fig. 10.3). In Whiteriver, the main settlement of the White Mountain Apache, the price level of forty-two different articles at the Lee Mercantile Company, one of the two local trading posts,[3] was on average almost 10% higher than in the tribal supermarket established there in 1977 (see fig. 10.4). For some items, including fruit and vegetables, the prices were on average 17% higher. Compared with the supermarkets in Pinetop, the nearest non-Indian settlement, the average difference in price for fruit and vegetables came to 32%.

2. For example, at the end of the Second World War there were still 146 trading posts on the Navajo Indian Reservation (Kluckhohn and Leighton 1962, 79), but only 60 were still open in 1986. In the case of the others, either the tribal government did not renew the licenses authorized and issued by the BIA, or the trading posts lost so many customers that they had no interest in remaining open. On the Fort Apache Reservation the number of trading posts declined from 10 to 6, and on the Gila River Indian Reservation from 8 to 3 <99, 31> (see fig. 9.2).

3. It was not possible to compare prices at the second trading post because brands and quantities were not always clearly marked.

Fig. 10.1: Hubbell Trading Post, Ganado, Navajo Indian Reservation, Arizona, 1992

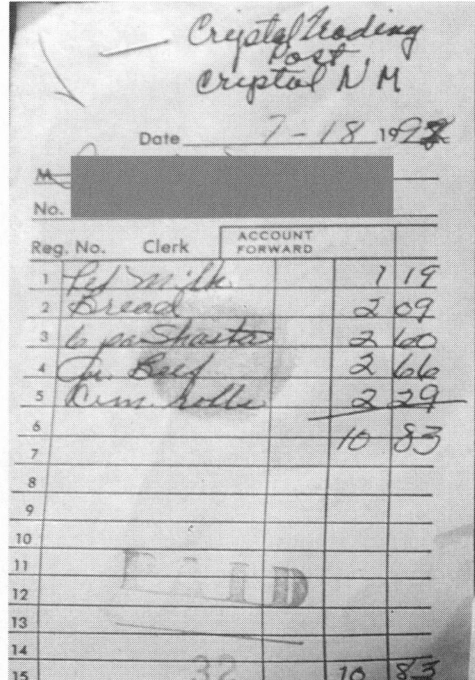

Fig. 10.2: Thumbprint as signature of a seventy-two-year-old customer at the Crystal Trading Post, Navajo Indian Reservation, New Mexico, 1992

According to the licensed owner of the Lee Mercantile Company trading post, as a result of competition the prices for goods in his store declined by 3% to 4% after the tribal supermarket opened <99>, drastically narrowing his profit margin. Furthermore, while in the past he had been able to count on immediate payment in cash for around 30% of his sales, this had dropped to 15% by the middle of the 1980s. He had allowed his Indian customers to postpone payment until they received wages or some other income and could pay off their debts, including the interest that had accumulated in the meantime. In the past ten years the owner of this trading post has been compelled to write off only a small part, around 2%, of the credit he had granted, because of inability to pay on the part of his debtors. Even so, because of the decline in the number of customers and the decrease in cash payments, as well as customer debts that could no longer be collected and, above all, almost impossible conditions that the tribal government attached to a renewal of the license, the Lee Mercantile Company finally had to close down in 1988.[4]

Compared to Whiteriver, the prices of the trading post at Peach Springs,

4. It was more than a hundred years ago that the first member of the Lee family came to the reservation to found a trading post. Like the traders on many other Indian reservations he came in the wake of the military.

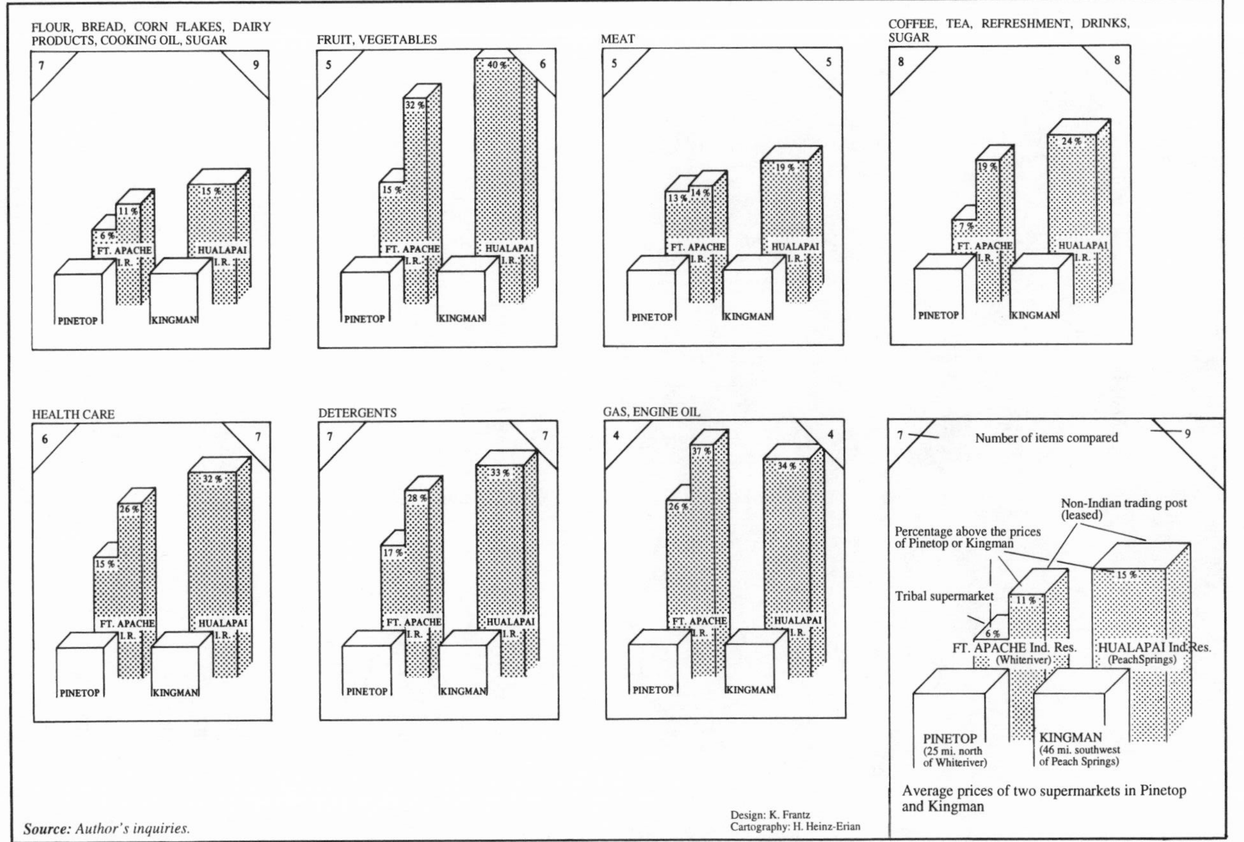

Source: *Author's inquiries.*

Fig. 10.3: The prices found in the supermarkets and trading posts in Whiteriver, Fort Apache Indian Reservation, Arizona; and Peach Springs, Hualapai Indian Reservation, Arizona, compared with the prices in the markets of the two nearest non-Indian settlements, 1986

Fig. 10.4: Tribally controlled shopping center of the White Mountain Apache, Whiteriver, Fort Apache Indian Reservation, Arizona, 1985

the only settlement of any importance on the Hualapai Indian Reservation, were considerably higher owing to the lack of other stores on the reservation. The only competitors of the licensed owner of the trading post were in Kingman, a small town about fifty miles to the southwest. On the basis of forty-six items that were compared, the prices here were, on average, more than 27% higher than in the supermarkets at Kingman, and 40% higher for fruit and vegetables (see fig. 10.3).

OTHER BUSINESSES IN THE SERVICE SECTOR FOR THE GILA RIVER, HAVASUPAI, AND FORT APACHE INDIAN RESERVATIONS

Except for the tribal supermarkets and the trading posts, which are almost exclusively in the hands of non-Indians or mixed-bloods, the private service sector on the reservations is, for reasons which have already been discussed, very small. Especially lacking are businesses owned by individual tribal members, despite the fact that initiatives from privately owned American Indian enterprises in particular can count on strong financial assistance from the federal government. The general lack of Indian entrepreneurs on the reservations is a consequence of prevalent disadvantages caused by their geographic location, their cultural beliefs and values, and their lack of investment capital. An important additional factor is that very few reservation Indians have experience or training in how to run a business.

For example, my fieldwork showed that on the Gila River Indian Reservation in Arizona, which had more than 8,000 residents in 1986 and covered an area of 574 square miles, there were only three retail shops and four additional private businesses in the service sector in that year, other than the trading posts already mentioned (see fig. 9.2). These facilities included a branch bank (fig. 6.11) and a laundromat, both leased to non-Indians; two stores, one for video rentals and one for leather goods; two smoke shops and another laundromat, with American Indian leaseholders. Various other businesses were tribally owned, including two supermarkets, a small grocery and convenience store, a restaurant, a laundromat, two gas stations, and a store for auto replacement parts attached to a repair shop. At the time all these service sector businesses employed thirty-three tribal members, only 1.8% of the gainfully employed.

In addition, the Gila River Indian Reservation accommodated a number of other facilities in the service sector, although with the exception of rental income, the tribal members derived hardly any benefit from them. These were a satellite tracking station operated by the U.S. Navy, nine radio and television transmission stations, and, above all, a recreational park used chiefly to entertain the residents of the Phoenix metropolitan area. This 440-acre recreational park, the Firebird International Raceway Park, is located near the reservation boundary nearest Phoenix (see figs. 9.2, 10.5). Managed by non-Indians, this facility appears to be a rather unusual component for an Indian reservation, with not only an outdoor amphitheater to accommodate 20,000 people for big events such as pop concerts, but also various courses for drag and off-road racing, all with freeway access. In the middle of this desert area there is also a 120-acre constructed lake, fed with groundwater which has to be pumped from a considerable depth for motor boat races and waterskiing tournaments in a region where water is very scarce.

To the north of the Firebird International Racing Park a National Football League stadium and a large complex for pari-mutuel wagering are planned (see fig. 10.6). The latter facility, to be created by a Miami consortium at a cost of an estimated $25 million, is strongly opposed by Arizona's attorney general, and has divided the Interior and Justice departments. It is said that this thirty-five-acre complex will provide approximately 500 jobs when completed, and that one-third of the workforce will be made up of members of the Gila River Indian Community.[5] The project will yield a 7.5% gaming tax to be collected by the tribe, which, it is hoped, will amount

5. This gambling hall would be the third of its kind in the immediate vicinity of a large city in Arizona to begin operation since the early 1980s. The same company that wanted to

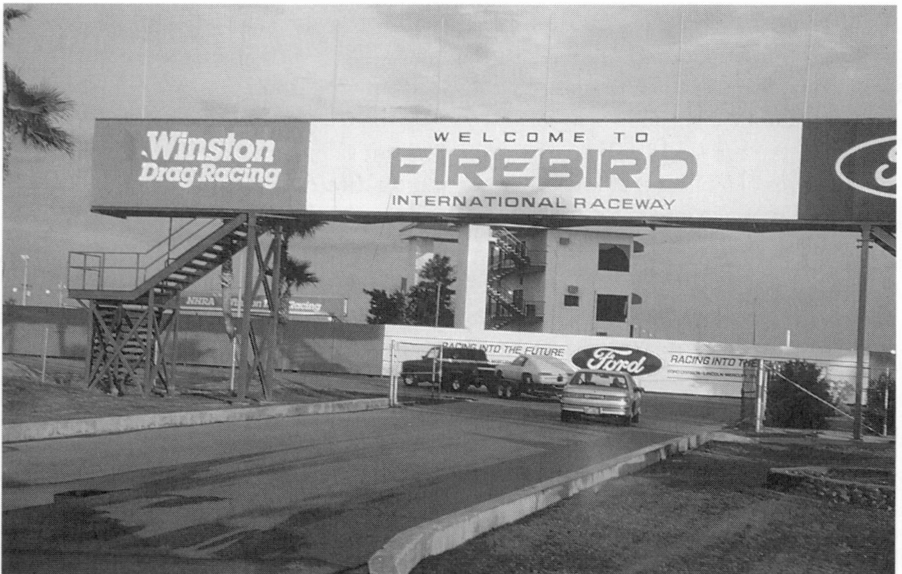

Fig. 10.5: Firebird International Raceway Park, Gila River Indian Reservation, Arizona, 1986

to $7 million in the first year, and an annual lease rental of $315,000. This company has promised to donate $1 million a year to a special fund for higher education and health care for the Pimas (Arizona Republic 1985a and 1985b) in order to make it attractive for the tribal government to approve their plans. The Florida investors seem to have found a gold mine with their projected establishment. (With its proximity to Phoenix in mind, the company is planning a complex that will accommodate 5,000 people.) Whether this facility will also be a gold mine for the Pimas remains to be seen. The managers of the casino on the Fort McDowell Indian Reservation have kept only a few of the financial and employment promises made to the Yavapai-Apache Indians.

Service facilities such as the three companies mentioned above are not usually found on reservations. They manifest the special location of Indian country bordering major urban areas, but as one can see from the example of some traditionally minded reservations in the vicinity of Albuquerque, New Mexico, the potential given by such an ideal geographic location is not necessarily made use of by the local people.

---

establish gaming on the Gila River Indian Reservation in 1986 had previously opened a casino on the small Pasqua Yaqui Indian Reservation close to Tucson. Still another casino was built on the Fort McDowell Indian Reservation northeast of Phoenix with investors from Kansas City. Another gaming facility is planned for the Papago Indian Reservation.

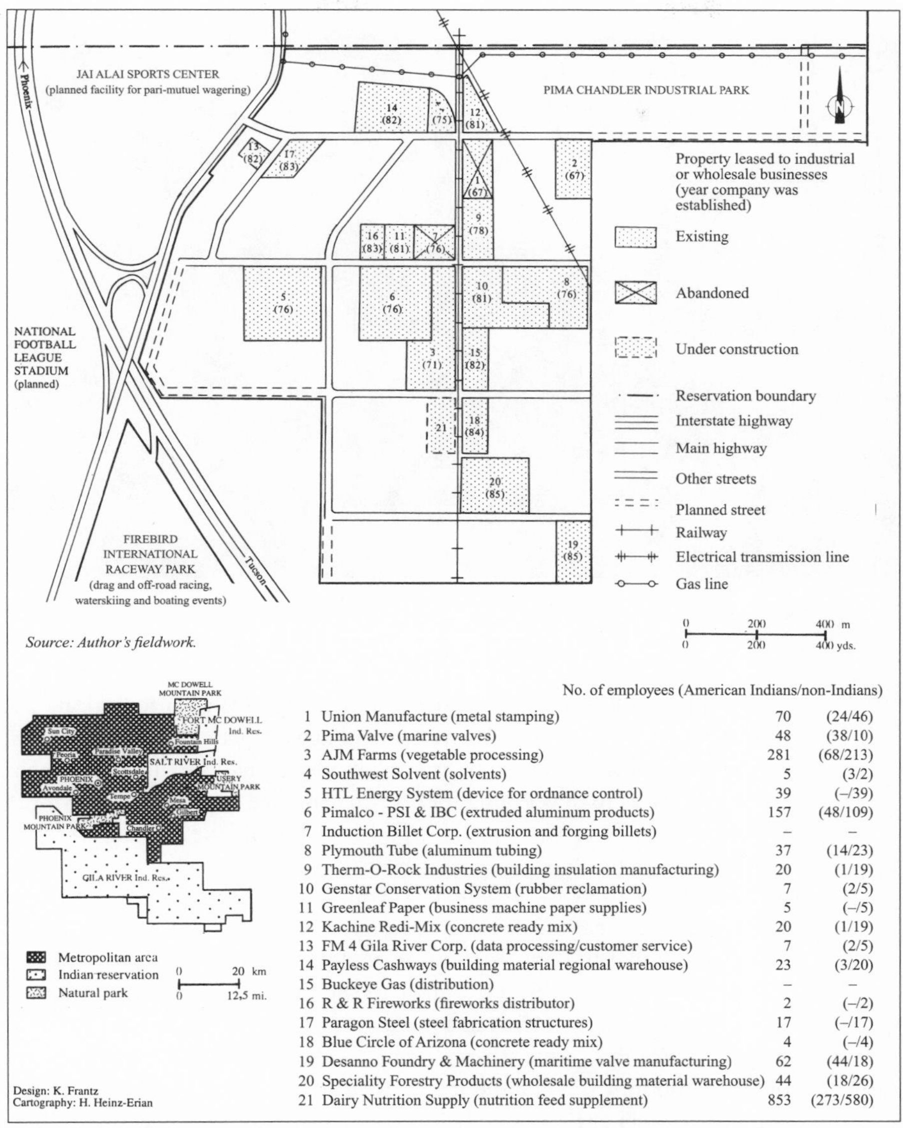

		No. of employees	(American Indians/non-Indians)
1	Union Manufacture (metal stamping)	70	(24/46)
2	Pima Valve (marine valves)	48	(38/10)
3	AJM Farms (vegetable processing)	281	(68/213)
4	Southwest Solvent (solvents)	5	(3/2)
5	HTL Energy System (device for ordnance control)	39	(–/39)
6	Pimalco - PSI & IBC (extruded aluminum products)	157	(48/109)
7	Induction Billet Corp. (extrusion and forging billets)	–	–
8	Plymouth Tube (aluminum tubing)	37	(14/23)
9	Therm-O-Rock Industries (building insulation manufacturing)	20	(1/19)
10	Genstar Conservation System (rubber reclamation)	7	(2/5)
11	Greenleaf Paper (business machine paper supplies)	5	(–/5)
12	Kachine Redi-Mix (concrete ready mix)	20	(1/19)
13	FM 4 Gila River Corp. (data processing/customer service)	7	(2/5)
14	Payless Cashways (building material regional warehouse)	23	(3/20)
15	Buckeye Gas (distribution)	–	–
16	R & R Fireworks (fireworks distributor)	2	(–/2)
17	Paragon Steel (steel fabrication structures)	17	(–/17)
18	Blue Circle of Arizona (concrete ready mix)	4	(–/4)
19	Desanno Foundry & Machinery (maritime valve manufacturing)	62	(44/18)
20	Speciality Forestry Products (wholesale building material warehouse)	44	(18/26)
21	Dairy Nutrition Supply (nutrition feed supplement)	853	(273/580)

Fig. 10.6: The Pima Chandler Industrial Park on the Gila River Indian Reservation in Arizona, 1986

Fig. 10.7: The idyllic seclusion of the Havasu Falls, Havasupai Indian Reservation, Arizona, 1992

The same tendency to reject development applies to reservations far from large urban centers, which are particularly attractive and sought after by recreational users because of their isolation. Here, too, the local people are not interested in developing their infrastructure or allowing non-Indians to add businesses that would utilize and exploit the recreational potential of their land. Moreover, the projects envisaged are often not in the best interests of the local people, and would not satisfy any urgent need. As a result, most projects devised by outside investors are never begun.

The Havasupai Indians, for example, elected not to have their idyllic seclusion at Havasu Creek, a tributary of the Colorado River at the Grand Canyon (see fig. 10.7), destroyed by a road or a cable railway, although they did agree to a campsite and a forty-eight-bed hotel, constructed with funds from the Economic Development Administration. Although this 1970s investment attracts approximately 15,000 visitors to the Havasupai Indian Reservation every year <100>, resulting in a mild form of tourism, the responsible tribal council members do not consider this a threat to the sociocultural fabric of their community or their way of life. Aside from these facilities which bring income to the tribe but from which mainly tourists benefit, the reservation has only one small cafe. There is no grocery store or any other kind of store to be found, and it takes the local people at least one day to leave the reservation to do their shopping.

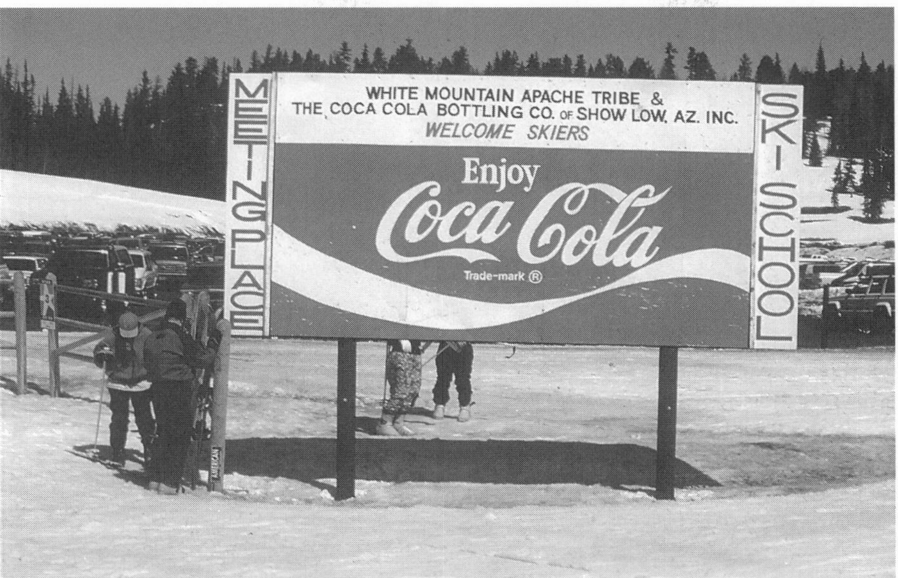

Fig. 10.8: Tribally owned ski center of the White Mountain Apache, Fort Apache Indian Reservation, Arizona, 1992

In contrast to the Havasupai Indians, the White Mountain Apaches have decided to open a portion of their reservation to mass tourism, namely the area northeast of the 11,350-foot Ord Peak. Here, at the instigation of outside investors, the Sunrise Park Ski Resort (see fig. 8.5) was built using federal, tribal, and private funds. With its eleven lifts and forty-five ski runs, this resort has become Arizona and New Mexico's largest skiing area (see fig. 10.8). Of the approximately twenty American Indians employed there, all except for two Apache board members appointed by the tribe have subordinate, auxiliary posts,[6] and earned a total of only about $250,000 in 1985 <101>. Except for these wages, this ski center has actually brought only losses to the tribe so far, and the individual Apache tribal members make hardly any use of the facility. Without doubt, Pinetop and Show Low, two non-Indian communities about thirty miles away which border the reservation to the north, benefit most from this resort. This is where the overwhelming majority of the skiers spend the night, since there is only a small, tribally owned hotel at Sunrise Park Ski Resort with ninety beds. Many of the business owners in the area maintain that without the winter sport activities on Apache lands the

6. Fifteen years after the Sunrise Park Ski Resort was opened there was not a single Apache among the approximately two dozen ski instructors working there at the height of the season.

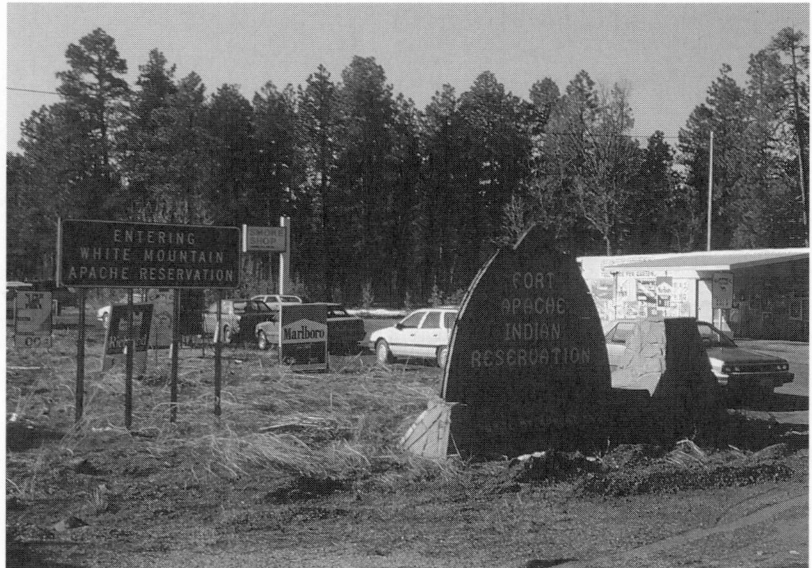

Fig. 10.9: Smoke shop in the northern borderland of the Fort Apache Indian Reservation, Arizona, 1985

lodges, restaurants, and tourist-oriented stores in Pinetop and Show Low could not stay open during the winter season <102–105>.

Apart from the ski resort, the service sector on the Fort Apache Indian Reservation, including the shopping facilities, is, as on most other reservations, quite insufficient. Except for a store which sells clothes, two smoke shops (see fig. 10.9), a pizzeria that is usually closed, and three trading posts leased by mixed-bloods, all business establishments and all other services are either tribally owned or belong to non-Indians. There is also the Commercial Center Enterprise in Whiteriver. The supermarket described earlier in this chapter is located there as part of a tribally owned complex that includes a motel, a restaurant, a movie theater, and a gas station, all operated by the tribe, as well as a branch bank and a video rental shop leased by someone off the reservation.

COMMERCIAL AND INDUSTRIAL ENTERPRISES IN INDIAN COUNTRY: A STOCKTAKING

Except for a few sawmills which were established by special arrangement with the BIA during the First World War, manufacturing is a rather recent phenomenon on Indian reservations. Apart from these sawmills, only four other industrial plants were located on the reservations before 1960 (Sorkin

1971, 81), despite the fact that the federal government had already created the legal framework for such enterprises under the Indian Reorganization Act in 1934 <29>. Before the 1960s, therefore, almost all Indians looking for a job in manufacturing had to work off the reservation, a move that frequently led to the loss of status as federally recognized American Indians. As we saw above, thousands of American Indians actually took this step during the Second World War and the postwar period..

A new political initiative for a selective industrialization of Indian country came in 1965 with the Public Works and Economic Development Act, which funded the Economic Development Administration (EDA) and provided the financial basis to create industrial ventures in chronically underdeveloped regions of the United States, including the Indian reservations. EDA funds were to create jobs for the great numbers of unemployed tribal members on reservations, and helped to establish many companies and a number of industrial parks. On the basis of this new policy, which may be considered a part of President Johnson's War on Poverty, the number of industrial enterprises increased from 24 in 1964 to 110 in 1968 (BIA, Washington Headquarters 1968, 8). According to the available records, around 8,500 people were employed in these plants in 1968, though only a little more than 48% of the workers were American Indians. It is no longer possible to ascertain how many of these enterprises belonged to individual Indians or to what extent the tribes were involved, but it may be assumed that even by this time non-Indian enterprises played a major role.

Precise information regarding the number of industrial enterprises on the reservations today are unfortunately not available. A task force on Indian Economic Development, set up by the Department of the Interior, however, came to the conclusion that the reservations and the Indian trust lands including the historic areas of Oklahoma in 1985 had 103 businesses with more than ten employees (see fig. 10.10). On the other hand, a BIA survey identified 124 such businesses by working through the classified telephone directories (U.S. Department of the Interior, BIA 1983). More than 8,400 people were employed in these businesses, but it is not known how many of these were American Indians.

The actual number of manufacturing establishments on reservations is surely somewhat higher. According to the 1980 census, in 1976 almost every tenth gainfully employed reservation Indian, at least 8,300 people, had jobs in the secondary sector (see tables 5.7 and 5.8). However, these figures cannot obscure the fact that the Economic Development Administration, which was intended to start a nationwide process of industrialization on the reservations, has not achieved its goal. The fact is that with 9.3% of reservation Indians working in the industrial sector, this percentage is still significantly lower than the U.S. average, which was 22.1% in 1980.

What are the characteristics of these reservation enterprises in the secondary sector? Most of the 103 previously mentioned businesses are small by American standards: 61% have fewer than 50 workers and almost 80% fewer than 100 employees. Only the three biggest businesses can be considered large-scale or at least medium-size industrial plants. These are a non-Indian clock factory on the Choctaw Indian Reservation in Mississippi with 900 employees, a forest products enterprise on the Navajo Indian Reservation in New Mexico with 450 employees, and a supply plant for electronic components and equipment on Choctaw land in Oklahoma with 400 employees. The two last-mentioned enterprises were joint ventures between the respective tribes and outside entrepreneurs (see fig. 10.10).

The extent to which the industrial sector on the reservations differs from everywhere else in the United States becomes clear when the question of ownership is considered. Whereas private business is the basis for the production of goods and services in the United States, and accounts for about 88% of the gross national product (Presidential Commission 1984, 37), on the reservations government-run tribal, welfare-oriented businesses predominate. One-quarter of reservation enterprises in the manufacturing sector are tribally owned. The tribes, with partners from outside, also participate in numerous joint ventures, either with their own funds or with federal funds, and an additional one-third of the businesses are in non-Indian hands. Private ownership by American Indians is hard to find. My inquiries showed that on Arizona's reservations, for example, where 24% of the BIA-identified manufacturing businesses on all U.S. reservations are located, there was not a single business that was privately owned by an American Indian.

Another striking fact is the uneven geographical distribution of manufacturing facilities in Indian country. Twenty-seven percent of all industrial companies on reservation land and 42% of all reservation Indian workers in the secondary sector (see fig. 10.10) are located on Choctaw land in Mississippi and Oklahoma, and on the Navajo and Pima Indian Reservations in the Southwest. The top fifteen tribes claim about 51% of all industrial companies on reservation land and 81% of all reservation Indian workers in the secondary sector. It is noticeable that on reservations near metropolitan areas the industrial plants are located in parts of the reservation immediately adjacent to the urban areas. This makes it possible for non-Indian employees and owners to reach their places of work faster, and they then often have no contact with the rest of the reservation.

A good many of the enterprises on the reservations either involve labor-intensive, low-wage business, or deal primarily with raw materials and their processing. The labor-intensive businesses include companies that produce electronic components and equipment, leather articles, textiles, and

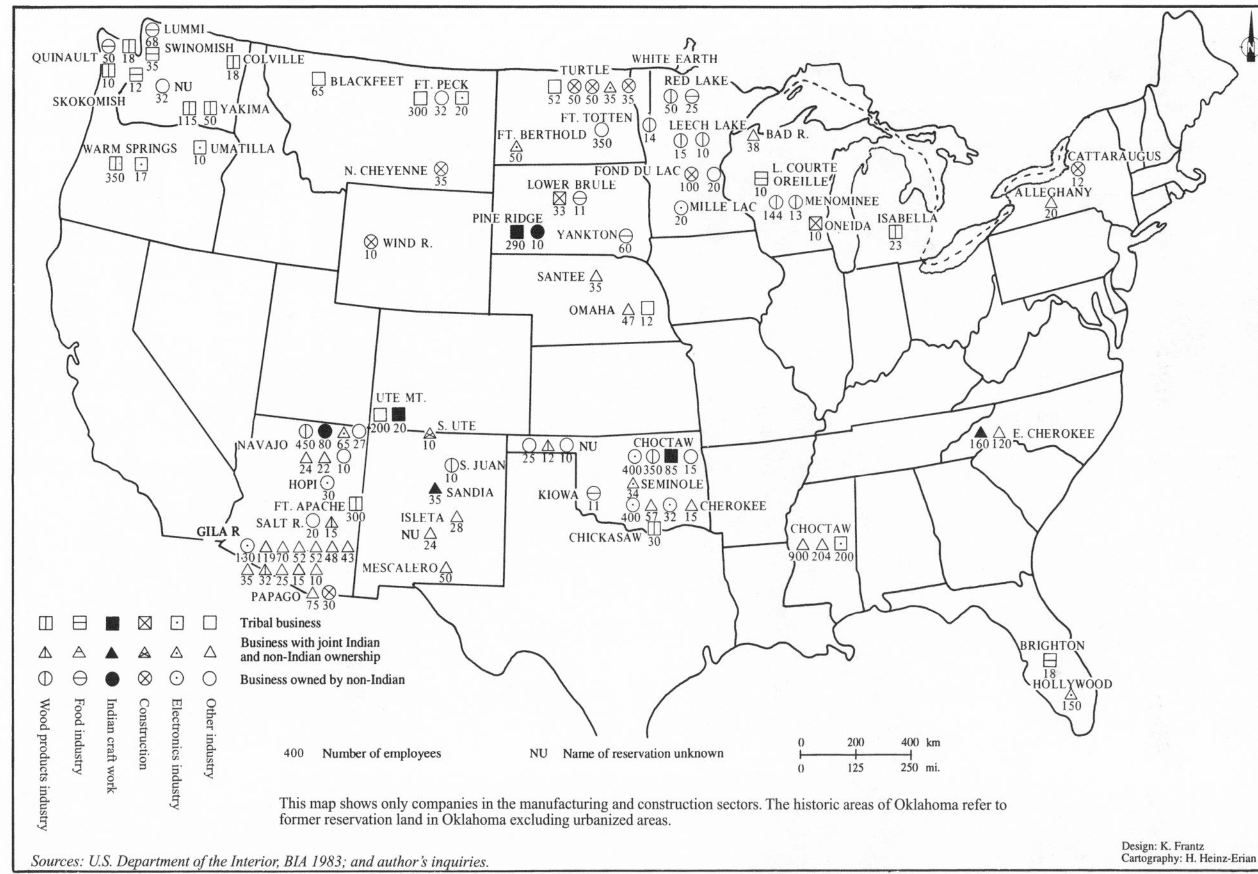

Fig. 10.10: Industrial enterprises with more than ten employees on Indian reservations and the historic areas of Oklahoma, 1983

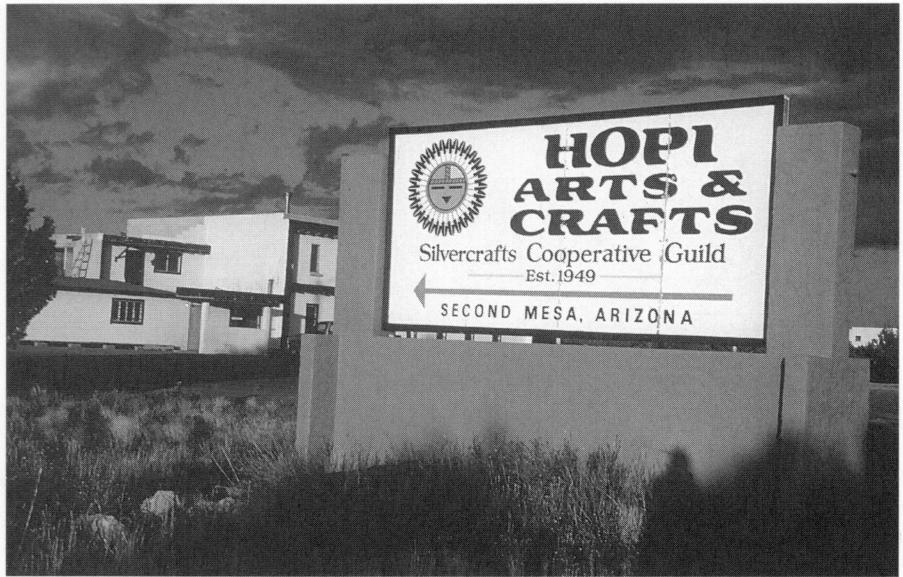

Fig. 10.11: Arts and crafts cooperative, Hopi Indian Reservation, Arizona, 1992

clocks; and some firms which produce articles of Indian craftsmanship that are partly machine-made, such as, jewelry, pottery, moccasins, and Indian dolls, as well as quilts and teepees (see fig. 10.11). These industries, which benefit from the fact that workers on reservations are rarely unionized, constitute more than 27% of the companies on the reservations, and 45% of the reservation Indians employed in the secondary sector work for them. The presence of large electronics plants is particularly striking. Since the Vietnam War these enterprises have started branch factories on reservation land with subsidies or direct funding through defense contracts from the Pentagon. Today there are branch businesses found on sixteen reservations where, as subcontractors,[7] they produce prewired switchboards, transistors, and other electronic components for the semiconductor industry (see fig. 10.12). The finishing stages in manufacturing which require highly skilled labor take place outside the reservations.

The enterprises most frequently found on reservations are those related to raw material production. Thus at the beginning of the 1980s, according to a BIA survey (U.S. Department of the Interior, BIA 1983), there were twenty-seven forest or wood products enterprises, seventeen food processing plants, and five companies which processed mineral resources on reser-

7. The largest of these manufacturing subsidiaries for electronic components, the Fairchild Semiconductor Plant on the Navajo Indian Reservation, still had around 1,200 Indian employees in 1971, but a few years after opening, this branch closed down.

Fig. 10.12: Subsidiary plant of an electronics company, Salt River Indian Reservation, Arizona, 1985

vations.[8] These companies comprised 40% of the enterprises on the reservations and employed 35% of all reservation Indians employed in the secondary sector. Considering how rich the reservations are in natural resources, however, the Indian raw materials are not processed to the extent they could be in Indian country by local enterprises, and thus do not meet the potential for value added to the product. Most of the mineral resources and agriculture, forestry, and fisheries products still leave the reservations before being refined.

In addition to hosting labor-intensive industries which require a great deal of raw materials, Indian lands are often used as locations for companies as testing grounds or for businesses that create pollution or have a negative effect on the environment. Such companies enjoy the benefit of a long-standing laissez-faire policy that was reinforced in the Reagan era Indian Economic Development Act (1987). This act stipulates, among other

8. Among them were seven large sawmills (Leech Lake, Red Lake, and White Earth Indian Reservations in Minnesota, and Navajo Indian Reservation in Arizona), five furniture or coffin manufacturers (two each among the Chickasaws in Oklahoma and the Yakima Indians in Washington, and one on the Gila River Indian Reservation in Arizona), a producer of prefabricated houses (Salt River Indian Reservation, Arizona), and seven companies which process seafood or freshwater fish (Annette Island in Alaska, the Lummi, Quinault, Skokomish, and Swinomish Indian Reservations in Washington, the Brighton Indian Reservation in Florida, and the Red Lake Indian Reservation in Minnesota).

things, that federal environmental policies and regulations shall not apply to certain areas on the reservations, including specifically designated enterprise zones (see Big Mountain Aktionsgruppe 1989, 30). Certain industrial plants and environmentally damaging facilities which non-Indian communities, particularly those in populous areas, did not want, have been established on reservations in accordance with this policy. The Gila River Indian Reservation (figs. 9.2 and 10.6), in proximity to the two million residents of Phoenix, is a good example for this trend. Since 1975 an urban sewage treatment plant has been set up there, as well as a recycling plant for ethylene; a hide tanning factory; a plant producing rocket propellants, fireworks, and solvents; and a rubber reclamation plant. In addition, testing grounds for jet engines, tractors, and farm equipment have been established on this reservation.

INDUSTRIAL PARKS, A NEW PHENOMENON OF THE INDIAN RESERVATIONS: THE CASE OF ARIZONA

The first industrial parks were created on some reservations in 1965 in connection with the selective industrialization of Indian lands. The federal government hoped that these parks would be incentives to stimulate the reservation economy and at the same time promote the creation of jobs to help relieve unemployment. These parks were constructed with substantial federal assistance on several reservations, and also in other economically distressed areas of the country, including downtown neighborhoods of large cities. Around forty industrial parks have since arisen on Indian reservations, without having stimulated any significant growth of tribal economies. The tribes have supplied not only their land but also a great deal of financial support to provide the necessary infrastructure for the parks, although these investments have not turned out to be profitable except in a very few cases. If anyone has benefited from these parks to any appreciable extent it has been the non-Indian entrepreneurs who have settled on the reservations. As has already been pointed out, they have been able to count on federal assistance from the various economic development programs and tribal funds, and have received loans on easy terms, tax relief, and often even construction of factory facilities on a turnkey basis. They also had access to a large and inexpensive labor pool, while their own investments were often very limited. Despite the opportunities presented by all these favorable conditions, relatively few non-Indian businesspersons have taken advantage of them, with the consequence that today many of these industrial parks are vacant or not used to their full capacity. The BIA is aware of this situation, since one of its surveys shows that barely

10% of these sites, created for the development of reservation economies, have actually been used <97, 106, 107>.

Of the twenty-one reservations in Arizona, only six had industrial parks in 1986.[9] The six reservations were the Ak-Chin, Colorado River, Gila River, Salt River, and San Carlos, as well as the San Xavier section of the Papago Indian Reservation. Two industrial parks on the Colorado River and San Carlos Indian Reservations had already been vacant for years at the time of my inquiries. On the San Xavier section of the Papago Indian Reservation there was only one tenant left in the industrial park, established in 1970, after several companies had come and gone. This was a non-Indian distributor of heavy equipment, whose business included repair and service facilities with seventy-five employees. There was only one company in the small industrial park of the Salt River Indian Reservation and two companies in the industrial park on the Ak-Chin Indian Reservation. The Salt River Indian Reservation had a factory for prefabricated homes and cabins with fifteen employees, and the Ak-Chin had a tribally owned grain storage and a meat packing company which employed eighteen people.

In comparison with the Ak-Chin, Papago (San Xavier), and Salt River Indian Reservations, and with most other U.S. reservations, the Gila River Indian Reservation has been exceptionally successful in its industrial parks (see fig. 9.2), yet even here, of the 1,625 acres set aside for development only about a quarter is used as intended. In the three parks of approximately equal size that were established on this reservation in the late 1960s and early 1970s there were twenty-eight business enterprises in 1986, eighteen of which were industrial in nature. They provided jobs to 976 employees, of whom barely one-third were American Indians. Even so, the industrial parks could claim second place on the Gila River Indian Reservation as employers of the Pima tribal members, first place being taken by the tribal administration, and third place by the BIA. Two of these parks, the San Tan and Pima Coolidge Industrial Park with respectively four and five small businesses, are not sufficiently utilized today and because of their geographical location bring in a rental payment of only $1,500 per acre. The Pima Chandler Industrial Park (see fig. 10.6) has great locational advantages because of its proximity to Phoenix, and these advantages manifest themselves in rental leases, which are from three to four and a half times higher than those of the two more isolated sites. These high rental leases on the Pima Chandler Industrial Park seem justified because the premises have direct access to a freeway and a state highway, and can be reached in twenty minutes from the center of Phoenix and in even less time from the

9. See chapter 9, note 3. The Navajo Indian Reservation is not considered here, because I was not able to include it in my inquiries.

communities of Chandler, Gilbert, and Mesa, which are rapidly expanding both economically as well as in population. Moreover, this industrial park, advantageously located in the midst of Maricopa County,[10] also has access to rail connections. The Pima Indians have made water, gas, and electricity available, as well as sewer systems, both with tribal and federal funding. In the future the park may even have access to a modern airport only three miles away, because the city of Phoenix hopes to obtain permission from the tribe to develop Memorial Airfield, constructed in the Second World War and hardly used since, into a large alternative airport to Phoenix's crowded Sky Harbor (see fig. 9.2).

In view of this advantageous geographical location, as well as its good infrastructure, it is understandable that the Pima Chandler Industrial Park is considered a highly attractive site for the thirteen industrial and six service enterprises which it contains today and which employ 853 people. This enterprise zone still remains attractive in spite of a tribal tax of 4% that is now levied instead of state property and sales taxes. If the entrepreneurs in this area, none of whom are American Indians and who mostly have leases for twenty-five years, were to relocate their businesses off the reservation they would have lease payments from 12% to 20% higher and their labor costs would increase from 30% to 45% <31, 70, 75, 86, 108>. It is not inconceivable that as a consequence of these unequal conditions, the Gila River Indian Reservation may yet become an outstanding industrial center alongside the reservation's function as the "vegetable garden" of Phoenix. It remains to be seen whether future tribal governments and the majority of Pima Indians would favor such a development, however. Many tribal members already feel that they are getting very little benefit from the industrial developments, which are not easily accessible from their communities unless they own cars. This attitude is not surprising when one considers that in 1985 the Pimas made a profit of only $300,000 from their three industrial parks, not counting unascertainable income from taxes and wages <31, 70, 108>. Nor has there been any appreciable decline in the unemployment rate of 31% in the past few years.

10. No less than 74% of all available jobs in Arizona's industrial sector are located in this county (Valley National Bank 1983, 16).

11 Epilogue

The geography of Indian country is unique to American geography. It is composed of unconnected land fragments from diverse cultures with diverse histories, yet all is one place with an overriding communality. That place is where the history, morals, economics, politics, and spirit of two cultures meet. It is within these cultural boundaries that Indian country is defined.

The intersection of those two cultures, one indigenous, one exotic, has been a severe test of the most basic values of each. Indian country has tested American commitment to the most fundamental democratic principle, equal rights under the law. For the same reasons the commitment of American Indians to preserve the integrity of community and homeland has been tested again and again. Indian country is as much a process of cultural definition as it is a place.

The last decade of the twentieth century has again brought a major shift in the boundary between the two cultures that shape Indian country. Federal Indian policy has been framed to seek a "government to government" relationship between each of the tribes and the United States of America (Public Law 103-413,1994) which reaffirms American commitment to coexistence with sovereign Native people.

The Tribal Self-Governance Demonstration Project Act was introduced in 1987 under President Bush, but implementation stalled until 1990 (Holmstrom 1995, 12). This act authorized the Department of the Interior to bypass the Bureau of Indian Affairs (BIA) and to provide block grants initially to ten tribal governments (Levitan and Miller 1993, 63). Subsequently, in 1994 the Tribal Self-Governance Act was passed (The Indian Self-Determination and Education Assistance Act, as amended by P.L. 103–413, P.L. 103–435, and P.L. 103–437, 1994), which expanded the number of tribes receiving block grants and significantly reduced the role of the BIA in tribal affairs. The BIA's capacity as manager of reservation resources and administrator of federal Indian programs and funding has decreased as more and more tribes assume direct control through self-governance and self-determination initiatives <121>.

At the same time, a substantial number of the 328 federally acknowledged tribes (U.S. Government Printing Office 1996, 58211–58216) have entered a new era of prosperity after a second affirmation of tribal sovereign powers came from the federal courts, which ruled that tribes could operate casinos on reservations if that type of gambling was legal in the states in which the reservations were located.

The "government to government" policy, coupled with the unprecedented wealth many tribes have derived from gaming profits, has opened a new chapter in the three-hundred-year-old drama of Indian country. The crippling obstacles of federal repression and chronic poverty suddenly have disappeared for some tribes owing to the Indian Gaming Regulatory Act of 1988 (Public Law 100-497, 1988), which opened the doors to numerous gambling operations. Many tribes now have a real possibility of achieving long-held goals of self-determination and self-sufficiency, and for the first time the achievement of these goals is largely in the hands of Native people themselves.

Prior to 1985, most gambling activity outside Indian land was restricted to two states, Nevada and New Jersey. Since then, gaming has mushroomed across the United States outside Indian reservations. Today all states have legalized gaming in some form, and some states permit casino gaming in limited locations, such as the riverboats on the Mississippi River and Gulf Coast. In 1996 there were thirty-one states in which gross wagering exceeded $1 billion, and throughout the United States the amount spent in 1996 on wagering exceeded $590 billion. The American Indian share of total money wagered in the same year was $65 billion, or 11 percent of the total. In 1986 there were 85 gambling facilities on Indian reservations, but by 1997 there were 142 tribes that operated a total of 281 Class III gaming facilities (casino gambling, slot machines, horse and dog racing, and jai alai), employing 180,000 people in twenty-four states (Winchell, D., Lounsbury, J., and Sommers, L., 1997, 1 and 3; <122>).

Despite the number of Indian gaming facilities, not all are highly profitable. Only 48 tribes earn more than $10 million a year from gaming operations (Washington Post Magazine, 1997). The wealthiest gaming tribe is the Mashantucket Pequot of Connecticut, which became a federally acknowledged tribe only in 1983. The tribe's Foxwood Casino employs 10,000 people, and is the country's largest casino. Under the terms of the tribe's compact with Connecticut, the tribe pays $150 million in revenue taxes on its gaming earnings to the state, a sum equal to 25 percent of the state's tax revenues from all sources (ibid., 5). With part of the profits from gaming and other enterprises, the Pequot have created a Museum of the American Indian, which opened on their reservation in August 1998. The museum reportedly cost $150 million, and has a research library on Native American subjects that contains 150,000 volumes in addition to educational exhibits on the tribe's history. Tribal leaders point to the new museum as an example of the vitality of their culture as they enter a new century <123>.

The federal "government to government" policy has supported tribal sovereignty in a number of interesting ways. The U.S. Environmental Protection Agency (EPA) was among the first federal agencies to implement this policy. When federal legislation dealing with water quality was passed in 1977, provision was made for states to receive technical and financial support from the federal government to carry out activities under the Clean Water Act (Public Law 95-217, 33 U.S.C. sec. 1344). Indian tribes, however, were omitted from the legislation, and therefore were not eligible for funding except in competition with cities and counties for money passed out by the states. The EPA addressed this omission by creating a process through which tribes could receive "treatment as a state" (33 U.S.C. 13442 Title 33, subchapter V, subsection 1377, Indian Tribes), with direct access to funds and technical support. "Treatment as a state" made it possible for tribes to set up their own environmental departments, and to set their own standards for water and air quality. For example, by choosing to establish their reservations as pristine air quality zones, tribes can affect the levels of pollution allowed on surrounding non-Indian lands.

Other federal agencies have followed the EPA's example. In cooperative efforts between tribes and the National Park Service, tribes have been responsible for planning and managing culturally significant places on federal land (http://www.NPS.gov, American Indian Liaison Office) even if they are outside reservation boundaries. Repatriation of Native human remains under federal law (Native American Graves Protection and Repatriation Act 1990) has affirmed both the human rights of Native people and their sovereignty. Formerly, human remains were classed as scientific objects and thousands were removed from their burial sites and maintained in museums and research institutions around the country. One tribal leader who protested this practice stated: "First you stole our land. Now you steal our dead." <124> Under present policy, remains are returned to requesting tribes with a cultural connection to the deceased. The decision on disposition of newly discovered remains is made in consultation between federal agencies and Indian communities. Research projects now include Native Americans as partners with scientists. In other examples, the U.S. Fish and Wildlife Service agreed to accept a tribal plan for management of the endangered Mexican Spotted Owl on forests of the Fort Apache Indian Reservation in Arizona (Statement of Relationship between the White Mountain Apache Indian Tribe and the U.S. Fish and Wildlife Service 1994). The White Mountain Apache Tribe was the first tribe to assume responsibility for management of a species protected under the Endangered Species Act (16 U.S.C. 1531, 1973).

Another gain for tribes has resulted from significant events which have reversed the steady erosion of the tribal land base. First, the number of federally acknowledged tribes in the contiguous forty-eight states has risen from 309 listed in 1985 to 328 in 1996 (U.S. Government Printing Office 1996, pp. 58211–58216). Several of the newly recognized tribes had small reservations established, so that by 1996 there were 314 tribes with their own reservations. These newly created reservations represent, however, only a minute part of the area which has been added to Indian lands in recent years. A total of 2,444 square miles, representing a gain of 2.9 percent of the total land base of all reservations, has been restored to tribes since 1984. The largest gain was on the Navajo Indian Reservation, which expanded by 1,128 square miles (U.S. Department of the Interior, BIA, Office of Trust Responsibilities 1996).

The Penobscot and the Passamaquoddy, two small tribes in Maine, successfully restored a large tract of ancestral land to its reservation. They reached a settlement with the State of Maine (Maine Indian Claims Settlement Act 1980) which restored tens of thousands of acres to the tribe, from which they have created a new reservation on their ancestral homeland. Another court decision upheld the validity of a 1783 Spanish land grant to the Pueblo of Sandia in New Mexico. The west slope of Sandia Mountain, which contains major recreational facilities and several enclaves of very expensive homes, was removed from the U.S. Forest Service and returned to the Pueblo. One of the homes on the mountain belongs to the present governor of New Mexico, who is now a trespasser living on tribal land (Albuquerque Journal 1998).

The tribe that owns the Agua Caliente Indian Reservation in California has reasserted decision-making power over land uses in the city of Palm Springs, which is wholly located on leased reservation land. The tribal council must approve zoning changes, and it is preparing a new general land-use plan and zoning ordinance to control growth in the city <125>. Until a few years ago, the tribe took an absentee landowner's role and left these matters to the Palm Springs Zoning Commission and city council.

Despite these signs of progress, the internal land problems on reservations created by fractionated heirship have not yielded to an easy solution. Land pooling, or consolidation of landownership among allottees, has not occurred to any significant degree, and multiple ownership of parcels remains an obstacle to the management and development of tribal lands. A noteworthy exception is the Pavilions Shopping Center in the Salt River Pima-Maricopa Indian community in Arizona. Adjacent to the wealthy community of Scottsdale, and built mostly on allotted land, this shopping center is one of the most successful in metropolitan Phoenix. It contains dozens of shops, several restaurants, and theaters in a lushly landscaped desert oasis which has jogging trails ringing a series of lakes and fountains. An adjacent development, also on mostly allotted land, supports a golf course. The example of Salt River indicates that cooperative development of fractionated land is indeed a possibility (Frantz 1996a).

Tribal economies are generally still based on the primary and secondary economic activities of raw materials extraction, agriculture, and industry. This is a major disadvantage in the context of a national economy dominated by the service sector. For some tribes, agriculture holds promise as a means to maintain traditional economic activities that can also be profitable. As America urbanizes, farmland is disappearing under urban uses, while reservations are the exception to this trend. Tribal land in agricultural production has increased, largely as a result of settled water claims. The Gila River Indian Community, just south of Phoenix, for example, is undertaking a major expansion of its agricultural lands, and is well situated to profit from its proximity to a growing urban area with a population approaching three million. The Fort Mojave tribe has extensive farmlands along the Colorado River which are very profitable and a key element in the tribal economy. As farm leases to non-Indians expire, Fort Mojave adds the acreage to the tribal farming enterprise. Within the first decade of the twenty-first century, all reservation farmlands are projected to be under direct tribal management.

Three tribes have assumed control of mineral extraction on their reservations, reversing the old pattern of non-Indian domination of this economic activity. The Jicarilla Apache in New Mexico already had established a tribally operated mining company in the early 1980s, but since that time the Navajo Nation also has assumed management of mining operations on its reservation as old leases expire. The Navajos have created a vertically integrated petroleum products enterprise, the Navajo Oil and Gas Company, which produces, refines, and markets oil and gas products in the Southwest. The Southern Utes in Colorado also have a tribally controlled mining operation, exploiting reservation gas fields with their own drilling company, Red Willow (Johnson 1995, 38). Revenues from mineral extraction have shrunk significantly since the mid-1980s, however, and mineral extraction has shifted from oil and gas to coal (Cornell and Kalt 1995; Wilson and Manydeeds 1994, 3).

On the whole, tribes still depend on old economic bases, and on gaming. Few tribes have any significant enterprises based on technology or services, fields that have come to dominate the American economy as a whole. Two enterprises, however, where tribes have entered these new sectors are the profitable Indian-owned wireless telecommunications corporations on the Gila River Indian Reservation in Arizona and the Fort Mojave Indian Reservation in California, Nevada, and Arizona. These companies provide service to off-reservation customers as well as reservation residents. Many tribes are struggling to make a successful transition from an economy dependent on the depletion of natural resources to a diversified economy. A particular problem confronting Indian tribes in their economic expansion is the lack of a significant private sector. It will be difficult to sustain growing reservation populations in the face of shrinking federal support if tribes do not find alternative economic endeavors and encourage a larger role for the private sector to create and expand current economic activities.

The long-term future and profitability of gaming is not assured, so those tribes that are able to profit from gaming must look to a time when that income source comes to an end. There is a window of perhaps five to ten years in which tribes must reinvest gaming profits to create balanced reservation economies for the time when gaming competition in a saturated market erodes Indian casino revenues, and marginal Indian casinos go out of business (Hallock 1996, 204). It is foreseeable that more and more states will turn to gaming as a revenue source because of voter resistance to new taxation. Gaming yields an "invisible" tax which is willingly tolerated by the gambling public, who engage in gaming as a recreational activity that rarely is perceived as producing revenue for governments.

It is inevitable that some tribes will succeed in making the transition into contemporary economies, and that others will fail. Some tribes have excellent economic opportunities, while others have dismal prospects. The situation is changing from one in which most tribes were have-nots, to one in which some tribes are wealthy, while others remain in poverty. As a result of this state of economic disparity, some wealthy tribes have begun "foreign aid" programs. The casino-rich Sycuan Indians in Southern California award grants on a competitive basis to tribal projects on other reservations that do not have profitable gaming operations <70>.

The new wealth from gaming has changed relations with neighbors off the reservation. Tribal representatives are now courted by politicians, consulting firms, and philanthropic concerns, and are invited to sit on boards and councils where membership by a Native American would have been rare fifteen years ago. Wealth has brought acceptance in non-Indian communities in the service of self-interest. Some reservations have state-of-the-art health care facilities, water and sewer infrastructure, and recreational facilities which are far better than those of their off-reservation neighbors. The Fort McDowell and Gila River Indian reservations in Arizona, and Sandia Pueblo in New Mexico, for example, have new health care facilities responsive to the unique needs of their tribes. Sandia Pueblo has also created a wetlands for sewage treatment which is used as an ecological study site. These sudden shifts in decades-old relationships have upset the rural social order in some areas, and there is resentment that American

Indians are no longer at the bottom rung of the ladder. This resentment often expresses itself in various schemes to tax Indian revenue sources to redistribute the wealth throughout the local economy. Congress has felt pressure from constituents and has put forward several bills proposing to tax casino profits or to cut federal funding levels in relation to casino revenues, for example, under the Indian Gaming Tax Reform Act (U.S. Congress 1997–1998).

Water rights remain one of the thorniest issues in Indian country. Some tribes, like Fort Mojave, have perfected rights which clearly define the water allocated to them. With this certainty about water resources, that tribe has prepared a reservation-wide water budget in conjunction with the reservation land use plan, assuring that there is adequate water to support tribally favored land uses.

The Salt River Pima-Maricopa Indian community, on the other hand, negotiated a settlement of its water claims (Salt River Pima-Maricopa Indian Community 1989) which resulted in the lease of a majority of its water to the nearby cities of Phoenix, Scottsdale, and Mesa for ninety-nine years. In reality, this water will never come back under tribal control. It is unlikely that any court would agree to cut off the water at the end of the lease term, for to do so would destroy a century of urban expansion which depends on tribal waters. The tribe, however, is in a strong position to continue to earn revenue from this resource.

The leasing of tribal water to non-Indians, a new development since the mid-1980s, was the result of a federal policy that encouraged tribes to seek negotiated settlements of water rights claims rather than continue with lengthy and costly litigation. Another example of the settlement of water rights is the Ak-Chin tribe in Arizona, which has agreed to lease some of its Central Arizona Project water allocation to fulfill the state's requirement that a one-hundred-year water supply must be in place in order for new residential development to be approved. The Ak-Chin Indian Community would lease water adequate to support a planned community of about 20,000 people, the equivalent of a city the size of Flagstaff, Arizona.

Despite tribal and federal emphasis on developing the economy on Indian reservations, economic development is not an end in itself. It is a means by which tribes may secure, in varying degrees, political self-determination, economic self-sufficiency, social well-being, and cultural vitality. Political self-determination has made real advances under the federal "government to government" policy. Economic self-sufficiency might also follow for some tribes, and is a real possibility for tribes with casino profits from which to build a diverse economy. As in the past, however, social well-being and cultural vitality may in the end be less affected by economics than by communal will and cohesion.

In this respect, the homogenizing effects of American culture may help to determine the nature of tribal society and culture in the future. Television, radio, and computer access to the Internet bring American Indians on even the most remote reservations into constant contact with non-Indian culture. The choice of many Indians to leave their reservations and live in urban areas also increases the intensity of contact with the outside culture. The effects of expanded contact can be positive. An urban setting has allowed many creative American Indians such as N. Scott Momaday, R. C. Gorman, and Allan Houser to achieve widespread acclaim and financial success in mainstream society.

Although poverty can destroy culture, so can sudden wealth. Just as the world at large is dividing into two sharply separated groups, rich and poor, so tribes are creating haves and have-nots within their own membership. On the Salt River Indian Reservation, payments for a freeway right-of-way by the State of Arizona created more than thirty instant millionaire families. Those allottees who hold land near the neighboring city of Scottsdale have similarly profited from the Pavilions Shopping Center, while allottees with land far removed from a customer base have little prospect of profitable development (Frantz 1996a). Tribal members whose allotments are located in more remote parts of the reservation have little hope that they will someday receive similar nouveau riche status. Thus, the system created by the General Allotment Act continues to work as a divisive force within the tribes. Although most American Indian communities maintain a traditional view of land as a community resource for the common good, some tribal members have become zealous property-rights advocates as real-estate values have soared in choice reservation locations. A policy attempting to equalize the benefit to tribal individuals from land development is unlikely because on many reservations it would be political suicide for its advocates. Even so, many tribes are expanding their tribally owned land base.

External forces of other kinds can have destructive effects on indigenous people. In addition to alcohol and drug addictions, a high-fat and high-sugar diet borrowed from America's fast-food and junk-food eating habits have exacerbated genetic predisposition to some diseases, notably diabetes. The Pima and Maricopa Indians on the Gila River Indian Reservation teach the benefits of a traditional Native diet to children at all grade levels as a means of combating endemic diabetes among the Pima people. In spite of these efforts, diabetes is increasing among all American Indians. In 1985, 24.4 American Indians per 100,000 were affected by this illness, compared to 9.6 per 100,000 for the entire U.S. population. By 1992, the diabetes rate had increased to 31.7 per 100,000 compared to 11.9 per 100,000 for the total U.S. population (U.S. Department of Health and Human Services, IHS 1996, 81).

Self-determination has placed the responsibility for the education of American Indians in the hands of tribes or locally controlled public school systems. The traditional role of the BIA as provider of American Indian education will soon be gone. Many BIA schools have been replaced with public and tribal schools. The BIA still operated forty-three on-reservation boarding schools nationwide in 1996, of which thirty-two were on the Navajo Indian Reservation, and four off-reservation schools (U.S. Department of the Interior, BIA, Office of Indian Education Programs 1997). When the Navajo Nation completes the takeover of education responsibilities on its reservation, only a handful of on- and off-reservation BIA schools will remain.

Through the control of education, Indian tribes are successfully guiding their youth to rediscover their roots and preparing them to meet the future. Throughout Indian country there are examples of tribally directed education programs that support tribal traditions, culture, and language (Frantz 1996b). At the same time, the number of tribally controlled community colleges has increased to twenty-six today. In 1996 these colleges enrolled almost 25,000 students (ibid., 6). The tribal community college system and three federal government-supported institutions of higher learning have met the needs of Indian students responsively and successfully. For example, the Southwestern Indian Polytechnic Institute in New Mexico has instituted a comprehensive water resources management curriculum to meet tribal health and resource management needs Not only does the program prepare students for well-paid technical jobs such as wastewater treatment plant operators, but it also meets all coursework criteria to allow students to transfer to four-year colleges for bachelor's degrees in science <126>. Education is undoubtedly a key to tribal survival, and some tribes have taken dramatic steps to assure that they will have educated leaders for the future. The Pueblo of Sandia in New Mexico, for example, requires all tribal members under the age of twenty-four to achieve a high-school diploma or its equivalent (Pueblo of Sandia Tribal Council 1998, 5); this "zero tolerance" policy toward dropping out is probably the only one in the United States.

Various American Indian tribes in the past have produced memorable individuals who conveyed their visions to both their own tribal members and to the non-Indian world. Chief Joseph, Sequoyah, and Tecumseh are notable examples of such memorable leaders, but the circumstances from which they emerged generally have disappeared. Today, such leaders have largely been replaced with tribal chairpersons the equivalent of corporate presidents. Perhaps this is the result of the Indian Reorganization Act, which artificially imposed an Anglo-inspired government structure foreign to tribal traditions of governance and decision making. Tribal leadership is often torn between traditional skills and practices and the bureaucratic demands of Anglo-style government.

American Indian tribes remain separate and diverse, yet there are glimmers of forces that could coalesce the many diverse tribes into a focused effort to address common concerns. Environmentalism is one such force where traditional concern for one's tribal homeland, and the knowledge that each reservation is virtually fixed in size and must sustain tribal members now and for all the generations to come, has given birth to a strong commitment to environmental protection on reservations. Positive effects of these common interests include information-sharing among tribes, the opening of challenging new career opportunities directed by trained tribal members, and the assertion of sovereign control over resource management.

Other indications of the coalescence among the many tribes are the National Congress of American Indians, which puts out position papers reflecting the consensus of member tribes after each annual congress, and almost 100 other national American Indian organizations, which range from political and trade groups to science and technology associations (Snyder 1997). One regional organization, the Affiliated Tribes of Northwest Indians, which has fifty-four tribal government members, was founded as far back as 1937 both to render technical assistance to member tribes and to provide a forum that determines the agenda for Northwestern tribal political action through lobbying and other means.

American Indians have also moved onto the larger, international stage, where they have joined with other indigenous peoples all over the world to lobby for human rights, environmental justice, land reform, improved health, and other issues of concern. Perhaps participation in this wider realm will bring greater unity among the tribes, serving to reinforce tradition, commitment to community, and commitment to homeland.

One area of sovereignty that has been crippled compared to the strength of the other areas of self-government is the tribal judicial system, which Congress has deliberately kept weak through the single provision that tribal courts do not have jurisdiction over non-Indians, or over members of other tribes, for major crimes committed on reservations. There is no indication that federal policy will correct this imbalance, which has important implications for land use and other issues. Some tribes, for instance, refuse to allow non-Indian housing development on the reservation, even though it would be very lucrative, because they refuse to allow people to live on the reservation who are not accountable for their criminal acts to the landowner, the tribe. Each tribe wrestles with this issue individually. The Fort Mojave Indians, for instance, have proceeded with non-Indian housing developments. The Tribal Council of the Gila River Indian Reservation has refused to allow them, even though thousands of homes crowd this reservation's northern border. An ironic twist to the tribes' lack of juridical power over non-Indians on the reservation is that some resident non-Indians have demanded representation on tribal councils. Again, each tribe decides the issue for itself: the Salt River Pima-Maricopa Indian Community has so far refused to allow political representation to non-Indians living on the reservation.

Another problem facing the tribes is how to define who is an American Indian and who will be an Indian in the future. Although raw numbers indicate an increase in the American Indian population, the numbers are deceiving. The product of improved health care has been lower mortality rates which, combined with continued high birth rates, contribute to the overall increase in population. The U.S. Census recorded 1.9 million American Indians in 1990. Population estimates for 1997 show an increase to almost 2.3 million <15>, or 63 percent since 1980. The proportion of American Indians who live on reservations is shrinking, however, and in 1990, 22.6 percent lived on reservations, down from 24.9 percent in 1980.

The most significant factor in the population increase is growing pride in Indian heritage and resulting self-identification as American Indian. The U.S. Census allows people to choose how they wish to be classified. The popularity of American Indians in American culture may cause people to choose to identify themselves as American Indians even though they have little or no Indian ancestry. People may also choose to classify themselves as American Indians from the perceived possibility that they may be eligible to share in the newly gained wealth of some tribes.

Despite the apparent increase in the American Indian population, it is foreseeable that there will come a time when few people meet the legal tribal and BIA blood quantum requirements presently in force for tribal enrollment. Blood quantum requirements have already been diluted by several tribes in order to maintain enrollment eligibility through successive generations. The Fort Sill Apache Tribe in Oklahoma lowered its requirement from one-eighth to one-sixteenth (Tempe Daily News 1997, 5). In another approach to answering the question of who is an Indian, the Pequot tribe abandoned the quantum requirement and replaced it with a determination of eligibility based on having an ancestor who was listed on the tribal roll of 1900 or 1910 (ibid.). There is an obvious limit to the number of times this can occur. Dilution of official tribal roles through marriage with non-Indians or Indians from other tribes, not assimilation, may bring the final dissolution of tribes across America.

Will Indian country, and the hundreds of separate tribes that represent Indian country, persist into the future? Will these tribes respond to their new opportunities and unresolved issues of the past in ways that expand

and strengthen their communities and homelands? Or are they the human equivalent of redwood trees, a relic culture formed in a distant time and place which can survive only on a shrunken scrap of familiar territory, unable to make the adaptations that would allow them to move into the changed ecosystem that has engulfed their native land?

The unique relationship between American Indians and the U.S. government, seen from a critical perspective, is well developed when compared to the status of indigenous people in Canada, Mexico, South America, and Australia, for instance. The U.S. example is worth noting as tribal people all over the world are reasserting their tribal affiliations and breaking up the nineteenth-century legacy of forcibly imposed nationalism. The survival of American Indian tribes attests to the unequaled status of minority people under the laws of the United States and the unique sovereign powers granted to, and maintained by, them. Despite its deeply uneven application over time, viewed on the whole, no other nation has so consistently upheld the rights of its indigenous people, who have been among the poorest and weakest groups in its society. American Indians have survived three centuries of efforts by invading European cultures to extinguish life, language, culture, economy, religion, and family. Their persistence is one of the great dramas of the human spirit and continues to represent a unique and dynamic legacy for the future.

Bibliography

BOOKS AND JOURNAL ARTICLES

Abbot, L. 1885. Address Given to the Lake Mohonk Conference of Friends of the Indian. Proceedings of the Third Annual Meeting, pp. 50–54.

Aberle, D. 1975. The Sun Dance and Reservation Underdevelopment: A Review Essay. JES, 3/1, pp. 66–73.

Aberle, D. 1983. Navajo Economic Development. In: Ortiz, A., ed., pp. 641–658.

Ablon, J. 1964. Relocated American Indians in the San Francisco Bay Area: Social Interaction and Indian Identity. HO, 23, pp. 296–304.

Adam, L., and Trimborn, H. 1958. Lehrbuch der Völkerkunde. Stuttgart, F. Enke Verlag, 303 pp.

Adams, W. 1963. Shonto: A Study of the Role of the Trader in a Modern Navajo Community. Washington, DC, Smithsonian Institution, Bureau of American Ethnology Bulletin 188, 329 pp.

Albuquerque Journal. 1998. Tribe Wins Sandias Suit. Aug. 1, A, pp. 1 and 5.

The American Indian Lawyer Training Program. 1982. Indian Water Policy in a Changing Environment: Perspectives on Indian Water Rights, n.p. Manuscript. 163 pp.

American Indian Policy Review Commission. 1976. Urban and Rural Non-Reservation Indians. Washington, DC, GPO, 23 pp.

Anders, G. 1980. Theories of Underdevelopment and the American Indian. Journal of Economic Issues, 14/3, pp. 681–696.

Anderson, R., ed. 1994. The Oil and Gas Opportunity on Indian Lands: Exploration Policies and Procedures. Golden, CO, U.S. Department of the Interior, BIA, Division of Energy and Mineral Resources (Publication G–94–3), 148 pp.

Andress, J., and Falkowski, J. 1980. Self Determination: Indians and the United Nations; The Anomalous Status of America's "Domestic Dependent Nations." ARIL, 8, pp. 97–116.

Arizona Republic. 1980. Like Giving Heroin to an Addict. Dec. 7, B, p. 6.

Arizona Republic. 1985a. Indian Stakes: Gaming Boom Worries Lawmen. April 14, B, pp. 9–10.

Arizona Republic. 1985b. Work Starts on Indians' Jai Alai Arena. May 7, B, pp. 2–3.

Arizona Republic. 1987. Fraud in Indian Country: A Billion Dollar Betrayal, Oct. 4–11, 36 pp.

Arizona Republic. 1989. War Waged in Forests of Arizona: Timber Industry Future at Issue, April 23, p. 13.

Arizona Republic. 1990. Inside Metro Phoenix 1990, 57 pp.

Arthur, H. 1978. Preface. In: Jorgensen, J., et al., eds., pp. 1–2.

Aschmann, H. 1970. Athapaskan Expansion in the Southwest. Yearbook of the Association of Pacific Coast Geographers 32, pp. 79–97.

Bähr, J. 1983. Bevölkerungsgeographie. UTB-Taschenbuch, 427 pp.

Bahr, H., Chadwick, B., and Day, R., eds. 1972. Native Americans Today: Sociological Perspectives. New York, Harper and Row, 547 pp.

Baldwin, G. C. 1973. Indians of the Southwest. New York, Capricorn Books, 192 pp.

Ballas, D. 1960. Notes on the Population, Settlement, and Ecology of the Eastern Cherokee Indians. Journal of Geography, 59, pp. 258–267.

Ballas, D. 1966. Geography and the American Indian. Journal of Geography, 65, pp. 156–168.

Ballas, D. 1970. A Cultural Geography of Todd County, South Dakota, and the Rosebud Sioux Indian Reservation. Ph.D. diss. (Geography), University of Nebraska, 245 pp.

Ballas, D. 1973a. Early Agriculture and Livestock Raising among the Teton Dakota Indians. Bulletin of the Illinois Geographical Society 15, pp. 53–62.

Ballas, D. 1973b. Cultural Traits Related to Socio-Economic Development on the Rosebud Sioux Indian Reservation. Geographical Survey, 2, pp. 17–26.

Ballas, D. 1985. Changing Ecology and Land-Use among the Teton/Dakota Indians, 1680–1900. Bulletin of the Illinois Geographical Society 27, pp. 35–47.

Barsh, R., and Diaz-Knauf, K. 1983. The Structure of Federal Aid for Indian Programs in the Decade of Prosperity, 1970–1980. N.p. Manuscript. 46 pp.

Barsh, R., and Henderson, J. 1980. The Road: Indian Tribes and Political Liberty. Berkeley, University of California Press, 301 pp.

Basso, K. 1983. Western Apache. In: Ortiz, A., ed., pp. 462–488.

Baumann, P., and Uhlig, H. 1973. Kein Platz für "wilde" Menschen. Vienna, Molden Verl., 303 pp.

Beaglehole, E. 1978. Notes on the Hopi Economic Life. 2d ed. New Haven, Yale University Press, 88 pp.

Bencherifa, A., and Popp, H. 1990. L'oasis de Figuig: Persistance et changement. Passauer Mittelmeerstudien 2 (special issue), 109 pp.

Bennett, R. 1961–1962. Building Indian Economics with Land Settlement Funds. HO, 20, pp. 159–163.

Berkman, R., and Viscusi, W. 1973. Indians Sold Down the River. In: Ralph Nader's Study Group ed., Damming the West. New York, Grossman, pp. 151–196.

Big Mountain Aktionsgruppe. 1989. Der Indian Economic Development Act oder die Lüge vom wirtschaftlichen Wohlstand. Coyote-Nachrichten und Berichte zur Situation nordamerikanischer Indianer, 1/1, pp. 28–33.

Biegert, C. 1983a. Indianerschulen. Als Indianer überleben—von Indianern lernen. Survival Schools. Reinbeck, Rowohlt Taschenbuch Verl., 295 pp.

Biegert, C., ed. 1983b. Die Wunden der Freiheit. Der Kampf der Indianer gegen die weiße Eroberung und Unterdrückung. Selbstzeugnisse, Dokumente, Kommentare. 5th ed. Reinbeck, Rowohlt Taschenbuch, 296 pp.

Bigart, R. 1972. Indian Culture and Industrialization. American Anthropologist, 74, pp. 1180–1188.

Birle, S. 1976. Irrigation Agriculture in the Southwest United States. Regional Variations of Crop Patterns. Marburger Geographische Schriften 67, 217 pp.

Blenck, J. 1974. Endogene und exogene entwicklungshemmende Strukturen, Abhängigkeiten und Prozesse in den Ländern der 3. Welt; dargestellt am Beispiel Liberia und Indien. In: Eichler, H., and Musall, H., eds., Heidelberger Geographische Arbeiten 40 (= H. Graul Festschrift), pp. 395–418.

Blindermann, A. 1978. Congressional Social Darwinism and the American Indian. The Indian Historian, 11/2, pp. 15–17.

Blume, H. 1975. USA. Eine geographische Landeskunde, vol. 1: Der Großraum in strukturellem Wandel. Darmstadt, Wissenschaftliche Länderkunde 9, 346 pp.

Blume, H. 1979. USA. Eine geographische Landeskunde, vol. 2: Die Regionen der USA. Darmstadt, Wissenschaftliche Länderkunde 9, 499 pp.

Bodine, J. 1972. Acculturation Process and Population Dynamics. In: Ortiz, A., ed. New Perspectives on the Pueblos. Albuquerque, University of New Mexico Press, pp. 257–285.

Boesch, H. 1973. USA. Werden und Wandel eines kontinentalen Wirtschaftsraumes. Bern, Kümmerly and Frey, 255 pp.

Bolton, H. 1921. The Spanish Borderlands: A Chronicle of Old Florida and the Southwest. New Haven, Yale University Press, 320 pp.

Bolton, H. 1986. The Padre on Horseback: A Sketch of Eusebio Francisco Kino S.J., Apostle to the Pimas. Chicago, Loyola University Press, 84 pp.

Bolz, P. 1986. Ethnische Identität und kultureller Widerstand. Die Oglala-Sioux der Pine Ridge-Reservation in South Dakota. Frankfurt, Campus Verlag, 277 pp.

Bowden, M. 1969. The Perception of the Western Interior of the United States, 1800–1870: A Problem in Historical Geography. AAAG, 59, pp. 16–21.

Boyce, G. 1974. When Navajos Had Too Many Sheep: The 1940's. San Francisco, Indian Historian Press, 273 pp.

Boyer, P. 1997. Native American Colleges: Progress and Prospects. Princeton, NJ, Carnegie Foundation for the Advancement of Teaching, 125 pp.

Brand, D. 1935. Prehistoric Trade in the Southwest. New Mexico Business Review, 4, pp. 202–209.

Brandes, R. 1960. Frontier Military Posts of Arizona. Globe, AZ, Dale Stuart King, 179 pp.

Bregman, S. 1982. Blessing or Curse? Uranium Mining on Indian Lands. Environment, 24/7, pp. 6–35.

Brophy, W., and Aberle, S. 1976. The Indian: America's Unfinished Business. 6th ed. Norman, Oklahoma University Press, 242 pp.

Brown, D., ed. 1982. Biotic Communities of the American Southwest: United States and Mexico. Tucson, University of Arizona Press, 315 pp.

Brown, R. 1948. Historical Geography of the United States. New York, Harcourt, Brace and World, 596 pp.

Brugge, D. 1983. Navajo Prehistory and History to 1850. In: Ortiz, A., ed., pp. 489–501.

Burrus, E., S.J. 1954. Kino Reports to Headquarters: Correspondence of Eusebio F. Kino, S.J., from New Spain to Rome; Original Spanish Text of Fourteen Unpublished Letters and Reports with English Translation and Notes. Rome, Institutum Historicum Societatis Jesu, 135 pp.

Buskirk, W. 1949. Western Apache Subsistence Economy. Ph.D. diss. (Anthropology), University of New Mexico, 305 pp.

Canby, W. 1981. American Indian Law. St. Paul, West Publishing Co., 288 pp.

Carlson, A. 1972. A Bibliography of the Geographical Literature on the American Indian, 1920–1971. Professional Geographer, 24, pp. 258–263.

Carlson, L. 1981. Indians, Bureaucrats, and Lands: The Dawes Act, and the Decline of Indian Farming. Westport, CT, Greenwood Press, 219 pp.

Castile, G. 1974. Federal Indian Policy and the Sustained Enclave: An Anthropological Perspective. HO, 33, pp. 219–228.

Champagne, D. 1994. Native America: Portrait of the Peoples. Detroit, Visible Ink Press, 786 pp.

Chaudhuri, J. 1974. Urban Indians of Arizona: Phoenix, Tucson, and Flagstaff. Tucson, University of Arizona Press, 68 pp.

Churchill, W. 1986. American Indian Lands: The Native Ethic amid Resource Development. Environment, 28/6, 6 pp.

Churchill, W., and Hill, N., Jr. 1979. Indian Education at the University Level: An Historical Survey. JES, 7/3, pp. 43–58.

Clemmer, R. 1978. Black Mesa and the Hopi. In: Jorgensen, J., et al., eds., pp. 17–34.

Cohen, F. 1982. Handbook of Federal Indian Law. 2d ed. Albuquerque, University of New Mexico Press, 662 pp.

Comeaux, M. 1981. Arizona: A Geography. Boulder, Westview Press, 336 pp.

Cook, J. 1981. New Hope on the Reservation. Forbes, 128, pp. 108–115.

Cornell, S., and Kalt, J. 1990. Pathways from Poverty: Economic Development and Institution-Building on American Indian Reservations. American Indian Culture and Research 14, pp. 89–125.

Cornell, S., and Kalt J., eds. 1995. What Can Tribes Do? Strategies and Institutions in American Indian Development. Los Angeles, American Indian Studies Center (American Indian Manual and Handbook Series 4), 336 pp.

Correll, J., et al. 1969 and 1973. Navajo Bibliography with Subject Index. 2 vols. Window Rock, AZ, The Navajo Tribe, Parks and Recreation Research Section. 326 pp. and 68 pp.

Cummings, H., and Harrison, M. 1972. The American Indian: The Poverty of Assimilation. Antipode, 4, pp. 77–99.

Cyrus, I. 1988. Die Indianer Nordamerikas (kurz & bündig). Vienna, hpt-Verlagsges., 124 pp.

De Cook, K. 1980. Arizona's Water Resources. (Conference: Toward the Year 2000: Arizona's Future, Oct. 26–29). Phoenix, Thirty-seventh Arizona Town Hall, pp. 87–102.

De Langé, N. 1983. Einkommensdisparitäten in den USA. Datenaufbereitung mit einfachen statistischen Verfahren im Hinblick auf die Verwendung im Geographieunterricht der Sek. II. GR, 35, pp. 189–194.

DeMallie, R. 1978. Pine Ridge Economy: Cultural and Historical Perspectives. In: Stanley, S., ed., pp. 237–312.

Debo, A. 1967. The Road to Disappearance: A History of the Creek Indians. 2d ed. Norman, Oklahoma University Press, 399 pp.

Deloria, V., Jr. 1972. Custer Died for Your Sins: An Indian Manifesto. 9th ed. New York, Avon Books, 272 pp.

Deloria, V. 1979. Self-Determination and the Concept of Sovereignty. In: Ortiz, R., ed., pp. 22–28.

Deloria, V., and Lytle, C. 1983. American Indians, American Justice. Austin, University of Texas Press, 262 pp.

Denevan, W., ed. 1976. The Native Population of the Americas in 1492. Madison, University of Wisconsin Press, 353 pp.

Denver Post. 1983. The New Indian Wars (Special Report), Nov. 20–27, 72 pp.

Despain, D. 1985. History and Results of Prescribed Burning of Pinyon:Juniper Woodland on the Hualapai Indian Reservation in Arizona. BIA, Truxton Cañon Agency. Valentine, AZ. Manuscript. 7 pp.

Dietrich, B., et al. 1933. Nord- und Mittelamerika. Die Arktis. In Natur, Kultur und Wirtschaft. Potsdam, Akad. Verlagsges. Athenaion (Handbuch d. Geog. Wissenschaft), 578 pp.

Dobyns, H. 1966. Estimating Aboriginal American Population: An Appraisal of Techniques with a New Hemispheric Estimate. Current Anthropology, 7, pp. 395–416.

Dobyns, H. 1976a. Scholarly Transformation: Widowing the Virgin Land. Ethnohistory, 23, pp. 95–104.

Dobyns, H. 1976b. Native American Historical Demography (Bibliography). Chicago, Newberry Library Center for the History of the American Indian, 95 pp.

Dobyns, H., and Euler, R. 1976. The Walapai People. 2d ed. Phoenix, AZ, Indian Tribal Series, 106 pp.

Driver, H. 1969. Indians of North America. 2d ed. Chicago, University of Chicago Press 632 pp.

Edwards, N. 1961–1962. Economic Development of Indian Reserves. HO, 20, pp. 197–202.

Estes, C. 1982. Indian Water Rights. In: The American Indian Lawyer Training Program, pp. 25–42.

Farnsworth, J. 1976. Native Employment in a Frontier Region. In: McDonald, J., and Lazewski, T., eds., pp. 73–88.

Federal Agencies Task Force. 1979. American Indian Religious Freedom Act. Washington, DC, Department of the Interior, n.p.

Feest, C. 1976. Das Rote Amerika. Nordamerikas Indianer. Wien, Europa Verlag, 463 pp.

Feest, C. 1978. Virginia Algonquians. In: Trigger, B., ed., Handbook of North American Indians, vol. 15: Northeast. Washington, DC, Smithsonian Institution, pp. 253–270.

Feest, C. 1980. Zur Domestikationsgeschichte der nordamerikanischen Indianer. Wiener Beiträge zur Geschichte der Neuzeit, 7, pp. 95–119.

Feest, C. 1981. Notes on Native American Alcohol Use. In: Hovens, P., ed., pp. 201–222.

Feest, C. 1986. Indianer Nordamerikas—Heute & Gestern. Vienna, Museum für Völkerkunde, 36 pp.

Feest, C., ed. 1987. Indians and Europe: An Interdisciplinary Collection of Essays. Aachen, Rader, 643 pp.

Flad, H. 1972. The Urban American Indians of Syracuse, New York: Human Exploration of Urban Ethnic Space. Antipode, 4, pp. 88–99.

Fliedner, D. 1974. Der Aufbau der vorspanischen Siedlungs- und Wirtschaftslandschaft im Kulturraum der Pueblo-Indianer. Eine historisch-geographische Interpretation wüstgefallener Ortsstellen und Feldflächen im Jemetz-Gebiet, New Mexico (USA). Arbeiten aus d. Geog. Inst. d. Universität d. Saarlanden 19, 63 pp.

Fliedner, D. 1975. Pre-Spanish Pueblos in New Mexico. AAAG, 65, pp. 363–377.

Fonaroff, L. 1963. Conservation and Stock Reduction on the Navajo Tribal Range. GRev, 53, pp. 200–223.

Fonaroff, L. 1964. Aid and the Indian: A Case Study in Faulty Communication. California Geographer, 5, pp. 57–68.

Fontana, B. 1983. History of the Papago. In: Ortiz, A., ed., pp. 137–148.

Forbes, J. 1981. Determining Who Is an Indian. Amerikastudien, 26, pp. 404–415.

Foreman, R. 1981. Indian Water Rights: A Public Policy and Administrative Mess. Danville, IL, Interstate Printers and Publishers, 233 pp.

Forest Industries. 1985. Leading Lumber Producers: U.S. and Canada. July, pp. 26–27.

Frantz, K. 1979. Review of Cherokees in Transition: A Study of Changing Culture and Environment prior to 1775, by G. C. Goodwin. Mitteilungen d. Österreichische Geographischen Gesellschaft, 121/1, pp. 166–167.

Frantz, K. 1987. Die Großstadt Angloamerikas im 18. und 19. Jahrhundert. Strukturwandlungen und sozialräumliche Entwicklungsprozesse anhand ausgewählter Beispiele der Nordostküste. Stuttgart, F. Steiner Verlag (Erdkundl. Wissen 77), 200 pp.

Frantz, K. 1989. Vom Stammesterritorium zur heutigen Indianerreservation. Idealtypische Überlegungen im Zusammenhang mit der territorialen Entwicklung des Reservationslandes innerhalb der USA und seiner Restsouveränität. Mitteilungsblatt –Arbeitskreis USA im Zentralverband der Deutschen Geographen, 10, pp. 15–16.

Frantz, K. 1990a. Zur Frage der territorialen Entwicklung und Souveränität der U.S.-amerikanischen Indianerreservationen. Mitteilungen der Österreichischen Geographischen Gesellschaft, 131, pp. 27–46.

Frantz, K. 1990b. Counting the Uncountable? Bevölkerungsstatistische Fragen im Zusammenhang mit den U.S.-amerikanischen Indianern. Vechtaer Arbeiten zur Geographie und Regionalwissenschaft, 10, pp. 67–74.

Frantz, K. 1992. Pottwalöl aus der Wüste: Zur Domestizierung und kommerziellen Nutzung des Jojoba-Strauches in Arizona. Mitteilungsblatt des Arbeitskreises USA im Zentralverband der Deutschen Geographen, 17, pp. 25–37.

Frantz, K. 1993. Kampf um die Wasserressourcen Arizonas. Der Konflikt zwischen Phoenix und der Fort Apache Indianerreservation. In: Santler, H., and Bogenreiter, H., eds., pp. 146–155.

Frantz, K. 1994. Little White Man Schools, Washington Schools, and Indian Schools: bildungsgeographische Fragestellungen dargestellt am Beispiel der Navajo Indianerreservation. Die Erde, 125, pp. 299–314.

Frantz, K. 1995. Die Indianerreservationen in den USA. Aspekte der territorialen

Entwicklung und des sozio-ökonomischen Wandels. 2d ed. Stuttgart, Franz Steiner Verlag (= Erdkundliches Wissen 109), 298 pp.

Frantz, K. 1996a. Education on the Navajo Indian Reservation: Aspects of the Maintenance of Cultural Identity as Seen from a Geographical Point of View. In: Frantz, K., and Sauder, R., eds., pp. 223–245.

Frantz, K. 1996b. Die Salt River Indianerreservation—Landnutzungskonflikte und sozioökonomischer Wandel am Rande der Großstadt Phoenix. Geographische Rundschau, 48, pp. 206–212.

Frantz, K., ed. 1996c. Human Geography in North America: New Perspectives and Trends in Research. Innsbruck, Innsbrucker Geographische Studien 26, 366 pp.

Frantz, K., and Sauder, R., eds. 1996. Ethnic Persistence and Change in Europe and America: Traces in Landscape and Society. Innsbruck, Austria, Veröffentlichungen der Universität Innsbruck 213, 270 pp.

Friese, H., and Hofmeister, B. 1983. Die USA. Wirtschafts- und Sozialgeographische Grundzüge und Probleme. 2d ed. Frankfurt, Moritz Diesterweg und Sauerländer, 182 pp.

Gaile, G., and Willmott, C., eds. 1989. Geography in America. Columbus, Merrill Publishing Co., 840 pp.

Gates, M. 1896. Address Given to the Lake Mohonk Conference of Friends of the Indian. Proceedings of the Fourteenth Annual Meeting, pp. 11–12.

Geisler, C., and Popper F., eds. 1984. Land Reform, American Style. Totowa, NJ, Rowan and Allanheld, 353 pp.

Getty, H. 1961–1962. San Carlos Apache Cattle Industry. HO, 20, pp. 181–186.

Gilbreath, K. 1973. Red Capitalism: An Analysis of the Navajo Economy. Norman, Oklahoma University Press, 157 pp.

Glazer, N., and Moynihan, D. 1963. Beyond the Melting Pot. Cambridge, MA, MIT Press, 360 pp.

Godwin, L., ed. 1994. Annual Report for the Fourth National Indian Energy/Minerals Conference. Golden, CO, BIA General Publication G-94-2, 104 pp.

Goodey, B. 1970. The Role of the Indian in North Dakota's Geography: Some Propositions. Antipode, 2, pp. 11–24.

Goodman, J. 1978. Resolution of the Hopi-Navajo Joint Use Area Dispute: A Progress Report. Geographical Perspectives, 41, pp. 1–9.

Goodman, J. 1982a. The Navajo Atlas: Environments, Resources, People, and History of the Dine' Bikeyah. Norman, Oklahoma University Press, 109 pp.

Goodman, J. 1982b. Resource Development and Its Significance to the Future of the Navajo. Journal of Cultural Geography, 272, pp. 101–109.

Goodman, J. 1985. The Native American. In: McKee, J., ed., pp. 31–53.

Goodman, J., and Elam, W. 1970. The Institute for Regional Geography of the American Southwest. Journal of Geography, 69, pp. 493–497.

Goodman, J., and Thompson G. 1975. The Hopi-Navajo Land Dispute. AILR, 3/2, pp. 397–417.

Goodwin, G. 1935. The Social Divisions and Economic Life of the Western Apache. American Anthropologist, 37, pp. 55–64.

Goodwin, G. 1969. The Social Organization of the Western Apache. 2d ed. Tucson, University of Arizona Press, 701 pp.

Goodwin, G. C. 1977. Cherokees in Transition: A Study of Changing Culture and

Environment prior to 1775. Geography Research Paper no. 181. Chicago, University of Chicago Department of Geography, 207 pp.

Graf, W. 1984. Thinking Like a River: Reflections of the Gila. Arizona Waterline, Summer, pp. 1–4.

Graves, T. 1972. The Personal Adjustment of Navajo Indian Migrants to Denver, Colorado. In: Bahr, H., Chadwick, B., and Day, R., eds., pp. 440–466.

Grey, Z. 1953. The Vanishing American. Roslyn, NY, W. J. Black, 308 pp.

Haas, W. 1925. The American Indian and Geographic Studies. AAAG, 15, pp. 36–37.

Haase, M. 1968. Raiders to Ranchers: The Historical Geography of the Arizona Apache Adoption of Cattle. MA thesis (Geography), Arizona State University, 85 pp.

Hackenberg, R., and Wilson, R. 1972. Reluctant Emigrants: The Roll of Migration in Papago Indian Adaptation. HO, 31, pp. 171–186.

Hagan, W. 1971. Indian Policy after the Civil War: The Reservation Experience. In: Prucha, F., et al., eds., pp. 20–36.

Hahn, R. 1990. USA. 2d ed. Stuttgart, E. Klett (Länderprofile), 385 pp.

Haines, F. 1976. The Plains Indians: Their Origins, Migrations, and Cultural Development. New York, Crowell Co., 213 pp.

Hallock, D. 1996. Second Contact: Redefining Indian Country. In: Frantz, K., ed., pp. 195–208.

Harris, M. 1989. Kulturanthropologie (Lehrbuch). Frankfurt, Campus Verl., 502 pp.

Hartmann, H. 1979. Die Plains- und Prärieindianer Nordamerikas. 2d ed. Berlin, Veröffentlichungen des Museums für Völkerkunde, N.F. 22, 422 pp.

Hauser, J. 1974. Bevölkerungsprobleme der 3. Welt. Uni-Taschenbücher 316, 316 pp.

Heizer, F., and Almquist, A. 1971. The Other Californians. Berkeley, University of California Press, pp. 23–46.

Henkel, R. 1985. Policy Implications of Non-Indian Population Growth on Indian Land and Resources in Arizona. Papers and Proceedings of the Applied Geography Conferences 8, pp. 298–306.

Hewes, L. 1942a. Indian Land in the Cherokee Country of Oklahoma. Economic Geography, 18, pp. 401–412.

Hewes, L. 1942b. The Eastern Border of the Cherokee Country of Oklahoma as a Cultural Fault Line. AAAG, 32, pp. 120–121.

Hewes, L. 1942c. The Oklahoma Ozarks as the Land of the Cherokees. GRev, 32, pp. 269–281.

Hewes, L. 1943. Cultural Fault Line in the Cherokee Country. Economic Geography, 19, pp. 136–142.

Hewes, L. 1944. Cherokee Occupance in the Oklahoma Ozarks and Prairie Plains. Chronicles of Oklahoma, 22, pp. 324–337.

Hewes, L. 1978. Occupying the Cherokee Country of Oklahoma. University of Nebraska Studies, n.s. 57, 77 pp.

Higgins, B. 1982. Urban Indians: Patterns and Transformations. Journal of Cultural Geography, 2, pp. 110–118.

Hilliard, S. 1971. Indian Land Cessions West of the Mississippi. Journal of the West, 10, pp. 493–510.

Hilliard, S. 1972. Indian Land Cessions. AAAG, 60, pp. 374.

Hodge, W. 1980. The First Americans: Then and Now. New York, Holt, Rinehart and Winston, 551 pp.

Hofmeister, B. 1967. The Public Domain. Entwicklung und gegenwärtige Problematik der Staatsländereien in den USA. GR, 19, pp. 48–56.

Hofmeister, B. 1975. Vorläufige Ergebnisse einer Forschungsreise durch den Südwesten der USA. Die Erde, 106, pp. 201–214.

Hofmeister, B. 1976. Indianerreservationen in den USA. Territoriale Entwicklung und wirtschaftliche Eignung. GR, 28, pp. 507–518.

Hofmeister, B. 1978. Ernährungsgrundlagen und Einkommensquellen bei zunehmendem Bevölkerungsdruck in den U.S.-amerikanischen Indianer-Reservationen. Verhandlungen d. 41. Dt. Geographentages (Mainz), pp. 331–346.

Hofmeister, B. 1979. Bedeutung und Möglichkeiten des Tourismus für die wirtschaftlichen Grundlagen ausgewählter Indianerreservationen in den USA. International symposium (IGU—Working Group of Geography of Tourism and Recreation) Tourism-Factor for National and Regional Development (Sofia, 1975), pp. 301–306.

Hofmeister, B. 1980. Die Grenze von Indianerreservationen in den USA. Landschaftselement, rechtlicher Status, ökonomische Bedeutung, Veränderlichkeit. In: Kishimoto, H., ed., Geographie und ihre Grenzen (= Boesch-Gedächtnisschrift); Kümmerly and Frey, pp. 69–79.

Hofmeister, B. 1988. Nordamerika (Fischer Länderkunde, 6), 8th ed. 455 pp.

Holmstrom, D. 1995. Tribal Self-Governance. American Studies Newsletter, 35, pp. 12–14.

Hoover, J. 1929a. Modern Canyon Dwellers of Arizona. Journal of Geography, 28, pp. 269–279.

Hoover, J. 1929b. The Indian Country of Southern Arizona. GRev, 19, pp. 38–60.

Hoover, J. 1930. Tusayan: The Hopi Indian Country of Arizona. GRev, 20, pp. 425–444.

Hoover, J. 1931a. Geographic and Ethnic Grouping of Arizona Indians. Journal of Geography, 30, pp. 235–246.

Hoover, J. 1931b. Navaho Nomadism. GRev, 21, pp. 429–445.

Hovens, P., ed. 1985. North American Indian Studies: European Contributions. Göttingen, Edit. Herodot, 546 pp.

Hoxie, F. 1991. Native Americans: An Annotated Bibliography. Pasadena, CA, Salem Press, 325 pp.

Hoxie, F. 1992. Treaties: A Source Book. Chicago, Newberry Library, 165 pp.

Hoxie, F. 1996. Encyclopedia of North American Indians. Boston, Houghton Mifflin Co. 756 pp.

Hughes, J. D. 1983. American Indian Ecology. El Paso, TX, Western Press, 174 pp.

Hundley, N. 1978. The Dark and Bloody Ground of Indian Water Rights: Confusion Elevated to Principle. Western Historical Quarterly, 9, pp. 455–482.

Hunt, J. 1970. Land Tenure and Economic Development on the Warm Springs Indian Reservation. Journal of the West, 9, pp. 93–109.

Iadarola, A. 1979. Indian Timber: Federal or Self-Management? Third Annual National Timber Symposium ("Indian Timber: A Multi-Billion Dollar Trust"), n.p. Manuscript. 97 pp.

Iverson, P. 1976. The Navajos. Chicago, Newberry Library Center for the History of the American Indian, 64 pp.

Jacobs, W. 1974. The Tip of an Iceberg: Pre-Columbian Indian Demography and Some Implications for Revisionism. William and Mary Quarterly, 31, pp. 123–132.

Jett, S. 1964. Pueblo Indian Migrations: An Evaluation of the Possible Physical and Cultural Determinants. American Antiquity, 29, pp. 281–300.

Jett, S. 1977. History of Fruit-Tree Raising among the Navajo. Agricultural History, 51, pp. 681–701.

Jett, S. 1978. The Origins of Navajo Settlement Patterns. AAAG, 68, pp. 351–362.

Jett, S. 1980. The Navajo Homestead: Situation and Site. Yearbook of the Association of Pacific Coast Geographers, 42, pp. 101–117.

Jett, S. 1994. A Bibliography of North American Geographers' Works on Native Americans North of Mexico, 1971–1991. Lawrence, KS, Haskell Indian Nations University, Studies in the Geography of the American Indian, no. 1, 82 pp.

Jett, S., and Spencer, V. 1981. Navajo Architecture: Forms, History, Distribution. Tucson, University of Arizona Press, 289 pp.

Johnson, D. 1995. Economic Renewal in Indian Country. American Studies Newsletter, 35, pp. 36–40.

Johnston, D. 1966. An Analysis of Sources of Information on the Population of the Navaho. Smithsonian Institution, Bureau of American Ethnology Bulletin 197. Washington, DC, GPO.

Jorgensen, J. 1971. Indians and the Metropolis. In: Waddell, O., and Watson, M., eds., pp. 67–113.

Jorgensen, J. 1978. A Century of Political Economic Effects on American Indian Society, 1880–1980. JES, 6, pp. 1–82.

Jorgensen, J., Davis, S., and Mathews, R. 1978. Energy, Agriculture, and Social Science in the American West. In: Jorgensen, J., et al., eds., pp. 3–16.

Jorgensen, J., et al., eds. 1978. Native Americans and Energy Development. Cambridge, MA, Anthropology Resource Center, 89 pp.

Josephy, A., Jr. 1971. Toward Freedom: The American Indian in the Twentieth Century. In: Prucha, F., et al., eds., pp. 38–65.

Josephy, A., Jr. 1977. Concise History of United States–Sioux Relations. In: Ortiz, R., ed., pp. 19–28.

Josephy, A., Jr. 1982. Now the Buffalo's Gone: A Study of Today's American Indians. New York, A. Knopf, 300 pp.

Joyner, C. 1978. The Historical Status of American Indians under International Law. The Indian Historian, 11, pp. 30–35.

Kamber, P. 1989. Timber and Termination: The Klamath Case. European Review of Native Studies, 3, pp. 43–47.

Khera, S., and Mariella, P. 1983. Yavapai. In: Ortiz, A., ed., pp. 38–54.

Kinney, J. 1937. A Continent Lost—A Civilization Won. Baltimore, John Hopkins Press, 366 pp.

Klink, H. J., and Mayer, E. 1983. Vegetationsgeographie. Braunschweig, Westermann (Das Geographische Seminar), 278 pp.

Kluckhohn, C., and Leighton, D. 1962. The Navaho. 2d ed. Garden City, NY, Natural History Library/Anchor Books, 355 pp.

Kluckhorn, F. 1961. Variations in Value Orientations. Evanston, IL, Row and Peterson, 437 pp.

Knox, P., et al. 1988. The United States: A Contemporary Human Geography. London, Longman, 287 pp.

König, R. 1973. Indianer—wohin? Alternativen in Arizona. Skizzen zur Entwicklungssoziologie. Opladen, Westdt. Verlag GmbH, 244 pp.

König, R., ed. 1974. Das Interview. Formen—Technik—Auswertung. 9th ed. Köln, Verlag Kiepenheuer and Witsch, 422 pp.

König, R. 1980. Navajo-Report 1970–1980. Von der Kolonie zur Nation. Neustadt, Arca Verlag, 190 pp.

Kopp, H. 1978. Die Arabische Republik Jemen. Entwicklungsprobleme und Entwicklungsmöglichkeiten eines rohstoffarmen Landes der Vierten Welt. GR, 30, pp. 88–93.

Kramm, H., and Brunner, H. 1977. Geographie der USA. Leipzig, VEB Hermann Haack, 167 pp.

Kroeber, A. 1939. Cultural and Natural Areas of Native North America. American Archaeology and Ethnology 38. Berkeley, University of California Press, 242 pp.

La Fontaine, F. 1974. The Native American Credit Problem. AILR, 2, pp. 29–40.

Laird, M. 1979. Water Rights: The Winters Cloud over the Rockies; Indian Water Rights and the Development of Western Energy Resources. AILR, 7, pp. 155–169.

Laird, W. 1977. Hopi Bibliography. Tucson, University of Arizona Press, 735 pp.

Latorre F., and Latorre D. 1976. The Mexico Kickapoo Indians. Austin, University of Texas Press, 401 pp.

Lazewski, T. 1973. Geographic Research and Native Americans. Introductory Commentary to the Special Interest Session on Native Americans in Contemporary Society: Problems and Issues. Paper for the AAAG Meeting in Urbana, IL. Manuscript, 7 pp.

Lazewski, T. 1976. American Indian Migration to and within Chicago, Illinois. Ph.D. diss. (Geography), University of Illinois–Urbana, 185 pp.

Lazewski, T. 1982. American Indian, Puerto Rican, and Black Urbanization. Journal of Cultural Geography, 2, pp. 119–134.

Lévi-Strauss, C. 1981. Das Prinzip der Gegenseitigkeit. In: Lévi-Strauss, C., ed., Die elementaren Strukturen der Verwandtschaft (translated from the French). Frankfurt, Suhrkamp, pp. 107–127.

Levine, S., and Lurie, N., eds. 1968. The American Indian Today. 2d ed. Jacksonville, FL, Convention Press, 229 pp.

Levitan, S., and Hetrick, B. 1971. Big Brother's Indian Programs: With Reservations. New York, McGraw-Hill Co., 228 pp.

Levitan, S., and Miller, E. 1993. The Equivocal Prospects for Indian Reservations. Washington, DC, George Washington University (Center for Social Policy Studies), 69 pp.

Levy, J., and Kunitz, S. 1969. Notes on Some White Mountain Apache Social Pathologies. Plateau, 42, pp. 11–19.

Lindig, W. 1972. Die Kulturen der Eskimo und Indianer Nordamerikas. Frankfurt, Athenaion, 380 pp.

Lindig, W., and Münzel, M. 1985. Die Indianer. 3d ed. Vol. 1 (Nordamerika). Deutscher Taschenbuch Verlag, 352 pp.

Littmann, G. 1970. Alcoholism, Illness, and Social Pathology among Indians in Transition. American Journal of Public Health, 60, pp. 1769–1787.

Locke, P. 1978. A Survey of College and University Programs for American Indians. 2d ed. Boulder, CO, Western Interstate Commission for Higher Education, 210 pp.

Lupe, R. 1986. Interior Indian Water Policy 1985: Consummation of Fraud. Fort Apache Scout, Jan. 3, pp. 2–3.

Lurie, N. 1968. Historical Background. In: Levine, S., and Lurie, N., eds., pp. 25–45.

Lurie, N. 1971. The World's Oldest On-Going Protest Demonstration: North American Indian Drinking Patterns. Pacific Historical Review, 40, pp. 311–332.

Lurie, N. 1972. From Reservation to Colony. HO, 31, pp. 257–269.

Manuel, H., Ramon J., and Fontana, B. 1978. Dressing for the Window: Papago Indians and Economic Development. In: Stanley, S., ed., pp. 511–575.

Marken, J., and Hoover, H. 1980. Bibliography of the Sioux. New York, Scarecrow Press, 370 pp.

Matheny, R. 1976. Lumbering in the White Mountains of Arizona, 1919–1942. Arizona and the West, 18, pp. 237–256.

McDonald, J., and Lazewski, T., eds. 1976. Geographical Perspectives on Native Americans: Topics and Resources. AAG, Associated Committee on Native Americans, Publication 1. N.p. Manuscript. 193 pp.

McDonald, P. 1980. Navajo Natural Resources. In: Vecsey, C., and Venables, R., eds.: American Indian Environments: Ecological Issues in Native American History. Syracuse, NY, Syracuse University Press, pp. 162–170.

McGuire, T. 1980. Mixed Bloods, Apaches, and Cattle Barons: Documents for a History of the Livestock Economy on the White Mountain Reservation, Arizona. Tucson, Arizona State Museum, University of Arizona (Archaeological Series 142), 227 pp.

McGuire, T. 1983. Walapai. In: Ortiz, A., ed., pp. 137–148.

McIntire, E. 1967a. Central Places on the Navaho Reservation: A Special Case. Yearbook of the Association of Pacific Coast Geographers, 29, pp. 91–96.

McIntire, E. 1967b. Changes in an American Peasant Society: The Hopi Farmer. Oregon Geographer, 1, pp. 9–13.

McIntire, E. 1969. Hopi Colonization on the Colorado River. California Geographer, 10, pp. 7–14.

McIntire, E. 1971. Changing Patterns of Hopi Indian Settlement. AAAG, 61, pp. 510–521.

McIntire, E. 1982. First Mesa Hopi in 1900: A Demographic Reconstruction. Journal of Cultural Geography, 2/2, pp. 58–71.

McKee, J. 1971. The Choctaw Indians: A Geographical Study in Cultural Change. Southern Quarterly, 9, pp. 107–141.

McKee, J. 1987. The Choctaw: Self-Determination and Socioeconomic Development. In: Ross, T., and Moore, T., eds., pp. 173–187.

McKee, J., ed. 1985. Ethnicity in Contemporary America: A Geographical Appraisal. Dubuque, IA, Kendall, 284 pp.

McKee, J., and Murray, S. 1986. Economics Progress and Development of the Choctaw since 1945. In: Wells, S., and Tubby, R., eds., pp. 122–136.

McKee, J., and Schlenker, J. 1980. The Choctaws: Cultural Evolution of a Native American Tribe. Jackson, University Press of Mississippi, 227 pp.

McNickle, D. 1962. The Indian Tribes of the United States: Ethnic and Cultural Survival. London, Oxford University Press, 79 pp.

McNickle, D. 1973. Native American Tribalism: Indian Survivals and Renewals. London, Oxford University Press, 190 pp.

Medicine, B. 1981. American Indian Family: Culture, Change, and Adoptive Strategies. JES, 8, pp. 13–23.

Meinig, D. 1965. The Mormon Culture Region: Strategies and Patterns in the Geography of the American West, 1847–1964. AAAG, 55, pp. 191–219.

Meinig, D. 1986. The Shaping of America: A Geographical Perspective on Five Hundred Years of History. Vol. 1, Atlantic America, 1492–1800. New Haven, CT, Yale University Press, 500 pp.

Meister, C. 1978. The Misleading Nature of Data in the Bureau of the Census Subject Report on 1970 American Indian Population. Indian Historian, 11/4, pp. 12–19.

Merewether, G., et al. 1994. Resources of Oil and Gas on Indian and Native American Lands of the United States. In: Anderson, R., ed., pp. 5–19.

Meriam, L., et al. 1928. The Problem of Indian Administration. Washington, DC, Brookings Institution (Institute for Government Research), 872 pp.

Merrill, J. 1980. Aboriginal Water Rights. Natural Resource Journal, 20, pp. 45–69.

Mooney, J. 1928. The Aboriginal Population of America North of Mexico. In: Swanton, J., ed., Smithsonian Miscellaneous Collection, 80/7, Washington, DC, 40 pp.

Mose, I. 1987. "So lange das Gras wächst und das Wasser fließt." Indianer und Reservationspolitik in den USA. Praxis Geographie, Heft 7/8, pp. 34–37.

Mose, I. 1990. Zwischen Widerstand und Resignation. Probleme der politischen und sozialen Selbstbehauptung von Indianern in den USA. Vechtaer Arbeiten zur Geographie und Regionalwissenschaft 10, pp. 75–84.

Müller-Vogg, H. 1987. Auf "Kriegspfad" gegen staatliche Bevormundung und Bürokratie. Frankfurter Allgemeine Zeitung, Aug. 17, p. 10.

Murray, S., and Tweeten, L. 1981. Some Trade-offs: Culture Education and Economic Progress on Federal Indian Reservations. Growth and Change, 12, pp. 10–16.

Nafziger, R. 1976. Indian Uranium Profits and Perils. Albuquerque, NM, Americans for Indian Opportunity.

Nakane, C. 1979. Japanese Society. 5th ed. Harmondsworth, Penguin Books, 162 pp.

Neils, E. 1971. Reservation to City: Indian Migration and Federal Relocation. Geography Research Paper 131. Chicago, University of Chicago, Department of Geography, 198 pp.

Novak, M. 1973. The Rise of the Unmeltable Ethnics. 3d ed. New York, Macmillan Publishing Co., 376 pp.

Nurge, E., ed. 1970. The Modern Sioux: Social Systems and Reservation Culture. Lincoln, University of Nebraska Press, 352 pp.

Oglala Sioux Community College, ed. 1983. Tokatakiya: A Self Study of Oglala Sioux Community College (OSCC), Pine Ridge Indian Reservation. Piya-Wiconi, SD, 128 pp. (*Tokatakiya* is a Lakota word meaning "onward" or "forward.")

Opler, M. 1983. Chiricahua Apache. In: Ortiz, A., ed., pp. 401–418.

Orrin, F., et al. 1976. Consumptive Use of Water by Crops in Arizona. 3d ed. Technical Bulletin 169, University of Arizona, College of Agriculture, 45 pp.

Ortiz, A., ed. 1983. Southwest Handbook of North American Indians, Vol. 10. Washington, DC, Smithsonian Institution, 868 pp.

Ortiz, R. 1980. Wounded Knee 1890 to Wounded Knee 1973: A Study in United States Colonialism. JES, 8/2, pp. 1–15.

Ortiz, R. 1984. Land Reform and Indian Survival in the United States. In: Geisler, C., and Popper, F., eds., pp. 151–170.

Ortiz, R., ed. 1979. Economic Development in American Indian Reservations. Berkeley, Moon Books, 224 pp.

Oswalt, W. 1966. This Land Was Theirs. New York, J. Wiley and Sons, 560 pp.

Owens, N. 1978. Can Tribes Control Energy Development? In: Jorgensen, J., et al., eds., pp. 49–62.

Parker, M. 1976. Native American Traditional Economic Values and Systems: Some Dispersed Samples. In: McDonald, J., and Lazewski, T., eds., pp. 11–40.

Paterson, J. 1984. North America: A Geography of Canada and the United States. 7th ed. New York, Oxford University Press, 497 pp.

Peters, U. 1987. "Radioaktivität kennt keine Reservationsgrenzen." Zum Uranabbau auf indianischem Land in den USA. Progrom, 18/8, pp. 29–32.

Peyer, B. 1985. Probleme indianischen Wasserrechts. In: Lindig, W., and Münzel, M., eds., pp. 281–285.

Pfister, A. 1984. Resolution of Indian Water Claims. Arizona Waterline, Winter, pp. 2–6.

Pierce, M. 1977. The Work of the Indian Claims Commission. American Bar Association Journal, 63, pp. 227–232.

Pipenstem, B., and Rice, W. 1978. The Mythology of the Oklahoma Indians: A Survey of the Legal Status of Indian Tribes in Oklahoma. AILR, 6, pp. 259–328.

Pohl, I., and Zepp, D., eds. 1975. Amerika. 11th ed. Munich, P. List Verlag (Harms, Handbuch der Geographie), 512 pp.

Price, J. 1968. The Migration of American Indians to Los Angeles. HO, 27, pp. 168–175.

Provenzo, E., Jr., and McCloskey, G. 1981. Catholic and Federal Indian Education in the Late Nineteenth Century: Opposed Colonial Models. Journal of American Indian Education, 21, pp. 10–18.

Prucha, F. 1962. American Indian Policy in the Formative Years: The Indian Tradeland Intercourse Acts, 1790–1834. Cambridge, MA, Harvard University Press, 303 pp.

Prucha, F. 1973. Americanizing the American Indians: "Friends of the Indians," 1880–1900. Cambridge, MA, Harvard University Press, 358 pp.

Prucha, F. 1976. American Indian Policy in Crisis: Christian Reformers and the Indian, 1865–1900. Norman, Oklahoma University Press, 468 pp.

Prucha, F. 1977. A Bibliographical Guide to the History of Indian-White Relations in the United States. Chicago, University of Chicago Press (Publication of Newberry Library), 454 pp.

Prucha, F. 1981. Indian Policy in the United States: Historical Essays. Lincoln, University of Nebraska Press, 272 pp.

Prucha, F. 1982. Indian-White Relations in the US: A Bibliography of Works Published 1975–1980. Lincoln, University of Nebraska Press.

Prucha, F. 1984. The Great Father: The United States Government and the American Indians. 2 vols. Lincoln, University of Nebraska Press, 1, 302 pp.
Prucha, F., ed. 1975. Documents of United States Indian Policy. Lincoln, University of Nebraska Press, 278 pp.
Prucha, F., Hagan, W., and Joseph, A., Jr. 1971. American Indian Policy: The Indian Historical Society Lectures, 1970–1971. Indianapolis, Indiana Historical Society, 65 pp.
The Pueblo of Sandia Tribal Council. 1998. Sandia Pueblo Today and Tomorrow. Albuquerque, NM, 8 pp.
Quinn, F. 1968. Must the American West Be Won Again? GRev, 58, pp. 108–132.
Ranquist, H. 1975. The Winters Doctrine and How It Grew: Federal Reservation of Rights to the Use of Water. Law Review (Brighton University), 1975/3, pp. 640–724.
Remerowski, A., et al. 1982. Bibliography on Indian Economic Development. Boulder, CO, Native American Rights Fund, 58 pp.
Reno, P. 1981. Navajo Resources and Economic Development. Albuquerque, University of New Mexico Press, 183 pp.
Rinschede, G. 1984. Die Wanderwirtschaft im gebirgigen Westen der USA und ihre Auswirkungen im Naturraum. Eichstätter Beiträge, Abt. Geographie 10, 486 pp.
Robbins, L. 1978. Energy Developments and the Navajo Nation. In: Jorgensen, J., et al., eds., pp. 35–48.
Rosenthal, H. 1985. Indian Claims and the American Conscience: A Brief History of the Indian Claims Commission. In: Sutton, I., ed., pp. 35–70.
Ross, A. 1981. Water Rights: Aboriginal Water Use and Water Law in the Southwestern United States; Why the Reserved Rights Doctrine Was Inappropriate. AILR, 9, pp. 195–209.
Ross, T., and Moore, T., eds. 1987. A Cultural Geography of North American Indians. Boulder, CO, Westview Press, 331 pp.
Rostlund, E. 1951. Freshwater Fish and Fishing in Native North America. Ph.D. diss. (Geography), University of California–Berkeley.
Royce, C. 1900. Indian Land Cessions in the United States. Washington, DC, GPO, Extract from the Eighteenth Annual Report of the Bureau of American Ethnology to the Secretary of the Smithsonian Institution, 1896–1897 (reprinted Arno Press, 1971), 997 pp.
Ruffing, L. 1978. Dependence and Underdevelopment. In: Stanley, S., ed., pp. 91–113.
Rühl, A. 1922. Die Wirtschaftspsychologie des Spaniers. Zeitschrift d. Gesellschaft für Erdkunde zu Berlin, pp. 81–115.
Rühl, A. 1925. Vom Wirtschaftsgeist im Orient. Leipzig, Verlag Quelle & Meyer, 98 pp.
Rühl, A. 1927. Vom Wirtschaftsgeist in Amerika. Leipzig, Verlag Quelle & Meyer, 134 pp.
Rühl, A. 1968. Einführung in die allgemeine Wirtschaftsgeographie. 2d ed. Leiden, A. W. Sijthoff, 94 pp.
Russell, F. 1908. The Pima Indians. Washington, DC, Twenty-sixth Annual Report of the Bureau of American Ethnology for the Years 1904–1905, 389 pp.
Samet, M., et al. 1984. Uranium Mining and Lung Cancer among Navajo Men. New England Journal of Medicine, 310, pp. 1481–1484.

Santler, H., and Bogenreiter, H., eds. 1993. Land ist Leben. Bedrohte Völker im Kampf gegen die Zerstörung der Umwelt. Vienna, Austria, Verlag Jugend und Volk, 256 pp.

Sargent, C. 1988. Metro Arizona. Scottsdale, AZ, Biffington Books, 195 pp.

Sasaki, T., and Basehart, H. 1961–1962. Sources of Income among Many Farms: Rough Rock Navajo and Jicarilla Apache: Some Comparisons and Comments. HO, 20, pp. 187–190.

Sauer, C. 1936. American Agricultural Origins: A Consideration of Nature and Culture. In: Leighly, J., ed. 1969. Land and Life: A Selection from the Writings of Carl Sauer. Berkeley, CA, University of California Press, pp. 121–144.

Sauer, C. 1944. A Geographic Sketch of Early Man in America. GRev, 34, pp. 529–573.

Sauer, C. 1954. Gatherers and Farmers in the Greater Southwest. American Anthropologist, 56, pp. 563–566.

Sauer, C. 1969. Seeds, Spades, Hearths, and Herds. Cambridge, MA, MIT Press, 175 pp.

Sauer, C. 1971. Sixteenth Century North America. Berkeley, University of California Press, 319 pp.

Sauer, C., and Brand, D. 1930. Pueblo Sites in Southeastern Arizona. Publications in Geography (Berkeley) 3/7, pp. 415–459.

Schmeckbier, L. 1972. The Office of Indian Affairs: Its History, Activities, and Organizations. 2d ed. New York, AMS Press (reprint), 591 pp.

Schmieder, O. 1963. Die Neue Welt, II. Nordamerika. Heidelberg-Munich, Keysersche Verlagsbuchhandlung, 548 pp.

Schulze-Thulin, A. 1976. Weg ohne Mokassins. Die Indianer Nordamerikas heute. Düsseldorf, Droste Verlag, 366 pp.

Schwarzbauer, P. 1986. Der Lakota-Report. Ein Volk kämpft ums Überleben. Wyk auf Föhr, Verlag für Amerikanistik, D. Kuegler, 101 pp.

Schwarzbauer, P., ed. 1989. Indianer Nordamerikas. Gegenwart und Vergangenheit. Informationen für Lehrer und politisch Interessierte. Wien, Gesellschaft für Bedrohte Völker, 132 pp.

Schwind, M. 1951. Kulturlandschaft als objektivierter Geist. In: Schwind, M. 1964, pp. 1–26.

Schwind, M. 1964. Kulturlandschaft als geformter Geist. Drei Aufsätze über die Aufgaben der Kulturgeographie. Darmstadt, Wissenschaftliche Buchgesellschaft, 73 pp.

Service, E. 1979. The Hunters. 2d ed. Englewood Cliffs, NJ, Prentice Hall, 118 pp.

Shafer, S. 1985. Zweisprachige Erziehung für Schulkinder im Südwesten der USA. Bildung und Erziehung, 38, pp. 343–356.

Shumway, J., and Jackson, R. 1995. Native American Population Patterns. GRev, 85, pp. 185–201.

Sibley, G. 1977. The Desert Empire: In Its Desperate Search for Water, the American West Meets the Limits of the Technological Ideal. Harpers, October, pp. 53–68.

Siuruainen, E. 1979. The Native Indian in the United States of America: Some Development Trends and Comparisons. Nordia 13/2, 61 pp.

Siuruainen, E. 1980. Reservation Indians in the Pacific North-West. Terra, 92, pp. 8–18.

Smith, F., Kessell, J., and Fox, F. 1966. Father Kino in Arizona. Phoenix, Arizona Historical Foundation.

Snyder, F., ed., 1996. Native American Directory. 2 vols. (Alaska and Canada and United States). San Carlos, AZ, National Native American Cooperative. Alaska and Canada, 184 pp; U.S., 583 pp.

Sorkin, A. 1971. American Indians and Federal Aid. Washington, DC, Brookings Institution, 231 pp.

Sorkin, A. 1973. Business and Industrial Development on American Reservations. Annals of Regional Science, 7, pp. 115–129.

Sorkin, A. 1978. The Urban American Indian. Lexington, MA, D. C. Heath, 158 pp.

Sparks, J. 1984. What's This I Hear about Indian Water Rights? Arizona Waterline, Winter, pp. 3–10.

Spicer, E. 1980. The American Indians. Dimensions of Ethnicity series. Cambridge, MA, Harvard University Press, 210 pp.

Spicer, E. 1981. American Indians. In: Thernstrom, S., ed., pp. 58–122.

Stanley, S. 1978. Introduction and Conclusion. In: Stanley, S., ed. 1978. American Indian Economic Development. The Hague, Netherlands, Mouton Publ., pp.1–14, 579–591.

Stanley, S., and Thomas, R. 1978. Current Demographic and Social Trends among North American Indians. Annals of the American Academy of Political and Social Science 436, pp. 111–120.

Starkey, M. 1972. The Cherokee Nation. 2d ed. New York, Russell and Russell, 355 pp.

Steinmetz, R. 1968. Alfred Rühl's Lebenswerk. In: Rühl, A., pp. 7–11.

Sutton, I. 1967. Private Property in Land among Reservation Indians in Southern California. Yearbook of the Association of Pacific Coast Geographers, 21, pp. 69–89.

Sutton, I. 1975. Indian Land Tenure: Bibliographical Essays and a Guide to the Literature. New York, Clearwater Publishing Co., 290 pp.

Sutton, I. 1976. Sovereign States and the Changing Definition of the Indian Reservation. GRev 66, pp. 281–295.

Sutton, I. 1982. Indian Land Rights and the Sagebrush Rebellion. GRev, 72, pp. 557–559.

Sutton, I., ed. 1985. Irredeemable America: The Indians' Estate and Land Claims. Albuquerque, University of New Mexico Press, Native American Studies, 448 pp.

Taylor, B., and Beach, P. 1969. Indian Economic Development. Phoenix, Fifteenth Arizona Town Hall, Conference on Economic Planning and Development, pp. 133–148.

Taylor, B., and O'Connor, D. 1969. Indian Manpower Resources in the Southwest: A Pilot Study. Tempe, Arizona State University, Bureau of Business and Economic Research, 374 pp.

Taylor, T. 1972. The States and Their Indian Citizens. Washington, DC, GPO, 307 pp.

Taylor, W. 1975. Land and Water Rights in the Viceroyalty New Spain. New Mexico Historical Review, 50, pp. 189–212.

Tempe Daily News. 1997. Indian Bloodlines Lose Purity as Tribes Grow, Jan. 27, A, p. 5.

Templer, O. 1978. Texas Surface Water Law. Historical Geography, 8, pp. 11–20.

Thernstrom, S., ed. 1981. Harvard Encyclopedia of American Ethnic Groups. 2d ed. Cambridge, MA, Harvard University Press, 1,076 pp.

Trelease, F. 1971. Federal-State Relations in Water Law (Indian Reserved Rights). Arlington, VA, National Water Commission, 324 pp.

Trink, R. 1989. Solange radioaktive Flüsse fließen. Uranabbau in den Black Hills (South Dakota). In: Schwarzbauer, P., ed., pp. 92–95.

Tucker, G. 1956. Tecumseh: Vision of Glory. Indianapolis, Bobbs-Merrill Co., 399 pp.

Turner-Ruffing, L. 1978. Navajo Economic Development: A Dual Perspective. In: Stanley, S., ed., pp. 15–86.

Tyler, L. 1973. A History of Indian Policy. Washington, DC, U.S. Department of the Interior, GPO, 328 pp.

Udall, S. 1963. The Quiet Crisis. New York, Holt, Rinehart and Winston, 208 pp.

United Effort Trust. N.d. Indian Water Rights. Washington, DC, 3 pp.

Useem, R., and Eicher, C. 1970. Rosebud Reservation Economy. In: Nurge, E., ed., pp. 3–34.

Veeder, W. 1982. Indian Water Rights in the Concluding Years of the Twentieth Century. Chicago, Newberry Library, Occasional Paper Series 5, 73 pp.

Vinje, D. 1982. Cultural Values and Economic Development: U.S.—Indian Reservations. Social Science Journal, 19, pp. 87–98.

Vlasich, J. 1980. Transitions in Pueblo Agriculture, 1938–1948. New Mexico Historical Review, 55, pp. 25–46.

Vogt, E., and O'Dea, T. 1953. A Comparative Study in the Role of Values in Social Action in Two Southwestern Communities. American Sociological Review, 18, pp. 645–654.

Vollmar, R. 1982. Kartenanfertigung und Raumauffassung nordamerikanischer Indianer. GR, 34, pp. 302–307.

Waddell, J., and Watson, O., eds. 1971. The American Indian in Urban Society. Boston, Little, Brown and Co., 414 pp.

Wallace, E., and Hoefel, E. 1952. The Comanches: Lords of the South Plains. Norman, University of Oklahoma Press.

Washburn, W. 1968. Philanthropy and the American Indians. Ethnohistory, 15, pp. 43–56.

Washburn, W. 1971. Red Man's Land; White Man's Law: A Study of the Past and Present Status of the American Indian. New York, Scribner's, 280 pp.

Washburn, W. 1973. The American Indian and the United States: A Documentary History. 4 vols. Westport, CT, Greenwood Press. 3,119 pp.

Washburn, W. 1975a. The Indian in America. New York, Harper and Row, 210 p.

Washburn, W. 1975b. The Assault on Indian Tribalism: The General Allotment Law (Dawes Act) of 1887. Philadelphia, America's Alternative Series, 79 pp.

Washburn, W. 1984. A Fifty-Year Perspective on the Indian Reorganization Act. American Anthropologist, 86, pp. 279–289.

Washburn, W. 1985. Land Claims in the Mainstream of IndianWhite Land History. In: Sutton, I., ed., pp. 21–33.

Washington Post. 1985. Watt Pursuing Business on Indian Reservations: Former Interior Chief Urging Oil, Gas Deal, March 30, p. 27.

Washington Post Magazine. 1997. The Unfashionable, Feb. 23, 24 pp.

Watson, J. 1979. The Indian Question. In: Watson, J., A Social Geography of the United States. London, Longman, pp. 20–48.

Wax, M. 1971. Indian Americans: Unity and Diversity. Engelwood Cliffs, NJ, Prentice Hall, 236 pp.

Weatherhead, L. 1980. What Is an "Indian Tribe?" The Question of Tribal Existence. AILR, 8, pp. 1–47.

Weaver, T. 1975. Indians of Arizona: A Contemporary Perspective. Tucson, VAP, 169 pp.

Weber, M. 1920. Die protestantische Ethik und der Geist des Kapitalismus. In: Weber 1963, vol. 1, pp. 17–206.

Weber, M. 1963. Gesammelte Aufsätze zur Religionssoziologie. Vol. 1–3. 2d ed. Tübingen, J. C. B. Mohr, 573 pp., 378 pp., and 442 pp.

Wehmeier, E. 1975. Die Bewässerungsoase Phoenix/Arizona. Stuttgarter Geographische Studien 89, 176 pp.

Wehmeier, E. 1987. Die Region Phoenix. Entwicklung und Strukturwandel unter dem Aspekt der Wasserversorgung. GR, 39, pp. 435–442.

Weibel, J. C. 1976. The American Indian Family in Los Angeles. In: McDonald, J., and Lazewski, T., eds., pp. 121–146.

Weibel, J. C. 1977. Native Americans in Los Angeles: A Cross Cultural Comparison of Assistance Patterns in an Urban Environment. Ph.D. diss. (Cultural Anthropology), University of California–Los Angeles, 421 pp.

Weightman, B. 1976. Indian Social Space: A Case Study of the Musquean Band of Vancouver, British Columbia. Canadian Geographer, 20, pp. 171–186.

Wells, S., and Tubby, R., eds. 1986. After Removal: The Choctaw in Mississippi. Jackson, University Press of Mississippi, 153 pp.

Weltfish, G. 1965. The Lost Universe. New York, Basic Books, 506 pp.

Weppner, R. 1972. Socio-Economic Barriers to Assimilation of Navajo Migrants. HO, 31, pp. 303–314.

Wesel, U. 1985. Frühformen des Rechts in vorstaatlichen Gesellschaften. Umrisse einer Frühgeschichte des Rechts bei Sammlern und Jägern und akephalen Ackerbauern und Hirten. Frankfurt, Suhrkamp, 388 pp.

Weslager, C. 1972. The Delaware Indians. New Brunswick, NJ, Rutgers University Press, 556 pp.

Wiley, P., and Gottlieb R. 1982. Empires in the Sun: The Rise of the New American West. New York, G. P. Putnam's Sons, 332 pp.

Wilkinson, C., and Biggs, E. 1977. The Evolution of the Termination Policy. AILR, 5/1, pp. 139–184.

Williams, W. 1978. United States Indian Policy and the Debate over Philippine Annexation. San Francisco, American Historical Association Meeting. Manuscript. 17 pp.

Wilson, P. 1972. Farming and Ranching on the Wind River Indian Reservation, Wyoming. Ph.D. diss. (Geography), University of Nebraska–Lincoln, 433 pp.

Wilson, R., and Manydeeds, S. 1994. National Indian Issues. In: Godwin, L., ed., pp. 1–16.

Winchell, D. et al. 1989. Geographic Research on Native American. In: Gaile, G., and Willmott, C., eds., pp. 9–13.

Winchell, D. 1992. Tribal Sovereignity as the Basis for Tribal Planning. The Western Planner, 13/3, pp. 9–13.

Winchell, D. 1996. The Consolidation of Tribal Planning in American Indian Tribal Government and Culture. In: Frantz, K., ed., pp. 209–224.

Winchell, D., Lounsbury, J., and Sommers, L. 1998. Indian Gaming in the U.S.: Distribution, Significance, and Trends. Focus, 44/4, pp. 1–10.

Winchell, D., et al. 1989. Geographic Research on Native Americans. In: Gaile, G., and Willmott, C., eds., pp. 239–255.

Windhorst, H. 1978. Geographie der Wald- und Forstwirtschaft. Stuttgart, Teubner Studienbücher, 204 pp.

Windhorst, H. 1987. Konzentrationsprozesse in der US-americkanischen Agrarwirtschaft und ihre räumlichen Konsequenzen. Geographische Rundschau, 39, pp. 474–480.

Windhorst, H. 1989. Die Industrialisierung der Agrarwirtschaft als Herausforderung an die Agrargeographie. Geographische Zeitschrift, 77, pp. 136–153.

Wirth, E. 1956. Der heutige Irak als Beispiel orientalischen Wirtschaftsgeistes. Die Erde, 8, pp. 30–50.

Wishart, D. 1976. Cultures in Co-operation and Conflict: Indians in the Fur-Trade on the Northern Great Plains, 1807–1840. Journal of Historical Geography, 2, pp. 311–328.

Wishart, D. 1979. The Dispossession of the Pawnee. AAAG, 69, pp. 382–401.

Wishart, D. 1985. The Pawnee Claims Case, 1947–1964. In. Sutton, I., ed., pp. 157–186.

Wishart, D. 1994. An Unspeakable Sadness: The Dispossession of the Nebraska Indians. Lincoln, University of Nebraska Press, 309 pp.

Wishart, D. 1996. Indian Dispossession and Land Claims: The Issue of Fairness. In: Frantz, K., ed., pp. 181–194.

Wistisen, M., et al. 1975. A Study to Identify Potential Feasible Small Businesses for the Navajo Nation, Phase 1: An Evolution of Income and Expenditure Patterns. Provo, UT, Brigham Young University, Center for Business and Economic Research, 67 pp.

Witt, S. 1968. Nationalistic Trends among American Indians. In: Levine, S., and Lurie, N., eds., pp. 53–75.

Woerner, D. 1941. Education among the Navajo: An Historical Study. Ph.D. diss. (Education), Columbia University, 227 pp.

Young, R. A. 1976. Regional Development and Rural Poverty in the Navajo Indian Area. Ph.D. diss. (Geography), University of Wisconsin–Madison, 475 pp.

Zelinsky, W. 1973. The Cultural Geography of the United States. Englewood Cliffs, NJ, Prentice-Hall, 164 pp.

ARCHIVAL AND STATISTICAL MATERIAL

Annual Reports of the Commissioner of Indian Affairs (ARCIA) to the Secretary of the Interior. Yearly issues 1849–1954. U.S. Department of the Interior, Bureau of Indian Affairs (formerly Office of Indian Affairs). Washington, DC.*

* Except for a few large libraries in the United States such as the Library of Congress, the Annual Reports of the Comissioner of Indian Affairs are to be found only in the BIA archives in Washington, DC. This is where I inspected them. From 1825 to 1848 these re-

ARCIA. *See* Annual Reports of the Commissioner of Indian Affairs.

ARIL. *See* U.S. Department of the Interior, BIA, Office of Trust Responsibilities.

Arizona Crop and Livestock Reporting Service. 1983. Arizona Agricultural Statistics, 1982. Phoenix, University of Arizona, College of Agriculture.

Arizona vs. California. 1963. 373 U.S. 546.

Arizona Water Commission. 1977. Alternative Futures, Arizona State Water Plan, Phase II.

Arnold, J. 1962. Arizona Watershed Management Progress Report. Arizona State Land Department, Watershed Management Division. Phoenix, AZ. Manuscript. 11 pp.

BIA, Fort Apache Agency. 1963. Cibicue Watershed Progress Report. Whiteriver, AZ, Aug. 27, 27 pp.

BIA, Fort Apache Agency. 1985. White Mountain Apache Environmental Assessment Plan. Whiteriver, AZ. Manuscript. 262 pp.

BIA, Fort Apache Agency, Branch of Forestry. 1982. Forestry Program Management Review. Whiteriver, AZ. Manuscript. 152 pp.

BIA, Fort Apache Agency, Branch of Forestry. 1983. Fort Apache Indian Reservation Forest Management Plan. Whiteriver, AZ. Manuscript. 298 pp.

BIA, Fort Apache Agency, Grazing Files. 1910–1934. Whiteriver, AZ.

BIA, Fort Apache Agency, Grazing Files. 1952. John Crow's letter to Ralph Colvin, BIA, Area Director in Phoenix, AZ, Feb. 29.

BIA, Fort Apache Agency, Natural Resource Information System. 1983. Inventory and Production Report. Whiteriver, AZ, 11 pp.

BIA, Phoenix Area Office. 1962. Accomplishments in Brush Control, 1959–1961. Phoenix, AZ, 76 pp.

BIA, Phoenix Area Office. 1965. Report of the Area Director from June 1964. Phoenix, AZ, 23 pp.

BIA, Phoenix Area Office. 1978. Environmental Assessment (Draft). Fort Apache Indian Reservation. Baseline Data. Phoenix, AZ, 249 pp.

BIA, Phoenix Area Office, Branch of Roads. 1972. Fort Mohave Indian Reservation, Highway System Map. Phoenix, AZ.

BIA, Pima River Agency, Land Operations. 1981. Annual Irrigation Crop Report. Sacaton, AZ.

BIA, San Carlos Agency. 1962. San Carlos Brush Control Report. San Carlos, AZ. Manuscript. 7 pp.

BIA, Truxton Cañon Agency. 1986a. Hualapai Membership Roll (handwritten, current membership roll). Valentine, AZ.

BIA, Truxton Cañon Agency. 1986b. Hualapai Tribal Forestry. Peach Springs, AZ. Manuscript. 183 pp.

---

ports were made by the War Department; but since 1849 they have been made by the Department of the Interior. Until 1861 they were issued jointly by the Senate and the House of Representatives. They include the annual report of the commissioner of Indian affairs and the reports of the BIA superintendents of the regional headquarters and the directors of the individual BIA agencies on the respective reservations. These reports also include a comprehensive collection of letters and documents sent to various Indian agencies or composed by them (see Jones 1955, 58–66). After 1920 there was a notable decline in their quality and since the 1930s they have appeared only intermittently.

BIA, Washington Headquarters. 1968. Summary Record of Plants Established as Result of Indian Industrial Development Program. Washington, DC. Manuscript. 17 pp.

Board of Consultants. 1972. CAP Indian Water Project: Report on Allocation of Central Arizona Project for Five Applicant Tribes. Phoenix, AZ. Manuscript. 97 pp.

Bureau of Social Science Research, Inc. 1983. Native American Statistical Profiles. Pts. s1–4. Washington, DC. Manuscript.

CAP Indian Water Project. 1972. Report on Allocation of Central Arizona Project Water for Five Applicant Tribes. Phoenix, AZ. Manuscript. 95 pp.

Cherokee Nation vs. State of Georgia. 1831. 30 U.S. (5. Petition) 1. Peter's United States Supreme Court Reports, pp. 1–3, 16–17.

Chronicle of Higher Education. 1984. Minority Enrollments at More Than 3,100 Colleges and Universities, 1982. Fact-File. Washington, DC, Dec. 5, 24 pp.

Clean Water Act, 33 U.S.C. 13442 Title 33, 1972 as amended.

Clotts, H. 1917. Report of the Indians and Irrigation on Indian Reservations. District no. 204: Fort Apache Reservation, Phoenix, AZ, 37 pp.

Colorado River Water Conservation District vs. United States. 1976. 424 U.S. 800.

Crouse, C. 1903. Report of the Agent for the Fort Apache Agency. In: ARCIA, pp. 147–151.

Crouse, C. 1905. Report of the Agent for the Fort Apache Agency. In: Forestry Files of the BIA. Whiteriver, AZ. Manuscript.

Department of Economic Security, Arizona, Population Statistic Unit. 1986. A Demographic Guide to Arizona, 1985. Report no. 14. Phoenix, AZ, 68 pp.

Endangered Species Act, 16 U.S.C. 1531, 1966, as amended.

Forest Industries. 1985. Leading Lumber Producers: U.S. and Canada. July, pp. 26–27.

General Accounting Office. 1975. Improving Federally Assisted Business Development on Indian Reservations. U.S. Department of Commerce, Washington, DC, 29 pp.

Gibson vs. Anderson. 9th Cir. 1904. 131 F. 39.

Hayden, C. 1924. A History of the Pima Indians and the San Carlos Irrigation Project. Senate Document no. 11, 89th Cong., 1st sess. Washington, DC, GPO (reprinted 1965). 94 pp.

The Indian Self-Determination and Education Assistance Act (P.L. 93-638), as amended by P.L. 103-413, P.L. 103-435, and P.L. 103-437, 1994.

Jones, J. 1955. Key to the Annual Reports of the U.S. Commissioner of Indian Affairs. In: Ethnohistory, 2, pp. 58–62.

Jones, R. 1982. Federal Programs of Assistance to American Indians. Report by Senate Select Committee on Indian Affairs of the U.S. Senate, Washington, DC, GPO, 279 pp.

Kappler, C. 1904. Indian Affairs: Laws and Treaties. Vols. 1 and 2. 58th Cong., 2d sess., Sen. Doc. 319. Washington, DC, GPO, 1,162 pp. and 1,099 pp.

Ludlam, A. 1880. Report of the Agent for the Pima Agency to the Commissioner of Indian Affairs. In: ARCIA, pp. 3–4.

Macmillan and Co. 1964. The Statesman's Year-Book: Statistical and Historical Annual of the States of the World for the Year 1964–1965. Vol. 101. London, 1,716 pp.

Macmillan and Co. and Walter de Gruyter and Co. 1984. The Statesman's Year-

Book: Statistical and Historical Annual of the States of the World for the Year 1984–1985. Vol. 121. London, 1,692 pp.

McCarran Water Rights Suits Act. 1952. July 10, P.L. 82–495, Section 208, 82d Cong., 66 Stat. 549, 560, codified at 43 USCA 666.

Maine Indian Claims Settlement Act. 1980. 25 U.S.C. subsections 1721–1735.

National Water Commission. 1973. Water Policies of the Future: Final Report to the President and the U.S. Congress. Washington, DC, GPO, 579 pp.

Native American Graves Protection and Repatriation Act, 25 U.S.C. Chapter 32, 1990 as amended. Oklahoma Organic Act. May 2, 1890. Ch. 182, § 201, 26 Stat. 2081.

Presidential Commission on Indian Reservation Economics. 1984. Report and Recommendations to the President of the United States. Washington, DC, GPO, 158 pp.

Public Law 95-217,Clean Water Act of 1977, 33 U.S.C. sec. 1344.

Public Law 100-497, Indian Gaming Regulatory Act of 1988, 25 U.S.C. sec. 2701, et seq. and 18 U.S.C./sec.1166–1168.

Public Law 103-413, Tribal Self-Governance Act of 1994, 25 U.S.C 450.

The Salt River Pima-Maricopa Indian Community. 1989. The Salt River Pima-Maricopa Indian Community Water Settlement Agreement. Scottsdale, AZ.

Stadtwerke Innsbruck. 1988. Jahresbericht—Wasserwerk 43, 50 pp.

Staff of the American Indian Policy Review Committee. 1976. Report on Terminated and Nonfederally Recognized Indians. 94th Cong., 2d sess. Washington, DC.

Stout, J. 1872. Report of the Agent for the Pima Agency to General O. Howard. In: ARCIA, pp. 166–168.

Tee-Hit-Ton Indians vs. United States. 1955. 348 U.S. 272.

United Nations, Department of International Economic and Social Affairs, Statistical Office. 1985. Demographic Yearbook 35. New York, UN, Publishing Division, 1,083 pp.

United States vs. Cook. 1873. 86 U.S. 591.

U.S. Commission on Civil Rights. 1981. Indian Tribes: A Continuing Quest for Survival. Washington, DC, GPO, 192 pp.

U.S. Congress. 1997–1998. Indian Gaming Tax Reform Act. House of Representatives Bill 325.IH. 105th session. Washington, DC.

U.S. Department of Commerce. 1974. Federal and State Indian Reservations and Indian Trust Areas. Washington, DC, GPO, 604 pp.

U.S. Department of Commerce, BC. 1890. Eleventh Census: Moqui Pueblo Indians of Arizona and Pueblo Indians of New Mexico (Extra Census Bulletin). Washington, DC, GPO, 1893, 129 pp.

U.S. Department of Commerce, BC. 1890. Eleventh Census. Report on Indians Taxed and Indians Not Taxed in the United States (Except Alaska). Vol. 10. Washington, DC, GPO, 1894, pp. 21–26.

U.S. Department of Commerce, BC. 1909. A Century of Population Growth in the United States, 1790–1900. Series A 195–209. Washington, DC, GPO, pp. 24–37.

U.S. Department of Commerce, BC. 1910. Thirteenth Census. Indian Population in the United States and Alaska. Washington, DC, GPO, 1915, 285 pp.

U.S. Department of Commerce, BC. 1910. Thirteenth Census: Statistics of the

Indian Population, Number, Tribes, Sex, Age, Fecundity, and Vitality. Washington, DC, GPO, 1913, 223 pp.

U.S. Department of Commerce, BC. 1930. Fifteenth Census: The Indian Population of the United States and Alaska. Washington, DC, GPO, 1937, 238 pp.

U.S. Department of Commerce, BC. 1940. Sixteenth Census: Characteristics of the Nonwhite Population by Race. Washington, DC, GPO, 1943, pp. 74–84.

U.S. Department of Commerce, BC. 1950. Seventeenth Census: Population; Characteristics of the Population; Territories and Possessions. Vol. 2, pts. 51–54. Washington, DC, GPO, 1953–1954, pp. 1–81.

U.S. Department of Commerce, BC. 1950. Seventeenth Census: Population (Special Report); Nonwhite Population by Race. Washington, DC, GPO, 1953, pp. 31–78.

U.S. Department of Commerce, BC. 1960. Eighteenth Census: Characteristics of the Population. Vol. I, chap. B. Washington, DC, GPO, 1961.

U.S. Department of Commerce, BC. 1960. Eighteenth Census: Population (Subject Report); Nonwhite Population by Race. Washington, DC, GPO, 1961, pp. 202–243.

U.S. Department of Commerce, BC. 1970. Nineteenth Census: Population (Subject Report); American Indians. Washington, DC, GPO, 1973, 192 pp.

U.S. Department of Commerce, BC. 1979. Twenty Censuses: Population and Housing Questions, 1790–1980. Washington, DC, GPO, 91 pp.

U.S. Department of Commerce, BC. 1980. Twentieth Census: Population; American Indians, Eskimos, and Aleuts on Identified Reservations and in the Historic Areas of Oklahoma, Excluding Urbanized Areas (Subject Report). Pts. 1 and 2. Washington, DC, GPO, 1986, 1,190 pp.

U.S. Department of Commerce, BC. 1980. Twentieth Census: Population; General Population Characteristics (Arizona). Washington, DC, GPO, 1982, pp. 10–121.

U.S. Department of Commerce, BC. 1980. Twentieth Census: Population; General Population Characteristics (New Mexico). Washington, DC, GPO, 1982, pp. 9–124.

U.S. Department of Commerce, BC. 1980. Twentieth Census. Population; General Population Characteristics (U.S. Summary). Washington, DC, GPO, 1982, pp. 20–22, 50–133, 300–305.

U.S. Department of Commerce, BC. 1980. Twentieth Census: Population; General Social and Economic Characteristics (Arizona). Washington, DC, GPO, 1982, pp. 42–77, 140–145, 278–281.

U.S. Department of Commerce, BC. 1980. Twentieth Census: Population; General Social and Economic Characteristics (New Mexico). Washington, DC, GPO, 1982, pp. 40–75, 115–120, 261–265.

U.S. Department of Commerce, BC. 1980. Twentieth Census: Population; General Social and Economic Characteristics (U.S. Summary). Washington, DC, GPO, 1982, pp. 91–162, 385–401, 451–462.

U.S. Department of Commerce, BC. 1980. Twentieth Census: Population (Supplementary Report); American Indian Areas and Alaska Native Villages. Washington, DC, GPO, 1984, 38 pp.

U.S. Department of Health, Education, and Welfare, IHS. 1978. Indian Health Trends and Services. Washington, DC, GPO, 89 pp.

U.S. Department of Health, Education, and Welfare, IHS. 1979. Selected Vital Statistics for Indian Health Service Areas and Service Units, 1972–1977. Rockville, MD, Public Health Service, 123 pp.

U.S. Department of Health, Education, and Welfare, IHS. 1984. Chart Book Series. Washington, DC, GPO, 47 pp.

U.S. Department of Health, Education, and Welfare, IHS. 1984. A Comprehensive Health Care Program for American Indians and Alaska Natives. Rockville, MD, 22 pp.

U.S. Department of Health, Education, and Welfare, NCHS. 1981. Vital and Health Statistics: Data Systems of the National Center for Health Statistics. Series 1/16. Hyattsville, MD, Public Health Service, 37 pp.

U.S. Department of Health, Education, and Welfare, NCHS. 1982. Health: United States. Washington, DC, GPO, 191 pp.

U.S. Department of Health, Education, and Welfare, NCHS. 1984. Monthly Vital Statistics Report. Advance Report of Final Natality Statistics, 1982. Vol. 33/6. Hyattsville, MD (Public Health Service), 44 pp.

U.S. Department of Health, Education, and Welfare, NCHS. 1984. Monthly Vital Statistics Report. Advance Report of Final Mortality Statistics, 1982. Vol. 33/9. Hyattsville, MD (Public Health Service), 44 pp.

U.S. Department of Health and Human Services, IHS. 1996. Regional Differences in Regional Health, 1991–1994. Rockville, MD, Public Health Service, 96 pp.

U.S. Department of Health and Human Services, IHS. 1996. Trends in Indian Health, 1991–1995. Rockville, MD, Public Health Service, 158 pp.

U.S. Department of the Interior. 1977. Topographic Map 1:100,000, Fort Apache Indian Reservation, AZ (East Half).

U.S. Department of the Interior. 1986. Report of the Task Force on Indian Economic Development. Washington, DC, GPO, 265 pp.

U.S. Department of the Interior, BIA. 1912. Executive Orders Relating to Indian Reservations: 1854–1912. Washington, DC, GPO, 226 pp.

U.S. Department of the Interior, BIA. 1922. Executive Orders Relating to Indian Reservations: 1912 (July)–1922. Washington, DC, GPO, 80 pp.

U.S. Department of the Interior, BIA. 1923. A Message of the Indian Commissioner. Washington, DC, Feb. 24. Newsletter. 1 p.

U.S. Department of the Interior, BIA. N.d. Resident Population on Indian Reservations, 1950. Washington, DC, 5 pp.

U.S. Department of the Interior, BIA. 1955. Indian Population of Continental United States, 1950. Washington, DC. Manuscript. 7 pp.

U.S. Department of the Interior, BIA. 1960. United States Indian Population and Land, 1960. Washington, DC. Manuscript. 5 pp.

U.S. Department of the Interior, BIA. 1968. Economic Development of American Indians and Eskimos. 1930–1967: A Bibliography. Washington, DC, GPO, 263 pp.

U.S. Department of the Interior, BIA. 1982. Indian Service Population and Labor Force Estimates, 1981. Washington, DC, 28 pp.

U.S. Department of the Interior, BIA. 1983. Buyer's Guide to Products Manufactured on American Indian Reservations. Washington, DC, 62 pp.

U.S. Department of the Interior, BIA. 1984. American Indians. Washington, DC, GPO, 44 pp.

U.S. Department of the Interior, BIA. 1984–1985. Education Directory. Washington, DC, 32 pp.

U.S. Department of the Interior, BIA. 1985. Indian Service Population and Labor Force Estimates, 1984. Washington, DC, 31 pp.

U.S. Department of the Interior, BIA. 1987. Indian Service Population and Labor Force Estimates, 1986. Washington, DC, 30 pp.

U.S. Department of the Interior, BIA. 1992–1993. Education Directory. Washington, DC, 40 pp.

U.S. Department of the Interior, BIA. 1996. Indian Service Population and Labor Force Estimates, 1995. Washington, DC, 20 pp.

U.S. Department of the Interior, BIA, Branch of Acknowledgement and Research. 1997. The Official Guidelines to the Federal Acknowledgement Regulations, 25 CFR 83. Washington, DC. Manuscript. 80 pp.

U.S. Department of the Interior, BIA, Division of Education. N.d. Career Development Opportunities for Native Americans. Washington, DC, 56 pp.

U.S. Department of the Interior, BIA, Office of Education Programs. 1997. Education Directory 1996–1997. Washington, DC, 50 pp.

U.S. Department of the Interior, BIA, Office of Indian Education Programs. 1982. Alcohol and Drug Abuse in BIA Schools. Washington, DC. Manuscript. 76 pp.

U.S. Department of the Interior, BIA, Office of Indian Education Programs. 1997. Education Directory 1996–1997. Washington, DC, GPO, 50 pp.

U.S. Department of the Interior, BIA, Office of Trust Responsibilities. 1996. Annual Report of Indian Lands. Washington, DC, 44 pp.

U.S. Department of the Interior, BIA, Office of Trust Responsibilities. 1974–1985. Annual Report of Indian Lands. Washington, DC.

U.S. Department of the Interior, BIA, Phoenix Area Office. 1976. Information Profiles of Indian Reservations in Arizona, Nevada, and Utah. Phoenix, AZ, 186 pp.

U.S. Department of the Interior, Bureau of Reclamation. 1961. Salt River Project: Reclamation Project Data, pp. 662–677.

U.S. Department of the Interior, National Park Service, n.d. Tribal Preservation Program. Washington, DC. Internet: http://www2.cr.nps.gov/tribal/thpo.htm.

U.S. Fish and Wildlife Service. 1994. Statement of Relationship between the White Mountain Apache Indian Tribe and the U.S. Fish and Wildlife Service. Whiteriver, AZ.

U.S. Government Printing Office. 1996. Federal Register, 61/220. Washington, DC. Nov. 13, pp. 58211–58216.

U.S. House of Representatives. 1920. House Executive Document no. 1, 42d Cong., 3d sess., serial 1560. Washington, DC.

U.S. House of Representatives. 1953. Report with Respect to the House Resolution Authorizing the Committee on Interior and Insular Affairs to Conduct an Investigation of the BIA. Union Calendar no. 790, 82d Cong., 2d sess. Washington, DC, GPO, 1,593 pp.

U.S. House of Representatives. 1985. Indian Economic Development Act of 1985 (H.R. 3597, 99th Cong., 1st sess.). Washington, DC, GPO, 56 pp.

U.S. Public Law 93-531. 1974. An Act to Provide for the Final Settlement of the Conflicting Rights and Interests of the Hopi and Navajo Tribes. 88 Stat. 1718, codified at U.S.C. 64d–3.

U.S. Senate Committee on Interior and Insular Affairs. 1958. Analysis of the Problems and Effects of Our Diminishing Indian Land Base. 85th Cong., 2d sess. Washington, DC.

U.S. Senate Report. 1969. Indian Education: A National Tragedy, A National Challenge (no. 501, 91st Cong., 1st sess., serial 12831–7). Washington, DC, GPO, 220 pp.

U.S. Senate Report. 1975. Indian Water Rights of the Five Central Tribes of Arizona (Ak-Chin, Fort McDowell, Gila River, Papago, Salt River, Pima Maricopa). Hearings of the Committee on the Interior and Insular Affairs, 94th Cong., 1st sess. Washington, DC, GPO.

U.S. Senate and Select Committee on Indian Affairs. 1993. Indian Country and the Clinton Administration: Forging a Visionary Partnership to Build Reservation Economics and Sustainable Homelands. Washington, DC, American Indian Resources Institute, 37 pp.

Valley National Bank of Arizona. 1983. Arizona Statistical Review. Phoenix, AZ, 41 pp.

White Mountain Apache Tribe. 1985–1986. Overall Economic Development Plan: Fort Apache Indian Reservation Whiteriver, AZ. Manuscript.

Winters vs. United States. 9th Cir. 1908. 145 Fed. 740; 148 Fed. 684; 207 US 564.

INFORMANTS

<1> Standing Elk: Adult Education; and Sam Johnson: Chief, BIA, Office of Education, Department of the Interior (Washington, DC).

<2> Rev. Art Geunther: Lutheran Church, Whiteriver (Fort Apache Indian Reservation, AZ).

<3> Philbert Whatahomigie: Teacher, Peach Springs School District (Peach Springs, Hualapai Indian Reservation, AZ).

<4> Bud Shapard: Chief, BIA, Branch of Acknowledgement and Research, Department of the Interior (Washington, DC).

<5> Robert Pennington: Former Chief, BIA, Branch of Acknowledgement and Research, Department of the Interior (Washington, DC).

<6> James Gilbert: Area Employment Officer, BIA, Phoenix Area Office (Phoenix, AZ).

<7> Joseph Johnston: Chief, BIA, Division of Energy and Minerals, Department of the Interior (Washington, DC).

<8> Wyman Babby: Superintendent, BIA, Crow Agency (Crow Indian Reservation, MT).

<9> Aaron Handler: Officer, IHS, Alcohol Program Statistics Branch, Department of Health and Human Services (Rockville, MD).

<10> Fred Deer: Officer, IHS, Alcohol Program Statistics Branch, Department of Health and Human Services (Rockville, MD).

<11> Hugo Kock: Officer, NCHS, Division of Vital Statistics, Department of Health and Human Services (Hyattsville, MD).

<12> Jeff Maurer: Statistician, NCHS, Nationality Statistics Branch (Mortality Data), Department of Health and Human Services (Hyattsville, MD).

<13> Selma Taffel: Statistician, NCHS, Nationality Statistics Branch (Natality Data), Department of Health and Human Services (Hyattsville, MD).

<14> Edgar Walema: Chairman, Hualapai Tribe (Peach Springs, Hualapai Indian Reservation); C.L. Henson: Superintendent, BIA, Truxton Cañon Agency (Valentine, AZ).

<15> Edna Paisano: Statistician, Bureau of the Census, Racial Statistics Branch, Department of Commerce (Washington, DC).

<16> George E. Smith: Chief, BIA, Division of Forestry, Department of the Interior (Washington, DC).

<17> Ronald Peake: Chief, BIA, Division of Housing Assistance, Department of the Interior (Washington, DC).

<18> Steve Permison, M.D.: Officer, IHS, Office of Program Operations, Department of Health and Human Services (Rockville, MD).

<19> Tony D'Angelo: Chief, IHS, Program Statistics Branch, Department of Health and Human Services (Rockville, MD).

<20> Donald Stewart, Sr.: Chairman, Crow Tribal Council (Crow Indian Reservation, MT).

<21> Rebecca Martagan: Area Educational Officer, BIA, Navajo Area Office (Window Rock, Navajo Indian Reservation, AZ).

<22> Philis Banashley: Secretary, BIA, Theodore Roosevelt School (Fort Apache Indian Reservation, AZ).

<23> John Reyhmer: Director, BIA, Havasupai School (Supai, Havasupai Indian Reservation, AZ).

<24> Earl Havatone: BIA, Education Officer (Peach Springs, Hualapai Indian Reservation, AZ).

<25> Gerald Scott: Officer, BIA, Office of Education, Department of the Interior (Washington, DC).

<26> Lucille Whatahomigie: Director, Hualapai Bilingual Program, Peach Springs School District (Peach Springs, Hualapai Indian Reservation, AZ).

<27> Kevin Theriot: Social Service Officer, Arizona Mesa Agency, Church of Jesus Christ of Latter-Day Saints (Mesa, AZ).

<28> Hazel Elbert: Deputy Director, BIA, Office of Indian Services, Department of the Interior (Washington, DC).

<29> John Fritz: Assistant Secretary of Indian Affairs, BIA, Department of the Interior (Washington, DC).

<30> Niphier Valandra: Superintendent, BIA, Office of Education; Myrna Hillyard: Principal, Elementary School; Wesely Bonito: Tribal Education Officer; Bret Hartzell: Teacher, Junior High School (all in Whiteriver, Fort Apache Indian Reservation, AZ).

<31> Colleen Moyah: Director, Tribal Economic Development (Sacaton, Gila River Indian Reservation).

<32> Leona Kakar: Chairwoman, Ak-Chin Community Council (Maricopa, Ak-Chin Indian Reservation, AZ).

<33> Art Converse: Vice President, First Interstate Bank-Pinetop (Pinetop, AZ).

<34> Charley Billie: Navajo Indian (at the time of the interview in Ganado, Navajo Indian Reservation, AZ).

<35> Ronnie Lupe: Chairman, White Mountain Apache Council (Whiteriver, Fort Apache Indian Reservation, AZ).

<36> Hal Butler: General Manager, Fort Apache Timber Company (Whiteriver, Fort Apache Indian Reservation, AZ). At the time of the interview Butler had held this position for more than thirty years with only a few brief interruptions.

<37> Taylor Satala: Director, IHS, Health Center (Peach Springs, Hualapai Indian Reservation, AZ).

<38> Ray Brady and Mark Nelson: Officers, Albuquerque Office, U.S. Geological Service (Albuquerque, NM).

<39> Floyd Correa: Member of the board of directors of the Atlantic and Richfield Company (Albuquerque, NM). This multinational concern is one of the largest mining cooperations in the world today. Mr. Correa is the former governor of the Laguna Tribal Council (Laguna Pueblo Indian Reservation, NM).

<40> Achmed Kooros: Officer, Council of Energy Resource Tribes (Denver, CO).

<41> Helfried Mostler: University professor, Department of Geology, University of Innsbruck (Austria); Wolfgang Klau: former employee of the Federal Institute of Geosciences and Mineral Resources (Hannover, Germany) and lecturer in the Department of Geology, University of Innsbruck (Austria).

<42> Cramer Bornemann: Geologist, BIA, Division of Energy and Mineral Resources, Department of the Interior (Washington, DC).

<43> Herb Voigt: Former planner of the Hualapai Tribe (Hualapai Mountains, Kingman, AZ). Mr. Voigt has a large collection of military correspondence (photocopies) having to do with the creation of the Hualapai Indian Reservation, and kindly placed this correspondence at my disposal.

<44> Name of person contacted unknown: Office of Minerals, Navajo Tribe (Window Rock, Navajo Indian Reservation, AZ).

<45> Kenneth Taylor: Attorney, Nordhaus, Haltom and Taylor (Albuquerque, NM).

<46> Richard Te Cube: Vice President, Jicarilla Apache Tribe (Dulce, Jicarilla Apache Indian Reservation, NM).

<47> Ron Solimon: Legal Assistant, Laguna Pueblo (Laguna Indian Reservation, NM).

<48> Robert Brauchli: Lawyer and former Tribal Attorney of the White Mountain Apache Tribe (Tucson, AZ).

<49> R. E. Johnson: Chief, BIA, Division of Tribal Government Services, Department of the Interior (Washington, DC).

<50> W. H. Veeder: Attorney of the White Mountain Apache Tribe (at the time of the interview in Whiteriver, Fort Apache Indian Reservation, AZ; otherwise in Washington, DC). Mr. Veeder represented the White Mountain Apache Tribe in the water rights conflict with the Salt River Project.

<51> Douglas Cole: Public Information Assistant of the Central Arizona Water Conservation District (Phoenix, AZ).

<52> Dick Jeffries: Hydrologist, BIA, Phoenix Area Office (Phoenix, AZ).

<53> Ron Moore: Director of the Planning Department, Colorado River Indian Tribes (Parker, Colorado River Indian Reservation, AZ).

<54> Bill Thompson: Director of the Havasu Pumping Plant, Central Arizona Water Project (Parker, AZ).

<55> Thomas Clark: General Manager, Central Arizona Water Conservation District (Phoenix, AZ).

<56> Raymus Albert: Tribal Land Office of the White Mountain Apache Tribe (Whiteriver, Fort Apache Indian Reservation, AZ).

<57> Mr. Green: Manager of the Tribal Irrigation Project, White Mountain Apache Tribe (Canyon Day, Fort Apache Indian Reservation, AZ).

<58> Mr. Grippin: BIA, Soil Conservationist (Whiteriver, Fort Apache Indian Reservation, AZ).

<59> Floyd Massey: BIA, Supervisor for Finances (Whiteriver, Fort Apache Indian

Reservation, AZ). Mr. Massey is a resident of Canyon Day (Fort Apache Indian Reservation, AZ).

<60> Josiah Moore: Chairman, Tohono O'Odham Council (Sells, Papago Indian Reservation, AZ).

<61> Russell Bradley: Superintendent, BIA, Fort Apache Agency (Whiteriver, Fort Apache Indian Reservation, AZ).

<62> Charles O'Hara: Former administrative aid to the Chairman of the White Mountain Apache Tribe (Whiteriver, Fort Apache Indian Reservation, AZ).

<63> John Nielsen: Tribal Planner, White Mountain Apache Tribe (Whiteriver, Fort Apache Indian Reservation, AZ).

<64> David Reinhold: Chief of Forestry, BIA, Fort Apache Agency (Whiteriver, Fort Apache Indian Reservation, AZ).

<65> Philipp Stago, Jr.: Director, Game and Fish Department, White Mountain Apache Tribe (Whiteriver, Fort Apache Indian Reservation, AZ).

<66> Bill Warskow: Supervisor, Salt River Project, Watershed Division (Phoenix, AZ).

<67> Bob Jones: Real Estate Service, BIA, Phoenix Area Office (Phoenix, AZ).

<68> Gert Pokorny: Project director, Lässer and Feizlmayr Engineers, Inc. (Innsbruck, Austria).

<69> Anthony Brazel: State Climatologist of Arizona and professor in the Department of Geography, Arizona State University (Tempe, AZ).

<70> Dorothy Hallock: Director, Office of Planning and Evaluation, Gila River Indian Community (Sacaton, Gila River Indian Reservation, AZ).

<71> William Talbow: Director, Department of Physical Resources, Gila River Indian Community (Sacaton, Gila River Indian Reservation, AZ).

<72> Jack Palmer: Manager, Gila River Farms (Sacaton, Gila River Indian Reservation, AZ).

<73> Bill Miller: Director, Amerind Agrotech Laboratories (Sacaton, Gila River Indian Reservation, AZ).

<74> Marian Miles: District #7 Coordinator, Maricopa Colony (Laveen, Gila River Indian Reservation, AZ).

<75> Mr. Levy: Real Estate Service, BIA, Pima Agency (Sacaton, Gila River Indian Reservation, AZ).

<76> Varnell Gatewood, Sr.: Manager, North Fork Cattle Association (Whiteriver, Fort Apache Indian Reservation, AZ).

<77> Ray Jackson: Soil Conservationist, BIA, Phoenix Area Office (Phoenix, AZ).

<78> Dale Bratcher: Manager, Tribal Farm of the Fort McDowell Tribe (Fountain Hills, Fort McDowell Indian Reservation, AZ).

<79> Theoria Lomahquahu: Employee of the Kaibab Band of Paiute government (Fredonia, Kaibab Indian Reservation, AZ).

<80> John Honeycutt: Natural Resources Manager, BIA, Fort Yuma Agency (Yuma, California).

<81> Kenny Guerrero: Councilman, Fort Mohave Indian Tribe (Needles, California).

<82> Bob McNichols: Natural Resources Officer, BIA, Truxton Cañon Agency (Peach Springs, Hualapai Indian Reservation, AZ).

<83> Frank Hunt: Rancher and freelance consultant to the BIA, Truxton Cañon Agency (Peach Springs, Hualapai Indian Reservation, AZ).

<84> Mayland Parker: Professor, Department of Geography, Arizona State University (Tempe, AZ). Prof. Parker was a VISTA coordinator for several Indian reservations in Arizona in the 1950s and 1960s.

<85> Nick Sunn: Co-founder of the Maricopa Indian Cooperative Association (Maricopa Colony, Gila River Indian Reservation, AZ).

<86> Robert Donlevy: Real Estate Service, BIA, Phoenix Area Office (Phoenix, AZ).

<87> Nicole Hulse: Employee of the Tribal Chairman, Ak-Chin Community Council (Maricopa, Ak-Chin Indian Reservation, AZ).

<88> Donna McCurdy: Realty Officer, BIA, Colorado River Agency (Parker, Colorado River Indian Reservation, AZ).

<89> Joycelyn Martinez: Realty Officer, BIA, Colorado River Agency (Parker, Colorado River Indian Reservation, AZ).

<90> Ron Moore: Director of the Planning Department, Colorado River Indian Tribes (Parker, Colorado River Indian Reservation, AZ).

<91> Father Hermann Zeller, S.J.: Professor at the Department of Apologetics, University of Innsbruck, and librarian of the Jesuit Monastery (Innsbruck, Austria).

<92> Carla Alchesay-Nachu: Service Unit Director, IHS, Hospital (Whiteriver, Fort Apache Indian Reservation, AZ).

<93> Fred Malroy: Former Chief of Forestry, BIA, Phoenix Area Office (Phoenix, AZ).

<94> John Philbin and Arch Wells: Employees of the Branch of Forestry, BIA, Phoenix Area Office (Phoenix, AZ).

<95> Scott Hilton: Former foreman of the Southwest Forest Industries (McNary, Fort Apache Indian Reservation, AZ).

<96> Sam Goodhope: Tribal consultant in forestry management for the White Mountain Apache Tribe (Whiteriver, Fort Apache Indian Reservation, AZ).

<97> Howard Pipenbrink: Chief, BIA, Branch of Land Use, Department of the Interior (Washington, DC).

<98> Peter Schwarzbauer: Associate Professor, Institute of Forest Sector Policy and Economics, University of Renewable Natural Resources (Vienna, Austria). Prof. Schwarzbauer is also a founding member of the Indians of North America "working circle," an Austrian branch of the Society for Threatened Peoples.

<99> Don Lee: Former owner of the licensed Lee Mercantile Company, a trading post closed in 1989 (Whiteriver, Fort Apache Indian Reservation, AZ). His forefathers came to the White Mountain Apache shortly after the reservation was established.

<100> Wayne Sinyella: Chairman, Havasupai Tribe (Supai, Havasupai Indian Reservation, AZ).

<101> Ron Malfara: General Manager of the tribally owned Sunrise Park Ski Resort (Fort Apache Indian Reservation, AZ).

<102> Mr. Natoli: Mayor (Pinetop-Lakeside, AZ).

<103> James Pettnadi: Board member, Chamber of Commerce (Pinetop-Lakeside, AZ).

<104> Kevin Dunlop: Town Manager (Pinetop-Lakeside, AZ).

<105> Mr. Hurlbut: Board member of the Lodge Owner Association (Pinetop-Lakeside and Show Low, AZ).

<106> William Buchholz: Officer, BIA, Division of Real Estate Services, Department of the Interior (Washington, DC).

<107> Lawrence Taylor: Geographic Planning Specialist, Department of Commerce (Washington, DC).

<108> Cecil Antone: Governor, Gila River Indian Community Council (Sacaton, Gila River Indian Reservation, AZ).

<109> Edgar Perry: Museum Director, White Mountain Apache Museum (Fort Apache, Fort Apache Indian Reservation, AZ).

<110> Susie Sato: Librarian, Arizona Historical Foundation, Arizona State University (Tempe, AZ).

<111> Amelia Flores: Librarian, Colorado River Indian Tribes Museum (Parker, Colorado River Indian Reservation, AZ).

<112> Santosham Mathuram: Assistant Professor, Johns Hopkins University (Baltimore, MD). At the time of the interview Mr. Mathuram directed a research project on infectious diseases for the White Mountain Apache Tribe (Whiteriver, Fort Apache Indian Reservation, AZ).

<113> Dave Blunt: Emergency Medical Services, White Mountain Apache Tribe (Whiteriver, Fort Apache Indian Reservation, AZ).

<114> Martha Bayer-Harvey: Community Health Educator, White Mountain Apache Tribe (Whiteriver, Fort Apache Indian Reservation, AZ).

<115> Kino Kane: Rainbow Center for Alcohol Rehabilitation (Whiteriver, Fort Apache Indian Reservation, AZ).

<116> Heinz Dorn: Director of the Criminal Investigation Department, Austrian federal police (Innsbruck, Austria).

<117> Bea Medicine: Associate Professor, Department of Anthropology, California State University–Northridge.

<118> Ardell Ruize: Officer, Human Resources (Sacaton, Gila River Indian Reservation, AZ).

<119> Jeff De Witt: Water Conservation Director, City of Phoenix (Phoenix, AZ).

<120> Kathy Jacobs: Director, Tucson Department of Water Resources (Tucson, AZ).

<121> Elizabeth Lohah Homer: Director, Office of American Indian Trust, U.S. Department of the Interior (Washington, DC).

<122> Tim Wapato: National Indian Gaming Association in Washington, DC, Director. Speech at the annual meeting of the Affiliated Tribes of Northwest Indians at Spokane, WA, August 1997.

<123> National Public Radio, "Morning Edition," August 11, 1998.

<124> Cecil F. Antone: Lieutenant Governor, Gila River Indian Reservation (Sacaton, AZ).

<125> Thomas J. Davis: Tribal Planning Director, Agua Caliente Indian Reservation (Palm Springs, CA). Speech at the Fifth Annual U.S. Environmental Protection Agency Conference (Region IX, Nov. 5, 1997, San Francisco, CA).

<126> Joseph L. Miller: Curriculum Advisor, Southwestern Indian Polytechnic Institute (Albuquerque, NM).

Index